Integrated Analytical Systems

Series Editor

Radislav A. Potyrailo, Niskayuna, NY, USA

This comprehensive and interdisciplinary series offers the most recent advances in all key aspects of development and applications of modern instrumentation for chemical and biological analysis on the microscale.

These key aspects will include (1) innovations in sample introduction through micro- and nano-fluidic designs, (2) new types and methods of fabrication of physical transducers and ion detectors, (3) materials for sensors that became available due to the breakthroughs in combinatorial materials science and nanotechnology, and (4) innovative data processing and mining methodologies that provide dramatically reduced rates of false alarms.

Clearly, a true multidisciplinary effort is required to meet objectives for a system with previously unavailable capabilities. This cross-discipline fertilization is driven by the expanding need for chemical and biological detection and monitoring and leads to the creation of instruments with new capabilities for new demanding applications. Indeed, instruments with more sensitivity are required today to analyze ultra-trace levels of environmental pollutants, pathogens in water, and low vapor pressure energetic materials in air. Sensor devices with faster response times are desired to monitor transient in-vivo events and bedside patients. More selective instruments are wanted to analyze specific proteins in vitro and analyze ambient urban or battlefield air. For these and many other applications, new features of modern microanalytical instrumentation are urgently needed. This book series is a primary source of both fundamental and practical information on both the current state of the art and future directions for microanalytical instrumentation technologies. This book series is addressed to the rapidly growing number of active practitioners and developers and those who are interested in starting research in this direction, directors of industrial and government research centers, laboratory supervisors and managers, students and lecturers.

Miguel Ángel Rodríguez-Delgado ·
Bárbara Socas-Rodríguez ·
Antonio V. Herrera-Herrera
Editors

Microextraction Techniques

Fundamentals, Applications and Recent Developments

Springer

Editors
Miguel Ángel Rodríguez-Delgado
Department of Chemistry
University of La Laguna
San Cristóbal de La Laguna, Spain

Bárbara Socas-Rodríguez
Department of Chemistry
University of La Laguna
San Cristóbal de La Laguna, Spain

Antonio V. Herrera-Herrera
Department of Chemistry
University of La Laguna
San Cristóbal de La Laguna, Spain

ISSN 2196-4475 ISSN 2196-4483 (electronic)
Integrated Analytical Systems
ISBN 978-3-031-50526-3 ISBN 978-3-031-50527-0 (eBook)
https://doi.org/10.1007/978-3-031-50527-0

This Springer imprint is published by the registered company Springer Nature Switzerland AG
The registered company address is: Gewerbestrasse 11, 6330 Cham, Switzerland

Paper in this product is recyclable.

Preface

Analytical chemistry constitutes a scientific area of great interest since it plays a key role in many other fields such as medicine, biology, food science, agriculture, environment, archaeology, etc. The main role of analytical chemists is the design, development, validation, and application of analytical processes that allow addressing the separation, identification, and quantification of many different types of compounds of interest within the studies carried out in the aforementioned fields, among others. In this sense, since its beginnings during the eighteenth century, the performance of reliable analytical procedures has been the principal goal, which has evolved towards the development of selective, sensitive, and rugged procedures supported by solid physico-chemical foundations.

In recent years the searching of reliable and effective analytical methodologies has been combined with the development of sustainable strategies based on the 12 principles of the green chemistry proposed by Anastas and Warner in 1998. This led to the development of an important working area at the beginning of this century and known as green analytical chemistry (GAC). It is focused on the performance of more environmentally friendly procedures, considering the main components of the analytical methodologies, i.e., operator, sample, reagents, instrument, method, and waste. The main goals of GAC are: elimination or reduction of the use of chemical substances, decrease of energy consumption, adequate management of analytical waste, and increase of the operator safety.

Different action lines have been proposed and developed to achieve those objectives. Among them, the use of miniaturised techniques has gained relevance. This approach provides important advantages and participates in the fulfilment of all established objectives. Particular attention has been given to the miniaturisation of sample preparation techniques since this constitutes one of the most important challenging steps in analytical procedures. The use of high volume of solvents and complex strategies to get proper analytical performance compromises the principles of GAC, which make necessary the searching of alternatives that allow addressing the problem and solve this limitations.

The design of efficient sorbents and solvents based on nanomaterials and natural components used in simple and effective extraction techniques have played a key role in the pursuit of green analytical procedures without hindering analytical performance.

This book has been designed and prepared with the aim of offering a general overview of the fundamentals, main applications, and novel developments of miniaturised extraction techniques developed so far. Editors have tried to compile the contribution of the most renown experts in this area to provide reliable and valuable information to graduates, postgraduates, and researchers who are immersed in the study, development, and application of sustainable extraction techniques.

San Cristóbal de La Laguna, Spain

Miguel Ángel Rodríguez-Delgado
Bárbara Socas-Rodríguez
Antonio V. Herrera-Herrera

Contents

Introduction

Introduction

Bárbara Socas-Rodríguez, Antonio V. Herrera-Herrera, and Miguel Ángel Rodríguez-Delgado

Abstract One of the most important lines of action in the area of Green Analytical Chemistry has been the development of miniaturised techniques that involves reducing the use of toxic solvents and hazardous substances and decreasing the complexity, cost and time of the procedures. Numerous advances have been reached in this fields and currently, there exists a great number of microextraction techniques thanks, in part, to the development of novel materials, such as nanomaterials and green solvents. In this sense, two big groups of microextraction techniques can be considered: sorbent-based microextractions and solvent-based microextractions. This book pretends to compile, from a general point of view (not focus on just one area of application), the fundamentals, main applications, and novel developments of all these techniques. The main audience will be graduates, postgraduates and researchers. It would be a very interesting option as academic book, especially for those that are working in the development of sustainable extraction techniques.

Keywords Green chemistry · Sorbent-based microextraction · Solvent-based microextraction · New materials

Abbreviations

DLLME	Dispersive liquid–liquid microextraction
GAC	Green Analytical Chemistry
GSP	Green sample preparation
HF-LPME	Hollow-fibre liquid-phase microextraction
HLLME	Homogeneous liquid–liquid microextraction

B. Socas-Rodríguez (✉) · A. V. Herrera-Herrera · M. Á. Rodríguez-Delgado
Departamento de Química, Unidad Departamental de Química Analítica, Facultad de Ciencias, Universidad de La Laguna (ULL), San Cristóbal de La Laguna, España
e-mail: bsocasro@ull.edu.es

A. V. Herrera-Herrera
Instituto Universitario de Bio-Orgánica Antonio González, Universidad de La Laguna (ULL), San Cristóbal de La Laguna, España

M. Á. Rodríguez-Delgado et al. (eds.), *Microextraction Techniques*, Integrated Analytical Systems, https://doi.org/10.1007/978-3-031-50527-0_1

LLE	Liquid–liquid extraction
LSE	Liquid–solid extraction
μ-dSPE	Micro-dispersive solid-phase extraction
MSPD	Matrix solid-phase dispersion
μ-SPE	Micro-solid-phase extraction
SBSE	Stir bar sorptive extraction
SDME	Single drop microextraction
SLE	Solid–liquid extraction
SPME	Solid-phase microextraction

1 Introduction

Green Chemistry has gained significant impetus in recent years due to growing concerns about environmental degradation and the adverse impact of traditional chemical practices [1]. Green chemistry is founded on a set of 12 principles [2] aimed at minimizing the generation of hazardous waste, reducing the use of toxic substances, and promoting sustainability in chemical processes. It proposes innovative scientific solutions throughout the entire life cycle (design, manufacture, use, and disposal) of a chemical product. Although this new way of conceiving chemical processes has skyrocketed from its appearance in the texts from Paul Anastas (1994–1998) [2–5], it has its roots in the US Federal Pollution Prevention Act of 1990.

Initially directed to industrial-scale processes, the 12 principles were also adapted to different chemical fields, including analytical chemistry. However, some of them are not directly applicable to the analytical field and certain fundamental aspects of analytical chemistry were not included in the general version of the requirements. In this regard, accuracy, sensitivity, and reproducibility should be cautiously considered. Thus, Green analytical chemistry (GAC) is defined as the discipline dedicated to develop cleaner and eco-friendly methodologies to analyse low concentrations of different molecules in complex sample matrices, without compromising the analytical parameters [6, 7]. In 2013, Gałuzska et al. [8] tailored and adapted the 12 principles in order to suit analytical chemistry requirements: (i) use direct analytical techniques, (ii) utilise minimal sample size and reducing the number of samples, (iii) conduct in situ measurements is advocated for in the analytical process, (iv) integrate analytical operation to reduce energy and reagent consumption, (v) opt for automated and miniaturised methods, (vi) avoid derivatisation steps, (vii) prevent the generation of a large volume of waste and proper management of it, (viii) choose multi-analyte or multi-parameter methods, (ix) minimise energy usage, (x) prioritise reagents derived from renewable sources, (xi) eliminate or replace toxic reagents, (xii) ensure the safety of the operator.

A chemical analysis consists of several sequential steps: sampling, sample preparation, analytical measurement, and data evaluation. Undoubtedly, with the progress in instrumentation, the sample preparation is one of the critical aspects that determines whether an analytical procedure can be labelled as "green". This preparation step encompasses not only the dissolution of target analytes in an appropriate solvent but also involves homogenisation, extraction, cleanup, and concentration processes. Recently, López-Lorente et al. [9] proposed the 10 principles of green sample preparation (GSP) based on the fact that the first principle of GAC is often misunderstood, leading to a mistaken notion that avoiding the sample preparation step is the ideal green approach. This interpretation overlooks the potential advancements in the analytical field and not consider these situations, in which direct analysis is not feasible. Analytical Sciences encounter intricate and interconnected challenges, both in on-site and laboratory situations, and sample preparation is frequently required to address these complexities. Therefore, 10 principles of GSP are [9]: (i) prioritise in situ sample preparation, (ii) opt for safer solvents and reagents, (iii) focus on sustainable, reusable, and renewable materials, (iv) minimise waste generation, (v) reduce sample, chemical, and material quantities, (vi) maximise sample throughput, (vii) encourage integration of steps and automation, (viii) minimise energy usage, (ix) select the greenest post-sample preparation configuration, (x) ensure operator safety. By incorporating the principles of Green Chemistry, GAC and GSP into extraction protocols, analysts can achieve remarkable reductions in solvent usage, hazardous substances, waste generation, and energy consumption.

First sample preparation protocols were laborious, time-consuming, and require large amounts of resources, generating hazardous waste. The appearance of analytical microextraction protocols exemplifies a harmonious alliance between environmental responsibility and analytical efficiency. As a result of the intensive research, analytical microextraction protocols frequently offer similar or enhanced selectivity and sensitivity. By minimising interferences and matrix effects, these protocols contribute to accurate and reliable results, reducing the need for reanalysis and further resource consumption. Although the first publications with the word "microextraction" appeared in the 1940s [10, 11], it was not until the 1990s when a real revolution occurred in this field with the development of solid-phase microextraction (SPME) by Arthur and Pawliszyn [12].

One of the primary concerns in traditional analytical methods is the excessive use of organic solvents, which can be harmful to both human health and the environment. Analytical microextraction protocols significantly reduce solvent consumption by employing miniaturised extraction techniques, such as sorbent-based microextraction and solvent-based microextraction. These approaches use reduced amounts of solvents or are solvent-free, thereby reducing the environmental impact and minimising waste generation. In this sense, the use of alternative, non-hazardous extraction phases disminishes toxicity and increases sustainability. These green alternatives offer comparable or improved extraction efficiencies while, the overall environmental footprint of the analytical process is substantially reduced.

It should be mentioned that analytical microextraction protocols, owing to their miniaturised nature, often require lower energy consumption compared to traditional sample preparation techniques, which not only contribute to sustainability but also lead to cost savings and faster analysis, thus promoting economic and environmental benefits. Also, while traditional sample preparation methods can generate substantial amounts of waste (including disposable extraction columns, cartridges, and excess solvents), miniaturised protocols can substantially minimise waste production by utilising reusable or disposable miniaturised extraction devices.

Based on their nature, extraction procedures can be categorised as exhaustive and non-exhaustive protocols. Exhaustive techniques fully extract analytes from the sample, while non-exhaustive methods do not transfer all compounds to the extraction phase. Exhaustive protocols often involve labor-intensive and costly procedures, and thus it is advisable to replace batch equilibrium techniques with flow-through techniques. On the contrary, non-exhaustive approaches can be employed under equilibrium or non-equilibrium conditions. In equilibrium non-exhaustive methods, the underlying principle is similar to equilibrium exhaustive techniques, but the main difference lies in the reduced capacity of the extraction phase, which is insufficient to completely extract analytes from the sample. The miniaturisation of the preparation steps of the analytical process faces its own challenges but, on contrast, it reduces dimensions of the whole analytical process, and advances to the design of portable analysers and on-site analysis. The progress in this field is not only oriented to the development of innovative solutions for isolating analytes, but also on the development of novel and alternative materials. These materials are distinguished by their heightened efficiency compared to traditional ones, leading to enhanced sensitivity and selectivity of the analytical process.

In this book, the fundamentals of each microextraction technique are exhaustively described and the main applications, the trends and the last developments are discussed in a general and didactic way. It is divided in two different big groups: solvent-based microextractions and sorbent-based microextractions (Fig. 1). The first one, composed of five chapters, comprise those sections devoted to micro-solid-phase extraction (μ-SPE), micro-dispersive solid-phase extraction (μ-dSPE) (also known as dispersive-micro-solid-phase extraction (d-μSPE)), solid-phase microextraction (SPME), stir bar sorptive extraction (SBSE), and matrix solid-phase dispersion (MSPD). The second part include those chapters dealing with single drop microextraction (SDME), hollow-fibre liquid-phase microextraction (HF-LPME), homogeneous liquid–liquid microextraction (HLLME), and dispersive liquid–liquid microextraction (DLLME). All of them include the fundamentals and general aspects of microextraction techniques, and novel developments (incorporating new materials and automation, if applicable). Moreover, certain particular aspects, such as the use of magnetic sorbents for μ-dSPE or the assistance by microwaves, ultrasounds or vortex in solvent-based microextractions, were also incorporated. The evolution of these techniques from the classical liquid–liquid extraction (LLE), solid–liquid extraction (SLE), and liquid–solid extraction (LSE) tries to achieve miniaturisation of the devices, enable multiclass analytes extraction, and implement automation.

This book could constitute a reference for novel researchers (including master, Ph.D. students and researchers working in this area) to learn the fundamentals of each technique, their advantages, and disadvantages. Also, it could serve as a guide to select the most suitable technique to solve each specific problem posed at the laboratory.

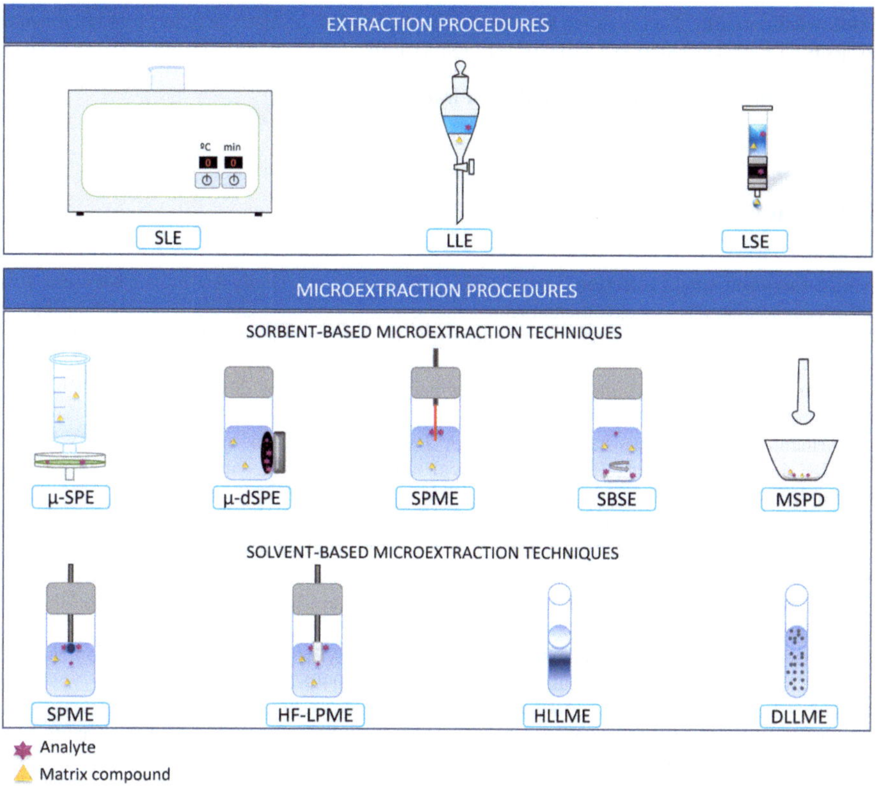

Fig. 1 Extraction and microextraction methods. Representative examples of each category are illustrated

2 Conclusions

Analytical chemists play a significant role in ensuring a sustainable future by incorporating green analytical protocols into their routines and research practices. The mutually beneficial collaboration between green chemistry and Analytical field not only contributes to a cleaner environment but also advances towards more sustainable and economically viable activities. Embracing this synergy offers a promising pathway for future advancements in analytical methodologies while safeguarding our planet's ecological balance.

This book explores the dynamic relationship between analytical microextraction protocols, highlighting how they complement each other and how they can help to achieve environmentally friendly and efficient analytical processes.

In the future, a growing utilisation of microextraction methods is anticipated. Due to their intrinsic advantages, we strongly advocate for the miniaturisation of standard techniques. It should be noted that similar or superior performance should be provided by these microextraction protocols to ensure an effective replacement.

Acknowledgements This work has been funded by the Spanish Ministry of Science and Innovation (projects CPP2021-009056 and TED2021-129480B-I00).

The authors have declared no conflict of interest.

References

1. US Environmental Protection Agency (2023) Basics of green chemistry. https://www.epa.gov/greenchemistry/basics-green-chemistry
2. Anastas PT, Warner JC (1998) Green chemistry: theory and practice. Oxford University Press, New York
3. Anastas PT (1994) Benign by design chemistry. In: Benign by design. American Chemical Society, pp 1–2.
4. Armenta S, de la Guardia M, Namiesnik J (2007) Green microextraction. In: Valcárcel M, Cárdenas S, Lucena R (eds) Analytical microextraction techniques. Bentham Science Publishers, pp 3–27
5. US Environmental Protection Agency (2022) Summary of the pollution prevention act. https://www.epa.gov/laws-regulations/summary-pollution-prevention-act
6. Armenta S, Garrigues S, Esteve-Turrillas FA, de la Guardia M (2019) Green extraction techniques in green analytical chemistry. TrAC—Trends Anal Chem 116:248–253. https://doi.org/10.1016/j.trac.2019.03.016
7. Anastas PT (1999) Green chemistry and the role of analytical methodology development. Crit Rev Anal Chem 29:167–175. https://doi.org/10.1080/10408349891199356
8. Gałuszka A, Migaszewski Z, Namieśnik J (2013) The 12 principles of green analytical chemistry and the SIGNIFICANCE mnemonic of green analytical practices. TrAC—Trends Anal Chem 50:78–84. https://doi.org/10.1016/j.trac.2013.04.010
9. López-Lorente ÁI, Pena-Pereira F, Pedersen-Bjergaard S, et al (2022) The ten principles of green sample preparation. TrAC Trends Anal Chem 148:116530. https://doi.org/10.1016/j.trac.2022.116530
10. Batt WG, Alber HK (1941) Systematic qualitative organic microanalysis. Ind Eng Chem Anal Ed 13:127–132. https://doi.org/10.1021/i560090a033
11. Tintrop LK, Salemi A, Jochmann MA, et al (2023) Improving greenness and sustainability of standard analytical methods by microextraction techniques: a critical review. Anal Chim Acta 1271:341468. https://doi.org/10.1016/j.aca.2023.341468
12. Arthur CL, Pawliszyn J (1990) Solid phase microextraction with thermal desorption using fused silica optical fibers. Anal Chem 62:2145–2148. https://doi.org/10.1021/ac00218a019

Sorbent-Based Microextractions

Micro-solid-phase extraction

Ahmad Reza Bagheri and Hian Kee Lee

Abstract The direct determination of compounds of interest is challenging and, in the majority of cases, impossible. This fact is related to the low concentration of analytes as well as the presence of possible interferences in many sample types. Hence, there is a need for a sample preparation step before the final analysis. To this end, great attention has been devoted to designing and using miniaturized extraction techniques (METs) as alternatives to conventional procedures. METs can address the main drawbacks of conventional sample preparation methods. These include excessive consumption of organic solvents or reagents, need for multiple operations or steps, inconvenience or difficulty of automation, etc. Amongst the METs, micro-solid-phase extraction (μ-SPE) has garnered much attention thanks to its demonstrable advantages. In μ-SPE the consumption of materials (solvents, reagents, etc.) is reduced. Also, sorbents with small (e.g., micro- and nanoscale) dimensions are utilized. Moreover, the procedure involves use of miniaturized extraction devices and apparatus, etc. The contents of this chapter are focused on the fundamentals, configurations and applications of μ-SPE. Novel materials that have been used as sorbents in μ-SPE are investigated and discussed. Finally, current trends and prospects concerning the application of the technique are assessed.

Keywords Sample preparation · Extraction methods · Miniaturized extraction techniques · Micro-solid-phase extraction · Novel sorbents

Abbreviations

β	Selectivity factor
μ-SPE	Micro-solid-phase extraction

A. R. Bagheri
Department of Chemistry, Yasouj University, Yasouj 75918-74831, Iran

H. K. Lee (✉)
Department of Chemistry, National University of Singapore, 3 Science Drive 3, Singapore 117543, Singapore
e-mail: chmleehk@nus.edu.sg; hiankeelee@u.nus.edu

M. Á. Rodríguez-Delgado et al. (eds.), *Microextraction Techniques*, Integrated Analytical Systems, https://doi.org/10.1007/978-3-031-50527-0_2

3D	Three-dimensional
BA	Benzoic acid
Bd	Benzidine
BTEX	Benzene, toluene, ethylbenzene, and xylenes
C_{18}	Octadecyl
COFs	Covalent organic frameworks
DAD	Diode array detection
DES	Deep eutectic solvent
DPX	Disposable pipette extraction
DSPE	Dispersive solid-phase extraction
FID	Flame ionization detection
FLD	Fluorescence detection
GAC	Green analytical chemistry
GC	Gas chromatography
GFAAS	Graphite furnace atomic absorption spectrometry
GO	Graphene oxide
Gr	Graphene
HG-AFLS	Hydride generation-atomic fluorescence spectrometry
HPLC	High-performance liquid chromatography
HVOCs	Halogenated volatile organic compounds
IF	Imprinting factor
IL	Ionic liquid
LC-APCI	Liquid chromatography atmospheric-pressure chemical ionization
LDHs	Layered double hydroxides
LOD	Limit of detection
LOQ	Limit of quantification
MB-μ-SPE	Membrane-based micro-solid-phase extraction
MEPS	Microextraction by packed sorbent
METs	Miniaturized extraction techniques
MICSM	Molecularly-imprinted chitosan microspheres
MOFs	Metal–organic frameworks
MS	Mass spectrometry
MS/MS	Tandem mass spectrometry
MSPD	Matrix solid-phase dispersion
MWCNTs	Multi-walled carbon nanotubes
NMBTEX	Non-metabolized benzene, toluene, ethylbenzene and xylene isomer compounds
NSs	Nanosheets
NTD	Needle trap device
OPPs	Organophosphorus pesticides
PAA	Poly(acrylic acid)
PAHs	Polycyclic aromatic hydrocarbons
PAN	Polyacrylonitrile
PCBs	Polychlorinated biphenyls
PT-SPE	Pipette-tip solid-phase extraction

PVA	Poly(vinyl alcohol)
QuEChERS	Quick, easy, cheap, effective, rugged and safe
rGO	Reduced graphene oxide
RSD	Relative standard deviation
SA	Salicylic acid
SBSE	Stir bar sorptive extraction
SBSEME	Sorbent-based sorptive extraction/microextraction
SC-SPE	Spin column solid-phase extraction
Si/PANI	Polyaniline silica
SPME	Solid-phase microextraction
SWCNT	Single-walled carbon nanotube
Tb	1,3,5-Triformylbenzene
TCS	Triclosan
TDMs	Two-dimensional materials
TFME	Thin-film microextraction
UHPLC	Ultra-high-performance liquid chromatography
UV	Ultraviolet
VWD	Variable wavelength detection

1 Introduction

There has always been a high demand for the analysis of compounds of interest in different (environmental, food, biological, etc.,) samples [1]. This is particularly so when new chemicals assume the status of emerging concern. This is not only because of the broad applicability of these compounds (e.g., as beneficial and health-promoting substances) but also due to their potential impact and effects as contaminants on humans and animal (especially wild) life [2]. Despite the advancements in the development of analytical instrumentation, however, the direct determination of compounds remains challenging. Direct determination is even almost impossible in many instances where the matrices are complex [3]. These challenges are related to the low concentration of compounds (at trace levels) and more significantly, due to the presence of possible interferences [4]. These interferences can cause problems for the instrumentation or accessories themselves or compromise the detection signals of the compounds of interest. For these reasons, the quantitative analysis of compounds in complex matrices usually has to be accompanied by a sample preparation step [5]. Sample preparation is a vital and inseparable step in almost all analytical procedures [6]. It can take up ca. 60–70% of the time and effort of the overall method. The main goals of sample preparation are the removal, or at least the minimization, of interferences, pre-concentration of the compounds and their delivery into an appropriate solvent system. Additionally, it is often necessary to convert recalcitrant compounds into more stable or volatile species suitable for, or compatible with, the chosen analytical technique [7].

Sample preparation methods can largely be divided into solvent-based (usually microscale or miniaturized formats in recent history) and sorbent-based sorptive extraction/microextraction (SBSEME) approaches. SBSEME is a versatile approach in which a sorbent is used for the isolation and pre-concentration of compounds [1, 8, 9], and is the general topic of the present chapter. SBSEME is arguably simple and easy to use and operate, and highly accessible (the intended materials can be designed and synthesized in-house or bought from commercial suppliers conveniently). Additionally, SBSEME provides reasonable extraction time, consumes low-to-moderate volumes of solvents, and is convenient for combining with different analytical techniques, even solvent-based ones [10, 11]. SBSEME can be applied in different configurations: Conventional formats (cartridge- and disk-mode solid-phase extraction (SPE)) [12] as well as miniaturized extraction techniques (METs) [13–16]. The conventional, especially commercially available formats of SBSEME are useful and popular in routine applications. As far as research activities on sample preparation method development are concerned, however, the main focus of analytical scientists has been to apply METs as alternatives to the conventional techniques. The main reasons for applying METs are to minimize, if not eliminate some of the main drawbacks of the conventional formats of SBSEME. Conventional SBSEME approaches generally consume large or moderate amounts of organic solvents and reagents. Also, it is possible for channeling to occur within the sorbent in cartridge-based SPE. Moreover, sorbent particles may also be trapped in the frits of the cartridge. Other limitations include the possible generation of high pressure in the cartridge, the need for processing large volumes (sometime >1 L) of aqueous samples for trace analysis and an excessive number of operational steps [17–19]. Further, concurrent with the emergence of green analytical chemistry (GAC), there is strong motivation and encouragement to develop environmentally friendly and sustainable analytical operations and actions. These serve to minimize health-and safety-related impacts and side effects on humans and the environment. Thanks to their attributes, SBEME techniques are often considered to subscribe to GAC principles [20]. GAC attempts to enable analytical laboratories to be sustainable in terms of costs, and energy. Another aspect of GAC is to reduce the scale of analytical operations, instrumentation, and apparatus [21]. Eliminating, or at least reducing the amounts of solvents, reagents, and sorbents themselves, and also downsizing the dimensions of analytical instrumentation, apparatus and devices, are also other goals of GAC. Hence, contemporary sample preparation researchers have been drawn to further the development of METs [22].

To date, different METs have been developed by analytical chemists. The following sample preparation techniques, micro-SPE (μ-SPE), pipette-tip SPE (PT-SPE), spin column SPE (SC-SPE), disposable pipette extraction (DPX), microextraction by packed sorbent (MEPS), extraction by needle trap device (NTD), stir bar sorptive extraction (SBSE), matrix solid-phase dispersion (MSPD), solid-phase microextraction (SPME), thin-film microextraction (TFME), and even conventional dispersive SPE (DSPE, in which loose sorbent are simply added to a liquid sample for extraction), etc., have been classified under METs [17–19]. The exclusive advantage of all of these techniques is their solventless or solvent-minimized character. Each of these techniques has its individual attributes and can be used for the (usually) chromatographic analysis of diverse compounds. The specific focus of this chapter is on μ-SPE. The fundamentals of μ-SPE are first introduced and the rest of the chapter covers its novel developments and recent main applications.

2 Fundamental Characteristics

The term "μ-SPE" has been widely used and reported under different definitions as to its formats or configurations [20]. One view defines μ-SPE as a technique that uses microgramme/low milligramme quantities of sorbent, in conjunction with microlitre volumes of solvents [1]. From another perspective, μ-SPE has also been described as a technique that uses sorbents with small dimensions like nanoparticles, nanomaterials or micro- (or sometimes meso-) spheres, etc. [19] (even though the technique itself may be at the conventional scale, like DSPE). The application of nanoparticles, nanomaterials, nanocomposites or spheres, etc., is mainly due to the primary features of these materials in terms of a large number of active sites, large surface area, high adsorption capacity, high chemical and physical stabilities, etc. [7, 19]. Sometimes, if hardware is involved such as columns, cartridges, etc., they are much smaller than conventionally scaled SPE devices, to warrant the "μ-" designation. Another definition of μ-SPE is based on the use of an extraction contrivance consisting of a sealed membrane bag (of low centimetre dimensions) containing micro- or nanoscale sorbent materials in small (microgramme/milligramme) quantities. In such a configuration, μ-SPE can be considered as a membrane-protected miniaturized version of DSPE. The filled bag is simply placed in a stirred sample for extraction [8]. Whatever the definition, μ-SPE is commonly considered to be at the leading edge of the trend toward the miniaturization of analytical chemistry in terms of the utilization of low volumes of solvents and reagents, small amounts of sorbent, and reduction of the size of the extraction apparatus or tools (where present). Another advantage of μ-SPE is that researchers can fabricate and use their own materials (especially homemade sorbents) and devices. Consequently, μ-SPE is considered an easy-to-implement and -use technique that has been widely applied in different formats. Decreasing the use of solvents, reagents, and also sorbents can consequently reduce the cost of analytical procedures. Reducing the scale of analytical instrumentation and accessories is another main aspect of GAC that is fulfilled by μ-SPE. Yet another characteristic of GAC that is met by μ-SPE is in the reduction of the side effects and impacts potentially caused by the practice of analytical chemistry that can adversely affect humans and the environment. Therefore, like all other METs, μ-SPE is inherently considered as complying with GAC principles, an attribute that normally need not be specifically advertised but is implicit.

Most of the μ-SPE formats have already been mentioned above (PT-SPE, DPX, MEPS, NTD, SC-SPE, and membrane-protected or -based μ-SPE (MB)-μ-SPE). In the following paragraphs, the fundamentals of these techniques and their attributes are described briefly.

PT-SPE is a simple and convenient MET that uses only a small amount (between 1 and 5 mg) of sorbent [19]. Due to the confinement of the sorbent to the tip of a pipette, PT-SPE consequently consumes low volumes of solvents (in conditioning, washing, and eluting steps) and also reagents, when and if needed [21]. Amongst PT-shaped configurations, the asymmetric (roughly conic) format is beneficial because its wider end is advantageous for sample loading and solvent elution. PT-SPE-based monolithic materials are examples of commercially available devices that are popular. ZipTip, NuTip, and MonoTip C_{18} are the names of some commercial PT-SPE products [22–24]. However, due to the affordability as well as the commercial availability of bare PTs, any researcher can prepare these devices that contain their own in-house-synthesized sorbents.

DPX is another MET that was introduced in 2003 [25]. DPX is essentially a combination of PT-SPE and DSPE, although the extraction device itself is a little more complicated than a normal PT. It consists of a tip filled with a small amount of sorbent material, which is freely dispersed between two filters contained inside the tip [26]. Commercially available tips range in capacities from 200 μL to 5 mL. No conditioning of the sorbent is necessary, as claimed by the manufacturer, which is an advantage that shortens the overall process. The sample is drawn in through the bottom of the device via air aspiration. The bubbles generated during aspiration mix the sample and sorbent more effectively [27]. After extraction, the spent sample is conveniently expelled and discarded. The elution solvent is then aspirated similarly to desorb the analytes. Like many other METs, in DPX, the amounts of sorbent material, sample and organic solvent are much reduced, as is the extraction time [28]. Apart from these, in-house sorbents can be conveniently used, and, importantly, there are several commercial autosamplers that allow the technique to be completely automated [19].

MEPS, which was introduced in 2004 [29], is a simple, fast, and purportedly user-friendly MET. In the technique, a syringe with volumes ranging between 100 and 150 μL is filled with a small amount of sorbent (usually 1–4 mg) [30]. In another format, a separate cartridge filled with a sorbent is inserted between the syringe barrel and the needle. MEPS can therefore be considered as a miniaturized version of cartridge-SPE. The sorbent can provide a selective medium suitable for sampling under a wide range of different conditions and for extraction of the target analytes [31]. The syringe plunger can be manipulated to allow better mixing of the sample and the sorbent, as in DPX [32]. Compared to cartridge-SPE, the packing is integrated directly into the syringe and not separately in a discrete column. Moreover, in MEPS, the sample and elution solvents are forced through the sorbent via the needle end. MEPS can be operated manually, semi-automatically or fully automatically. For the latter two modes, commercially available autosamplers can be programmed to undertake the required operations. Another useful feature of MEPS is that the amount of eluting solvent needed is usually in the same range as an extract introduced into the analytical system. Like DPX, no preconditioning of the sorbent is needed, and each device can be reused several times.

As a type of needle-based extraction technique, NTD resembles MEPS except that the sorbent is housed within the syringe needle itself [19]. Compared to the fibre (coating)-based SPME, NTD is considered to be more reliable and robust. Also, NTD has a lower cost and higher capacity [33]. The main component of an NTD is a narrow stainless-steel tube. The sorbent is packed inside near the open end of the NTD needle via mechanical pushing and compression. Then, by vacuum or pressure, a gas or liquid sample is forced through the sorbent. After adsorption of the analytes by the sorbent, they are usually desorbed into the gas chromatography (GC) injector using thermal desorption. Like SPME, NTD can be considered as a solvent-free technique [34]. Exploiting its attributes, NTD is most appropriate for the extraction and pre-concentration of volatile organic compounds (VOCs). NTD has been mainly used in the headspace mode for trapping such volatile and semi-volatile analytes [35]. Moreover, comparing with fibre-SPME, NTD can be applied for exhaustive extraction by appropriate selection of the experimental variables [36]. In some cases, NTD has also been coupled with liquid chromatography (LC) in which the analytes are desorbed by a small volume of an eluent, via an offline process [33, 37].

SC-SPE is a variant of PT-SPE in which a small amount (milligramme quantity) of a sorbent is tightly packed between two frits in a spin column (instead of a pipette) [38]. The main difference of SC-SPE compared to conventional SPE is that all the manipulations are accomplished in a centrifuge. The sample is loaded from the top and is drawn through the sorbent by a centrifugal force. Therefore, no pressure or vacuum pumps are needed [39]. Also, SC-SPE is reputedly easier and faster [40, 41]. SC-SPE decreases manual manipulation errors and inconsistencies and provides a much higher analyte trap capacity than PT-SPE. By optimizing centrifugation duration as a vital parameter, effective analyte trapping can be realized. However, SC-SPE needs repeated insertion of the column tip in a rotor of a centrifuge, and this is considered a drawback in comparison with PT-SPE [42]. Another drawback of SC-SPE is that it cannot be automated since a centrifuge cannot be integrated with chromatographic instrumentation. In particular, this drawback arises since after passing a sample solution through the SC by centrifugation, the column needs to be manually re-loaded onto the equipment for subsequent analyte elution [43]. The extract then has to be manually retrieved for analysis.

Membrane-protected or -based (MB)-μ-SPE was introduced in 2006 [44], and is essentially a variant of DSPE. It was originally termed simply as μ-SPE but is named MB-μ-SPE in this article to distinguish it from other μ-SPE techniques. The procedure is based on confining loose sorbent media within a membrane bag. The bag or envelope (usually made from polypropylene flat-sheet membrane) serves as a standalone extraction device that is simply added to a stirred sample. The bag is made by superimposing two membrane sheets (of low centimetre dimensions) whose edges are heat-sealed after the sorbent is introduced [45]. The membrane (usually of pore size of 0.2 μm) allows analytes to diffuse freely through it to be adsorbed by the sorbent, but can filter out extraneous substances in the sample matrix [46]. The main advantage of MB-μ-SPE is that it performs extraction, cleanup, and pre-concentration in a single step, whilst also providing protection of the enclosed sorbent from highly complex matrices. Moreover, this method is not only portable, robust, and durable but also that the bag, after appropriate rinsing, can be reused several times [47].

Having the flexibility of being able to select either commercially available or self-prepared sorbents is an additional advantage of μ-SPE techniques. The creativity and resourcefulness of analytical chemists in designing and applying new materials as sorbents is one of the most interesting and enduring features of μ-SPE.

3 Novel Developments, and Applications

In μ-SPE, in whatever format or configuration, the sorbent plays a critical role in the extraction/isolation, cleanup, pre-concentration of, and potential selectivity for, analytes from complex samples. Additionally, since these techniques, by definition, use only a small amount of sorbent, the choice of the most suitable sorbent is a vital task to maximize extraction capacity, efficiency, and selectivity [48]. On these points, the sorbent should have certain characteristics for effective operation. For instance, it should be stable under different conditions. It should have efficient interactions with the analytes as well as possess acceptable selectivity [49, 50]. One issue in many μ-SPE techniques particularly the home-prepared and -assembled approaches concerns the potential variability and other characteristics of the sorbent. There are also considerations of dispersibility in the sample, "stickiness" to membrane and glass, batch-to-batch inconsistencies or discrepancies, etc. [45]. This is why in most μ-SPE studies, the physical characterization of the sorbents and their properties with respect to surface area, adsorption capacity, selectivity, stability, durability, recyclability and adsorption kinetics and mechanism, etc., are of paramount importance, even as analytical performance metrics normally earn most of the attention [51]. Concurrent with developing new sorbents, a great deal of effort has also been dedicated to improving the efficiency and repeatability of μ-SPE techniques. To achieve these goals, the implementation and use of automated operations are gaining some attention, even if they are not widespread as yet [52, 53]. Automation of course not only reduces operator and indeterminate errors but also improves efficiency and sample throughput. It represents a natural progression towards practical routine laboratory applications [53]. In the following sections, recent developments of applications of new sorbents in, and automated, μ-SPE are discussed. As should be clear by now, there are various modes of sorbent-based METs that are classified as μ-SPE. In the discussion that follows, we make no judgment on the μ-SPE designation as reported by the authors concerned; so long as the procedures are termed as "μ-SPE," we abide by the respective authors' descriptions.

3.1 Two-Dimensional Materials

Graphene (Gr) and its derivatives (graphene oxide (GO) and reduced graphene oxide (rGO)) are materials with two-dimensional structures that have been widely used in different fields. The latter can not only be attributed to the simple synthesis and modification of these materials but also due to their useful properties [54, 55]. Two-dimensional materials (TDMs) have advantages like high porosity, surface area, adsorption capacity, and ease of functionalization. In addition, they show low toxicity, excellent biocompatibility, high physical, chemical, and thermal stability, high elasticity and flexibility, and high resistance and thermal conductivity [56]. Hence, given these properties, TDMs have been widely applied in μ-SPE. Notwithstanding its suitable features at first sight as a sorbent, Gr does have some shortcomings. These include defects and inter-sheet junctions which can reduce its surface area, and subsequently, its adsorption capacity.

The strong π-π interactions between the Gr layers can produce aggregation and consequently limit the utilization of native Gr as a sorbent. Balanced against this, however, is Gr's capability for facile functionalization and modification. For example, Sotolongo and co-workers constructed a three-dimensional (3D) Gr-nickel (Ni) foam functionalized with an ionic liquid (IL) as sorbent for online μ-SPE of mercury species in water samples with atomic fluorescence spectrometric detection [57]. The main reasons for using ILs are based on their properties of air and moisture stability, good thermal stability, very low volatility, relatively favourable viscosity and miscibility with water and organic solvents. Also, IL has good extracting ability that can be incorporated into new sorbents with wide applicability, as demonstrated by these authors [57]. The functionalization of IL onto 3D Gr-Ni foam was through van der Waals forces and π-π interactions. The authors claimed that the foam was free of defects and inter-sheet junctions. Another key advantage imparted by the foam was that agglomeration of the final sorbent was much reduced. A sensitivity enhancement factor for inorganic mercury was 180. The method also presented a limit of detection (LOD) of 3.6 ng L^{-1} with relative standard deviation (RSD) of 4.1%.

Similarly, Yuan and colleagues constructed a composition of Gr/multi-walled carbon nanotubes (Gr/MWCNTs) as a sorbent in the PT-SPE of 17β-estradiol in milk products prior to high-performance liquid chromatography-fluorescence detection (HPLC-FLD) [58]. The combination of Gr and MWCNT not only overcame the aggregation of Gr but also addressed the disordered structures of CNTs provided by the strong interactions between their layers. The adsorption experiments were carried out using only 1.0 mg of Gr/MWCNTs, showing the high applicability of the composite for the adsorption of the 17β-estradiol. Moreover, the sorbent produced by the combination of the two materials may also have the potential capability for extracting compounds with conjugated systems (such as aromatic rings) through π-π and hydrophobic interactions. This type of work has not been demonstrated as yet, however. The method had a linear range of 5–250 ng mL^{-1} with LOD and a limit of quantification (LOQ) of 0.7 and 2.3 ng mL^{-1}, respectively. The precision of the method was investigated via repeatability and intermediate precision. Applying this method, 17β-estradiol was detected in three out of eight real milk samples. This result confirmed the accuracy and applicability of the proposed method.

In another interesting study, novel semi-automated syringe infusion-pump-assisted Gr nanosheets (GNSs)-based PT-μ-SPE was developed [59]. The procedure was applied as an eco-friendly technique for the HPLC-ultraviolet (HPLC–UV) analysis of triclosan (TCS) as an emerging environmental pollutant in water. The GNSs were fabricated via a microwave-assisted method and packed into a 100-μL PT that was connected with a commercial plastic syringe (containing the water sample). The assembly was attached to a programmable auto-syringe infusion pump for the GNSs-PT-μ-SPE process. The GNSs have multidimensional structures and could provide high adsorption capacity due to their π-π interactions with the TCS. The PT-μ-SPE procedure was described as simple, low-cost, and efficient. Another claimed advantage of the work was the high reusability of the packed PT (twenty extraction cycles). Although the authors asserted that the GNSs had higher physical and chemical stability when compared to commercially available sorbent materials, they did not provide any supporting experimental data. The selectivity of the GNSs for TCS was also not investigated. Under the optimized conditions, the method showed a linear range of 2–250 ng mL^{-1} and an LOD of 0.5 ng mL^{-1}. Use of the method for determination of TCS in river and lake waters resulted in relative recoveries of TCS of between 94.6 and 102.4% with an RSD < 7.8%.

In another application of Gr-based materials, Seidi et al. prepared polyamide-GO-polypyrrole by electrospinning and used the nanofibres for the SC-SPE of parabens in milk samples, followed by HPLC–UV analysis [60]. The combination of GO and polypyrrole in the polymeric network of polyamide improved the extraction efficiency of the electrospun sorbent. This is because the composition of these materials can provide the possibility of various interactions with the target analytes such as hydrogen bonding, hydrophobic and π-π stacking. The method was claimed to address the drawbacks of conventional SPE in terms of labour-intensive steps, consumption of hazardous organic solvents, and high prices of commercial sorbents. The material could be used for seven times. The maximum adsorption capacities of the sorbent for methyl paraben, ethyl paraben, and propyl paraben were 1.827, 2.570, and 3.023 mg g^{-1}, respectively. It would seem possible that the composite can be modified with other materials to increase its adsorption capacities for the analytes in question. The method represented a linear range of 10–1000, 15–1000, and 20–1000 ng mL^{-1} for methyl paraben, ethyl paraben, and propyl paraben, respectively. The LOD values were < 7.0 ng mL^{-1}. Intra- and inter-assay RSDs were < 8.6% and 5.8%, respectively, which makes the method a good candidate to analyse target analytes in complex matrices.

In a DPX application that did not make use of a vendor-supplied sorbent, Tan and Lee prepared graphitic carbon nitride (g-C_3N_4) as a sorbent suitable for the extraction of polychlorinated biphenyls (PCBs) [61]. The emulsification-enhanced DPX of PCBs in environmental waters, followed by GC-mass spectrometry (GC–MS) analysis was fully automated. One key advantage of the work was that the synthesis of g-C_3N_4 was direct, solventless, and inexpensive. Moreover, g-C_3N_4 with a two-dimensional structure showed high dispersibility in the sample/solvent and strong affinity for PCBs. The large number of nitrogen functionalities in g-C_3N_4 not only improved its dispersibility but also increased its active surface area. Moreover, since g-C_3N_4 has π-layers in its structure, they could form strong π-electron analyte-extractant interactions with the PCBs and therefore, increased the extraction of the analytes significantly. Coupled with the pre-emulsification step, g-C_3N_4 could extract the analytes within 20 s of gentle turbulence within the DPX device. Besides the wide linear range, the method also exhibited reasonable enrichment factors of between 34 and 57 and RSDs of $\leq 8.95\%$ and $\leq 12.6\%$ for intra- and inter-day precision, respectively. Another key property of the method was its good resilience against matrix interferences. Some applications of TDMs and their composites in different modes of μ-SPE are presented in Table 1.

3.2 *Silica-Based Materials*

Silica is a versatile material that has attracted a lot of attention of analytical chemists in recent years. Silica not only has high stability, surface area and adsorption capacity, but is also easily modifiable. Considering these attributes, silica-based materials have been broadly used as extractant phases and of course, as the base material in HPLC stationary phases for decades [78, 79]. Given the experience gained from its use in stationary phases, the application of silica-based sorbents in μ-SPE techniques was a natural progression.

There is no doubt that silica-based octadecyl (C_{18}) materials are excellent sorbents. In a study conducted by Teo et al., an MB-μ-SPE method followed by LC-isotope dilution MS analysis was used for the fast and accurate analysis of carbamazepine in surface water [80]. The μ-SPE device (~0.8 × 1 cm) was constructed by folding and heat-sealing the edges of two overlapped polypropylene membrane sheets that housed the sorbent, C_{18}-modified silica (Fig. 1).

In the extraction procedure, the analytes (both carbamazepine and isotope-labelled carbamazepine) were first extracted by the μ-SPE device in the sample (10 mL) via agitation, then desorbed by an organic solvent (1 mL) via ultra-sonication. The μ-SPE device was easily fabricated (around 5–6 devices could be fabricated in only 30 min), and was much cheaper than commercially available SPE cartridges. The method in this work was faster and more time-efficient compared to that using cartridge-based SPE methods. The protection offered by the membrane meant that the method did not require sample pre-treatment, i.e., filtration of solid particles and sediments that are usually present in environmental waters. In addition, only 10 mL of water sample was needed compared to SPE which generally required 200–1000 mL. The μ-SPE procedure needed very little organic eluent (1 mL) making it more environmentally friendly. This work, making use of an easily accessible sorbent that can be purchased off-the-shelf, demonstrated that even for newly developed METs, well-established materials can still provide good extraction performance, and it is not always necessary to come up with new sorbents for existing analytes. The LOD and LOQ values of the method for carbamazepine were 0.5 ng L^{-1} and 1.6 ng L^{-1}, respectively. The RSD value of 0.7% indicated the high precision of the method.

In another study leveraging on the use of silica, poly(glycidoxypropylmethyl-co-dimethylsiloxane) (PGDMS) was thermally immobilized on the material (Si(PGDMS)) and used as a selective DPX sorbent for carbendazim residues in

Table 1 Some recent applications of TDMs and their composites in different modes of μ-SPE

Material	Extraction mode	Analyte(s)	Real sample	Linear range	LOD	LOQ	Detection	Ref.
EPETNGONPs[a]	μ-SPE	Tetracycline	Honey	50–5000 μg kg^{-1}	15.3 μg kg^{-1}	47.1 μg kg^{-1}	HPLC–UV	[62]
		Cefotaxime		10–5000 μg kg^{-1}	3.0 μg kg^{-1}	10 μg kg^{-1}		
GO-starch-based nanocomposite	μ-SPE	Antibiotic residues	Cow's milk	3–1000 μg kg^{-1}	0.8–1.5 μg kg^{-1}	2.7–5.0 μg kg^{-1}	HPLC–UV	[63]
DESFGO[b]	PT-SPE	Toluene and xylene exposure biomarkers	Urine	0.200–200 mg mL^{-1}	1.66–2.89 ng mL^{-1}	5.52–9.63 ng mL^{-1}	HPLC–UV	[64]
DESMGr[c]	PT-SPE	Sulfamerazine	River water	–	0.01 mg mL^{-1}	0.03 mg mL^{-1}	HPLC–UV	[65]
PEMPAFSCGO[d]	PT-SPE	Phenolic acids	Chinese Wolfberry drink	0.1–400.0 mg mL^{-1}	–	–	HPLC–UV	[66]
GO/polypyrrole foam	PT-SPE	Sulfonamide residues	Honey and milk	0.01–10.00 mg mL^{-1}	1.04–1.50 ng mL^{-1}	3.48–5.05 ng mL^{-1}	HPLC–UV	[67]
3D-IL-Fe_3O_4-GO[e]	PT-SPE	PAHs[f]	Human blood	0.007–0.013 mg L^{-1}	0.002–0.004 mg L^{-1}	0.007–0.013 mg L^{-1}	GC–MS	[68]
IL-Gr	PT-SPE	Auxins	Soybean	0.03–5.00 μg g^{-1}	0.004–0.026 μ g^{-1}	0.013–0.086 μ g^{-1}	HPLC–UV	[69]
GO/zinc oxide nanocomposite	PT-SPE	Rhodamine B	Seawater	5–250 μg L^{-1}	1.0 μg L^{-1}	–	Spectrophotometry	[70]
		Malachite green			1.2 μg L^{-1}			
GO	DPX	Herbicides	Sugarcane-derived foods	–	–	1.0–25.0 ng g^{-1}	LC–MS/MS[g]	[71]

(continued)

Table 1 (continued)

Material	Extraction mode	Analyte(s)	Real sample	Linear range	LOD	LOQ	Detection	Ref.
g-C_3N_4	DPX	Organochlorine pesticides	Lake and river water	10–100,000 ng L^{-1}	2.4–46.2 ng L^{-1}	7.5–163 ng L^{-1}	GC–MS	[72]
GNSs	MEPS	Pesticides	Coffee	–	–	–	GC–MS/MS[h]	[73]
GO reinforced polyamide nanocomposite	MEPS	OPPs[i]	Environmental aqueous samples	1–1000 μg L^{-1}	0.2–1 μg L^{-1}	–	GC-FID[j]	[74]
rGO	MEPS	Local anesthetics	Human plasma and saliva	5–2000 nmol L^{-1}	2–4 nmol L^{-1}	5–2000 nmol L^{-1}	LC–MS/MS	[75]
Polyacrylonitrile/ GO nanofibres	MEPS	Drugs and their metabolites	Human plasma	2–2000 nmol L^{-1}	0.25–2.50 nmol L^{-1}	2–10 nmol L^{-1}	LC–MS/MS	[76]
rGO-melamine formaldehyde	NTD	VOCs	Water	1–400 μg L^{-1}	0.2–1.5 μg L^{-1}	0.8–5.0 μg L^{-1}	GC-FID	[77]

[a]Electrospun polyethylene terephthalate nanofibres doped with graphene oxide nanoparticles
[b]Deep-eutectic-solvent-functionalized graphene oxide
[c]Deep-eutectic-solvent-modified graphene
[d]Polyethyleneimine modified porous aromatic framework and silane coupling agent grafted graphene oxide
[e]Three-dimensional-ionic liquid-magnetic graphene oxide
[f]Polycyclic aromatic hydrocarbons
[g]Liquid chromatography atmospheric-pressure chemical ionization tandem mass spectrometry
[h]Gas chromatography tandem mass spectrometry
[i]Organophosphorus pesticides
[j]Gas chromatography-flame ionization detection

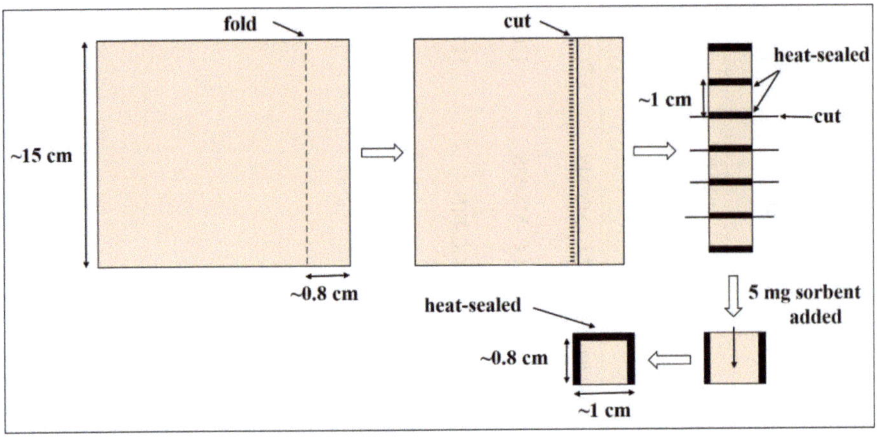

Fig. 1 Typical fabrication of μ-SPE device (not drawn to scale) (Modified from ref. [80] with permission from Elsevier)

orange juice. Analysis was by HPLC-diode array detection (DAD) [81]. The preparation of the Si(PGDMS) was simple, and therefore accessible to most laboratories, and did not require refined and expensive reagents. Another important property of the sorbent is that the glycidoxy group in the monomeric structure of PGDMS could provide more selective interactions with the carbendazim. Comparing some other methods used for the determination of carbendazim in orange juice, the presented method needed less sorbent, smaller sample volume and lower eluent volume. The sample preparation process itself was faster (<3 min) without the need of any cleanup step or equipment. Also, only one extraction cycle was enough for adequate retention of the carbendazim in the sorbent. The DPX method provided high sample throughput (ca. 20 samples per hour) [81]. The method showed recoveries of 93–110% with RSD < 16%.

Still on the use of silica, Dalvand and Ghiasvand synthesized a polyaniline silica (Si/PANI) organic–inorganic nanocomposite by a combination of electrospinning and in situ polymerization processes [33]. This approach prevented the aggregation of PANI during the polymerization. It also provided a higher synthetic yield and more uniformity of the composite. Si/PANI was then packed inside a stainless-steel needle to fabricate an NTD. The NTD was used for the headspace extraction and GC-FID analysis of PAHs, and benzene, toluene, ethylbenzene, and xylenes (BTEX) in polluted soil samples. The main advantage of the method was its capability for directly analyzing complex solid samples in a reasonable time and temperature, without significant matrix effects. Wide linear ranges (1–2000 ng g^{-1} for BTEX and 0.2–2000 ng g^{-1} for PAHs) and also LODs of 0.02–0.1 ng g^{-1} for BTEX and 0.001–0.01 ng g^{-1} for PAHs showed the suitability of the method for extracting and determining these analytes from contaminated soil samples.

As mentioned earlier, the MEPS technique is capable of being fully automated. This important feature was demonstrated by Khesina et al. [82]. The authors considered parabens in cosmetics which were analysed by HPLC–UV. In the work, the MEPS device was based on a digital analytical 50-μL syringe packed with 4 mg of C_{18} sorbent. The technique was fast, needed only minimal consumption of organic solvents and permitted processing of the sample from interfering components. The effect of the extraction factors was investigated and optimized by Plackett–Burman design. The true selectivity of the sorbent for other analytes was not studied in this work, although based on hundreds, if not thousands of literature reports, C_{18} is known to be a versatile sorbent, and should be suitable for extracting many other compound classes. The method presented LODs of 2–5 ng mL^{-1} and extraction recoveries of 89–105% for parabens in cosmetic samples.

Hakme and Poulsen also reported what they described as an automated μ-SPE clean-up system for the GC–MS analysis of pesticide residues in cereals [52]. Once again, C_{18} was applied as the sorbent. The μ-SPE step was conducted using a commercial robotic autosampler system, after an initial QuEChERS (quick, easy, cheap, effective, rugged and safe) cleanup. Only the μ-SPE step was automated. Overall, nevertheless, the method showed a high capacity for removing matrix-interfering components and in the recovery of pesticides. The authors suggested that some additional accessories such as a thermostatic autosampler, a larger size tray, and an automatic decapping and capping system would make the system even more functional and powerful. Some other applications of silica-based materials in various μ-SPE are listed in Table 2.

3.3 Layered Double Hydroxides

Layered double hydroxides (LDHs) are two-dimensional materials with high negative charges which can act as ligands [89]. LDHs have high surface area, thermal stability, tunability and flexibility in their interlayer spaces. They can readily undergo ion exchange reactions and are also capable of being easily modified [90, 91]. Due to these properties separately or in combination, LDHs have been successfully applied as sorbents in sample preparation methods.

In one report, Tang and Lee fabricated a dissolvable magnesium–aluminium LDH material and used it as a sorbent for the DSPE of aromatic acid anions [92]. The extraction of the analytes was based on a coprecipitation method. The method relied on dissolving the original LDH (the "sorbent") at pH < 4 and then reforming a new LDH after coprecipitation reaction with the analytes by adjustment of the pH. The analytes were coprecipitated and intercalated into the interlayers of the DHs. By manipulating the pH, the new LDH was dissolved, releasing the analytes into solution. Applying a dissolvable LDH enabled the removal of some of the discrete physical steps. These are normally associated with conventional DSPE (filtration, centrifugation or magnetic separation, and particularly analyte elution using a solvent). The potential for a fully

Table 2 Applications of silica-based materials in various μ-SPE techniques

Material	Extraction mode	Analyte(s)	Real sample	Linear range	LOD	LOQ	Detection	Ref.
Gr/silica nanospheres	PT-SPE	Antiviral drugs	Yellow catfish (*Pelteobagrus fulvidraco)*	–	≤0.8 ng mL^{-1}	–	UHPLC-MS/MS[a]	[83]
C_{18}	PT-SPE	Propofol	Human plasma	0.005–5 μg mL^{-1}	–	–	LC-APCI-MS/MS[b]	[84]
OFHSM[c]	PT-SPE	Cannabidiol and tetrahydrocanna-binol	Plasma	10–50 ng mL^{-1}	–	10 ng mL^{-1}	UHPLC-MS/MS	[85]
zIL@silica[d]	MEPS	PAHs	Brewed coffee	5–100 μg L^{-1}	1.0–1.5 μg L^{-1}	–	GC–MS	[86]
ASGONCC[e]	NTD	PAHs	Contaminated soil samples	0.001–2.0 μg g^{-1}	5–38 pg g^{-1}	–	GC–FID	[87]
Carbotrap/silica	NTD	HVOCs[f]	Air	0.01–100 ng mL^{-1}	0.01–0.03 ng mL^{-1}	0.05–0.09 ng mL^{-1}	GC–MS	[88]

[a]Ultra-high-performance liquid chromatography-tandem mass spectrometry
[b]Liquid chromatography atmospheric-pressure chemical ionization tandem mass spectrometry analysis
[c]Octyl-functionalized hybrid silica monolith
[d]Zwitterionic liquid-modified silica
[e]Aminosilica/graphene oxide nanocomposites covalently attached to cotton
[f]Halogenated volatile organic compounds

automated sample preparation procedure by making use of such a dissolvable sorbent was apparent in the work.

Thus, in a follow-up study, the authors constructed another dissolvable LDH, this time magnetic magnesium–aluminium LDH-iron oxide (Fe_3O_4) core–shell microspheres [93]. These were used for fully-automated DSPE of several pharmaceuticals and personal care products with HPLC–DAD determination. After extracting the analytes, and concomitantly forming a "new" LDH, the latter was immobilized by magnetic force, and isolated from the spent sample solution which was collected by a syringe. The LDH was then dissolved by acid to release the analytes. Therefore, due to the magnetic and dissolvable attributes of the sorbent, the extraction process was fully automatized. The procedure afforded manual labour-free convenience and seamless integration with HPLC analysis. Although due to its dissolvability, the sorbent could not be reused, only 0.1 mg of it was used each time. Besides a small amount of sorbent, only small volumes of both the sample (1 mL) and final solvent (10 μL) were needed. Thus, the extraction was claimed to be not only simple and effective but also economical. In addition to the low LODs (0.021–0.042 μg L^{-1}), the method also showed good repeatability (RSDs < 4.8%).

Exploiting the ease of modifying and functionalizing LDHs, Manouchehri et al. coated a magnesium–aluminum LDH on partially rGO NSs and used the composite for a home-made MEPS device [94]. MEPS of parabens in human breast milk followed by HPLC–UV determination was then conducted. The surface oxygen groups of the GO can be ionized in aqueous media. Therefore, the layers of GO can be negatively charged and can provide sites for anchoring the appropriate species on the GO surface. Hence, in this work, the GO/LDH composite had synergetic effects for extracting analytes. The maximum extraction recovery (87.2–104.4%) was obtained at a pH of 6.0 which can be related to the increase in the hydrophobicity of the parabens and the strong interaction between the neutral parabens and the sorbent through π-π stacking, and hydrogen bonding. Another mechanism for extracting analytes was indicated to be π-cation interaction due to the positive charge of the composite surface in acidic media and the electron cloud of the aromatic rings in the parabens. Although the work was aimed at improving the adsorption capacity by fabrication of a composite, the relevant experiments were not conducted, and the capacity was not calculated. The adsorption capacity was only surmised from the experimental results. Notwithstanding, a major feature of the material was its impressive reusability (25 cycles). The relative recoveries ranged from 87.2 to 104.4% for breast milk, with RSDs of between 4.2 and 9.5%.

GO was also composited with a zinc (II)-chromium (III) LDH and used as an efficient sorbent for the PT-SPE of lead (II) in hair samples followed with analysis by graphite furnace atomic absorption spectrometry (GFAAS) [95]. For decreasing possible interactions between lead (II) and the sample matrix as well as preparing solid samples for the PT-SPE procedure, the hair samples were first treated using acid digestion. The adsorption capacity of the composite was found to be 16.92 mg g^{-1}. The possible interaction mechanisms between the lead (II) and LDH were given as follows: (i) Interaction of LDH with lead (II) via the former's surface hydroxyl groups, (iii) isomorphic substitution, and (iii) the precipitation of metal hydroxides.

Investigating coexisting ions such as cobalt (II), cadmium (II), nickel (II), zinc (II) and manganese (II) indicated that the material had a higher selectivity for lead (II). Although copper (II), chromium (VI), and iron (III) showed similar characteristics as lead (II), they did not disrupt the adsorption of the latter. Thus, lead (II) measurement due to the high adsorption capacity of the composite and the inherent selectivity of AAS was deemed to be reliable. The main reasons for the selectivity of the composite towards lead (II) can be attributed to the (i) Lewis hard-soft acid–base principle, (ii) ionic radius (the isomorphic substitution caused by replacement of metals is related to their ionic radius and the ions must have approximately the same size and total ionic charge as those being replaced) [96–98], (iii) different hydration degrees of the cations and thus hydrated radius, which are affected by the sample pH, (iv) the composition of reagents used for the synthesis of LDH, and (v) electronegativity effects [95]. Since this LDH was not meant to be dissolvable like most conventional sorbents, it would be advantageous for it to be able to be reused, so recyclability would be a desirable property. However, despite its other useful properties, reusability of the LDH was not assessed in the work. If it turns out that this value is not appreciable, then this deficiency may limit its practical usefulness. The method showed a linear range of 0.5–15 ng mL^{-1} with LOD and LOQ of 0.1 $\mu g\ g^{-1}$ and 0.5 $\mu g\ g^{-1}$, respectively. Also, the obtained recoveries were from 92 to 104% with RSDs < 12.5%. More examples in applications of LDH-based materials in μ-SPE are given in Table 3.

3.4 Metal–Organic Frameworks

Metal–organic frameworks (MOFs) have been receiving a lot of attention due to their gas storage, separation and delivery capabilities. MOFs are already in commercial use in several industries (semiconductors, materials and power generation, etc.). These materials which are formed based on the coordination reactions between metal ions/clusters and organic ligands have a high surface area and adsorption capacity [103, 104]. Their other properties include structural diversity, pore structure uniformity, tunable porosity, extensive variety, and flexibility in network topology, geometry, dimension, and chemical functionality [105, 106]. Hence, it is not surprising that MOFs and their composites have enthused analytical chemists, especially those interested in sample preparation, to explore their applicability as sorbents.

As an example of using a MOF as a sorbent, Amini and co-workers synthesized electrospun polyacrylonitrile (PAN)/zinc-MOF-74@GO nanocomposite for online μ-SPE of chlorobenzenes in water, soil and food samples prior to HPLC–DAD determination [107]. Since GO has a large number of hydroxyl, carboxyl, and epoxide groups, its incorporation into the MOF network can increase the porosity, stability, and extraction efficiency of the MOF. Also, zinc-MOF-74@GO nanoparticles can be emplaced in the PAN nanostructure to produce a new composite. The PAN/zinc-MOF-74@GO composite can interact with analytes via hydrogen bonding and π-π stacking interactions. Online μ-SPE addresses some drawbacks of conventional

Table 3 Applications of LDH-based materials in various μ-SPE techniques

Material	Extraction mode	Analyte(s)	Real sample	Linear range	LOD	LOQ	Detection	Ref.
Nickel–iron LDH	μ-SPE	Nonsteroidal anti-inflammatory drugs	Human urine	10.0–1000 ng mL^{-1}	1.0–10.0 ng mL^{-1}	–	HPLC–UV	[99]
Dissolvable LDH	Dispersive-μ-SPE	Sulfonylurea herbicides	Wolfberry	–	0.01–0.5 ng g^{-1}	0.1–2.0 ng g^{-1}	LC–MS/MS	[100]
Immobilized LDH onto cotton fibre	PT-SPE	Fluoroquinolone drugs	Chicken eggs and pork liver	0.03–25 μg L^{-1}	0.01–0.02 μg L^{-1}	0.03–0.05 μg L^{-1}	HPLC-FLD	[101]
Electrospun nickel (II)-iron (III) LDH/Nylon 6 composite	MEPS	Naproxen	Whole blood	75–2000 ng mL^{-1}	20 ng mL^{-1}	75 ng mL^{-1}	HPLC–UV	[102]
		Diclofenac		75–2000 ng mL^{-1}	20 ng mL^{-1}	75 ng mL^{-1}		
		Mefenamic acid		50–2000 ng mL^{-1}	15 ng mL^{-1}	50 ng mL^{-1}		

METs in terms of time consumption and low repeatability. This is because it integrates all the steps of injection, extraction, and desorption which can subsequently reduce human errors, avoid tedious labour, and enhance reliability. The PAN/zinc-MOF-74@GO composite also had acceptable chemical stability and a relatively long lifetime (60 extraction cycles) [107]. The extraction procedure provided a linear range of 0.25–700.00 ng mL^{-1} and LODs of 0.08–1.10 ng mL^{-1} for chlorobenzenes.

In another study on these interesting class of materials, the MOF, MFU-4l, was used for SC-SPE of chlorophenols before GC–MS analysis [39]. The MOF has different amino groups in its linker structure that supposedly can improve extraction performance. The authors claimed that the sorbent had appropriate selectivity towards the target analytes. However, they did not investigate the selectivity of the sorbent towards different analytes in terms of adsorption capacities, etc. Also, as already mentioned above (Sect. 2), the SC-SPE method has one limitation: The adsorption and desorption steps both contained cycles during which the sample or eluent tubes had to be physically collected and reloaded onto the centrifuge. The method presented a linear range of 0.5–400 μg kg^{-1} for water samples, and 1.0–400 μg kg^{-1} for soil samples. Also, preconcentration factors were between 26.3 and 29.6 for aqueous samples.

MOFs have also been applied to MB-μ-SPE. One such study using zirconium (Zr)-based MOF, UiO-66(Zr), has been reported [108]. The UiO-66(Zr) sorbent was used for extracting the hormones, androgens and progestogens, in environmental water samples with LC-tandem MS (MS/MS) analysis [108]. Experimental and molecular simulation strategies were applied to evaluate the adsorption capability of the finally-selected water-stable MOF. A combination of computer simulations and experiments was conducted to select the appropriate sorbent amongst several (chromium-based MOF (MIL-101(Cr)), iron-based MOF (MIL-100(Fe)), aluminium-based MOF (MIL-53(Al)) and UiO-66(Zr)) that were prepared and to explain the possible extraction mechanisms. To this end, AutoDock 4 (a program primarily designed to predict how a small molecule binds to a 3D receptor) was used to calculate the possible interactions between the MOF and analytes. The results indicated that the main interaction mechanisms were hydrophobic effects between the phenyl rings in the MOF and the steroid ring systems in the hormones. Other possible mechanisms were intermolecular hydrogen bonds between the –COOH of the MOF and the –OH of testosterone, as well as π-σ interactions between the phenyl rings in MOF and the saturated carbon chains in the hormones. Another useful property of the UiO-66(Zr) was its reusability of over at least 10 cycles [108].

NTD extraction has also been applied that used a zinc-based MOF as sorbent [109]. The analytes were PAHs in air. The MOF was fabricated via an electrochemical method and then packed inside a 22-gauge needle to serve as the NTD. The preparative procedure was conducted in water and was based on the concurrent steps of synthesis and deposition of the sorbent on an electrode. The procedure was indicated to have key advantages such as green synthesis conducted at ambient temperature and pressure, short synthesis time (300 s), non-requirement of organic solvent, and high purity of the sorbent. Moreover, the storage capability of NTD was assessed at 4 °C, and according to the results, there was no significant reduction in the amounts of analytes extracted even after 60 days [109]. This means the sorbent can potentially function as part of a field-deployable extraction system that can be employed at remote locations where immediate analysis is not possible. The LOD and LOQ of the procedure were within the ranges of 0.011–0.021 and 0.03–0.07 mg m^3, respectively. Moreover, repeatability and reproducibility of the method were estimated in the ranges of 3.6–9.9% and 5.3–24.1%, respectively.

The MB-μ-SPE technique can be in a different format from the original configuration (i.e., sorbent held within a membrane bag [48]). Zhou et al. have developed a syringe-based MB-μ-SPE method based on MOF mixed-matrix membranes for the extraction and pre-concentration of five PAHs in tea infusion prior to HPLC-FLD [110]. Bare MOFs are often difficult to be retrieve and reused when used in extraction applications. Hence, application of mixed-matrix membranes incorporating MOFs can address this problem. The presence of MOFs dispersed on the membrane can provide some synergistic advantages of MOFs and membranes. This includes the increase in the specific surface area and improved hydrophilicity and fouling resistance. In the work mentioned, MOF nanoparticles were anchored on the surface of polyvinylidene difluoride membrane which was then inserted between two syringe filters and tightened into place at the syringe outlet. (This format appears to be similar to that of membrane disk-SPE (Empore) that was commercialized in the 1980s by the 3M Company.) The extraction procedure of PAHs by this MOF mixed-matrix membrane is shown in Fig. 2.

The post-synthesis fabrication of the composite was claimed to be simple so that task-specific and highly selective membranes could be easily enabled. Four different membranes (UiO-66(zirconium), MIL-53(aluminum), MIL-101(iron) and NH_2-MIL-101(iron)) were prepared, with the UiO-66(zirconium) membrane giving the best performance (extraction recovery of 74.4%). The UiO-66(zirconium) composite had a highest specific surface area (320.5 m^2 g^{-1}) compared to other MOFs. The composite-on-membrane could be applied for four cycles. Beyond this, however, as it is worth mentioning, the edge of the membrane began to degrade, which might

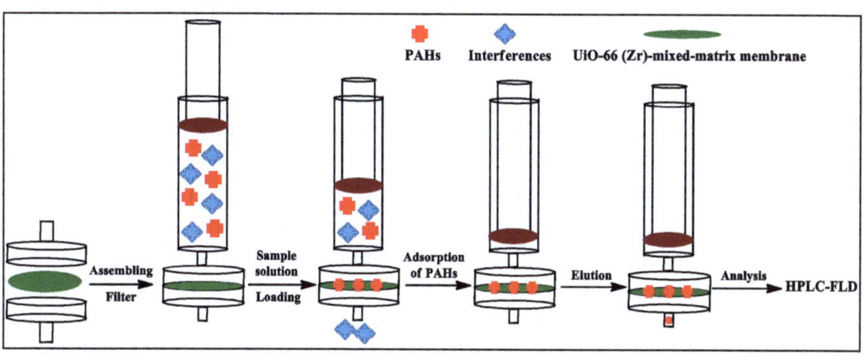

Fig. 2 Extraction procedure of polycyclic aromatic hydrocarbons by metal–organic framework mixed-matrix membrane (Modified from ref. [110] with permission from Elsevier)

be due to the friction between the membrane and the filters because of tightening actions. The combination of the syringe MB-μ-SPE with HPLC-FLD represented a relatively simple, and convenient method which eliminated the need for vortex-mixing or ultrasonication. The use of sorbent particles immobilized on membranes means that there is an alternative to the inconvenience of using conventional SPE where cartridges need to be packed or DSPE where the sorbent needs to be separated from the sample after extraction. Another main advantage of the work is that due to the easy manipulation and processability of MOF-MB-μ-SPE, it can potentially be developed into an automated approach. Under the most favourable conditions, the method provided LOD values as low as 0.02–0.08 μg L^{-1} with extraction recoveries of between 85.5 and 102.1%, and inter-day and intra-day precision < 8.4%, for the PAHs in tea infusions. Table 4 shows other examples of using MOF-based materials in μ-SPE.

3.5 Covalent-Organic Frameworks

Another group of emerging materials that have attracted a lot of interest are covalent-organic frameworks (COFs). COFs are fabricated by connecting covalent bonds between different building blocks [124, 125]. Due to these covalent bonds, these materials possess higher stability compared to MOFs. Like MOFs, COFs have porous structures with ordered channels, easily modifiable capability, large surface area and large adsorption capacity [126–128], all attributes that make them amenable as sorbents in extraction procedures.

Like several other types of materials already discussed, COF composite nanofibres have been used in PT-SPE. These were constructed via electrospinning and used for the extraction of tetracycline antibiotics in grass carp and duck prior to HPLC–DAD [129]. The COF was fabricated at ambient temperature. Moreover, the fibres showed high thermal (300 °C) as well as chemical stability. The composition of COF with electrospun nanofibres both increased the adsorption capacity of the electrospun nanofibres and prevented the leakage and high pressure caused by the sole use of nanosized COFs in PT-SPE. The interactions of analytes and sorbent were mainly related to their structures at different pH values. At lower pH (3.0), with the analytes in their cationic forms, the sorbent adsorbed them via electrostatic, hydrogen bonding and π-π interactions. Also, at pH 4.0–7.0, the adsorption was via hydrophobic interaction because of the tetracylines being in their intermediate forms. At higher pH (7.0–11.0) with the analytes in their anionic forms, the sorbent could not adsorb them effectively. Hence, maximum extraction was obtained at a pH of 7.0 [129]. The linear range of the method was from 4 to 70 ng mL^{-1}. The LOD and LOQ ranged from 0.6 to 3 ng mL^{-1} and from 2 to 10 ng mL^{-1}, respectively.

In another study on COFs, magnetic COF (1,3,5-triformylbenzene (Tb) and benzidine (Bd) as building blocks) (Fe_3O_4@TbBd) was fabricated for MEPS [130]. The analytes were BTEX's chief biomarkers (trans, trans-muconic acid, mandelic acid,

Table 4 Applications of MOF-based materials in various μ-SPE techniques

Material	Extraction mode	Analyte(s)	Real sample	Linear range	LOD	LOQ	Detection	Ref.
Chromium-based MOF	PT-SPE	Methyl paraben	Wastewater and shampoo samples	1.0–200.0 μg L^{-1}	0.25 μg L^{-1}	0.83 μg L^{-1}	Spectrophotometry	[111]
		Propyl paraben			0.24 μg L^{-1}	0.80 μg L^{-1}		
ZIFFGOS[a]	PT-SPE	Chlorophenols	Environmental and food samples	20–2000 ng mL^{-1}	0.2–20 ng mL^{-1}	–	HPLC–UV	[112]
CMOF-74CA[b]	PT-SPE	Triazole fungicides	Fruits and vegetable	0.098–200.0 mg kg^{-1}	0.033–0.065 mg kg^{-1}	–	GC-FID	[113]
Boronic acid grafted MOF	PT-SPE	*cis*-Diol-containing compounds	Human urine	0.01–50 μg mL^{-1}	0.005–0.012 μg mL^{-1}	–	–	[114]
ECA/PTMgMOFNFs[c]	PT-SPE	Anti-cancer drugs	Biological fluids	0.1–1500.0 μg L^{-1}	0.03–0.10 μg L^{-1}	0.10–0.33 μg L^{-1}	HPLC–UV	[115]
MOF	MEPS	PAHs	Water	–	6.7–27 ng L^{-1}	–	–	[116]
MCMOFCAMS[d]	MEPS	Nitroimidazoles	Water	–	8.250–16.33 ng L^{-1}	–	UPLC-MS/MS	[117]
MIL-100(iron)	NTD	Airborne organochlorine pesticides	Air	–	0.04–0.41 μg m^{-3}	0.21–1.82 μg m^{-3}	–	[118]
Zirconium-based MOF	NTD	Amphetamine derivatives	Urine	0.5–40.0 ng mL^{-1}	0.06–0.09 ng mL^{-1}	0.5–0.8 ng mL^{-1}	GC-FID	[119]
3D nickel/cobalt-based MOF	NTD	NMBTEX[e]	Urine	–	0.01–0.04 ng mL^{-1}	0.2–1.1 ng mL^{-1}	GC-FID	[120]

(continued)

Table 4 (continued)

Material	Extraction mode	Analyte(s)	Real sample	Linear range	LOD	LOQ	Detection	Ref.
Hybrid monoliths with MOFs	SC-SPE	Non-steroidal drugs	Human urine	–	0.1–7 μg L^{-1}	–	RPHPLC-UV[f]	[121]
MOF mixed-matrix membrane	MB-μ-SPE	Bisphenol A	Milk and milk packaging	0.1–50 μg L^{-1}	16 μg L^{-1}	–	HPLC-FLD	[122]
UMCM-1BMHFMOFDESMIPs[g]	MB-μ-SPE	Phthalate esters	Yogurt, water and edible oil	–	0.008–0.03 μg L^{-1}	0.028–0.12 μg L^{-1}	GC-FID	[123]

[a]Zeolitic imidazolate framework-8/fluorinated graphene coated silica composites
[b]Carbonized MOF-74/carbon aerogel
[c]Electrospun cellulose acetate/polyacrylonitrile/thymol/Mg-metal organic framework nanofibres
[d]Monolithic and compressible MIL-101(chromium)/cellulose aerogel/melamine sponge
[e]Non-metabolized benzene, toluene, ethylbenzene and xylene isomer compounds
[f]Reversed-phase-high-performance liquid chromatography-ultraviolet
[g]UMCM-1 based monolithic and hollow fibre-metal–organic framework deep eutectic solvents/molecularly imprinted polymers

hippuric acid, and 3-methylhippuric acid) in urine which were determined by HPLC–UV [130]. The COF-packed MEPS technique was claimed to be rapid, easy, user-friendly and environment-friendly. It was conjectured that these properties as well as the satisfactory recoveries (70–87%) obtained could form the basis of a method to monitor workers' exposure to these BTEX biomarker compounds in an occupational setting. A linear range of between 0.05 and 300 μg mL^{-1} and LOD ranging from 0.02 to 0.5 μg mL^{-1} were obtained under the optimized conditions.

An NTD procedure has also been developed in which a composite of a melamine-based COF (Schiff base network (SNW-1)) and single-walled carbon nanotube (SWCNT) was synthesized and applied for the sampling of phenolic compounds in air [131]. Analysis was by GC-FID. Being portable and solventless, the SNW-1/SWCNT-NTD technique was evaluated in the field (a plastics factory) after being first assessed in the laboratory. The NTD was taken back to the laboratory for further processing. When compared with the US National Institute for Occupational Safety and Health (NIOSH)-2546 Method, the results of the extraction and analysis of the phenolic compounds showed a strong correlation (R^2 value of 0.9812) between the two methods. The LOD of 0.002–0.011 ng mL^{-1} and LOQ of 0.008–0.037 ng mL^{-1} showed the high sensitivity of the procedure. Table 5. lists applications of COF-based materials in μ-SPE.

3.6 *Chitosan- and Cellulose-Based Materials*

Despite the vibrant research activities on the employment of diverse sorbents in μ-SPE, most of the materials reported still face some important drawbacks. These limitations are in terms of protracted reaction duration, relatively expensive synthesis conditions, and consumption of moderate volumes of organic and other hazardous solvents and reagents, etc. In recent years, a lot of efforts have been devoted to the preparation of relatively easy-to-produce, inexpensive, abundant, and generally regarded as 'green' or environmentally friendly sorbents [137]. One main aspect of this trend is in regard of the design, synthesis/modification and application of biological or natural materials (and their composites). Chitosan or cellulose, and modified chitosan-/cellulose-based materials are widely considered as cost-effective and environmentally friendly sorbents with high biocompatibility and low or negligible toxicity.

Considering the properties of chitosan and molecularly-imprinted polymers as selective sorbents, Li and Row fabricated deep eutectic solvents (DESs) cross-linked with molecularly-imprinted chitosan microspheres (MICSM) for the μ-SPE of p-hydroxybenzoic acid (PHBA) from pear rind [138]. To investigate the molecular recognition ability of the MICSM and CSM, imprinting factor (IF) and selectivity factor (β) values were measured. Benzoic acid (BA), salicylic acid (SA), and phenol were used for comparison with PHBA. The IF values for PHBA, BA, SA, and phenol toward MICSM were found to be 9.42, 1.48, 2.64, and 4.12, respectively. Also, the β values of MICSM for PHBA, BA, SA, and phenol were calculated to be 1.14, 0.05,

Table 5 Applications of COF-based materials in μ-SPE (PT-SPE)

Material	Extraction mode	Analyte(s)	Real sample	Linear range	LOD	LOQ	Detection	Ref.
COFBECNFMs[a]	PT-SPE	Inorganic arsenic	Rice	–	0.015 μg L^{-1}	–	HG-AFLS[b]	[132]
ECOF/PNF[c]	PT-SPE	Plant growth regulators	Food samples	5–500 ng mL^{-1}	0.24–3.19 ng mL^{-1}	1.65–5.72 ng mL^{-1}	HPLC–DAD	[133]
COF-MOF	PT-SPE	Sulfonamides	Environmental water, milk and meat samples	10–2000 ng mL^{-1}	1–10 ng mL^{-1}	–	HPLC-VWD[d]	[134]
COF Schiff base network-1	PT-SPE	Sulfonamides	Milk and honey	5–500 ng mL^{-1}	013–0.25 ng mL^{-1}	0.44–0.82 ng mL^{-1}	HPLC–DAD	[135]
COFIEN[e]	PT-SPE	Sulfonamides	Meat	5–125 ng mL^{-1}	1.7–2.7 ng mL^{-1}	–	HPLC–DAD	[136]

[a] COF-based electrospun composite nanofibre membranes
[b] Hydride generation-atomic fluorescence spectrometry
[c] Electrospun covalent organic framework/polyacrylonitrile nanofibre
[d] High-performance liquid chromatography-variable wavelength detection/>/>
[e] COF incorporated electrospun nanofibre

0.18 and 0.3, respectively. It was clear that the sorbent exhibited high selectivity for PHBA. The differences in IF values can be ascribed to the type of functional group and steric effects that affect the generation of suitable hydrogen bonds involved in the imprinting process. Besides the high selectivity of the sorbent, another main feature of the work was the use of DESs as alternatives to conventional organic solvents and the synthesis of the chitosan microspheres from natural, sustainable materials. However, the role of the chitosan in the extraction was not clarified in the work, only that its presence apparently enhanced the selectivity of the sorbent towards PHBA. The obtained recoveries ranged from 87.86 to 103.65% for PHBA with RSDs $\leq$ 4.32%.

In PT-SPE, the use of small particles can cause high backpressure that is unfavourable to the vacuum pump used for suction during the extraction process. One way to address this shortcoming is to use electrospun nanofibres which, apart from their other useful sorbent properties like high surface area and stability, are highly porous. In an example of the application of an electrospun sorbent, poly(vinyl alcohol)-(PVA)-poly(acrylic acid) (PAA)/CNT-cellulose nanocrystal (CNC) composite nanofibres were constructed and used for the PT-SPE of seven opioid analgesic drugs in biological samples with HPLC–UV analysis [139]. Despite the outstanding reusability (110 times) of the composite in standard analyte solutions, in plasma samples, reusability was more modest (20 times). The main interactions between the composite and analytes were indicated to be hydrogen bonding, and hydrophobic and π-π interactions. However, no specific experiments were conducted to provide supporting evidence of this claim. The authors also stated that, due to the porous structure of the composite with aromatic rings and many hydroxyl groups, it can also extract other compounds like dopamine, uric acid, herbicides, fungicides, etc., through the above-mentioned hydrogen bonding, hydrophobic, and π-π interactions. Considering that the possible interactions between sorbent and analytes can potentially indicate the universality of the sorbent, the authors' claims, however, were not definitively demonstrated or supported experimentally. Under the optimized conditions, the linear range was from 1.5 to 700.0 ng mL^{-1} for morphine, codeine, oxycodone, and tramadol, and 0.5 to 1000.0 ng mL^{-1} for nalbuphine, thebaine, and noscapine. Also, the method had LODs of 0.15–0.50 ng mL^{-1} and recoveries of between 87.3 and 97.8%.

In yet another PT-SPE application, chitosan-modified phosphoric acid-activated biochar was fabricated for extracting triazine herbicides from rice followed by HPLC-VWD determination [140]. The interaction of the composite with analytes was explained to be via electrostatic interaction, hydrogen bonding, pore-filling interaction, halogen-bonding interaction, and hydrophobic and π-π interactions (Fig. 3).

Besides the multi-interactions property of the composite, it also showed superior performance to three commercial single-mode sorbents, and comparable results as those of another commercial hydrophilic-lipophilic balance material [140]. Besides the low LODs (1.41–3.35 ng g^{-1}) and satisfactory linear range (0.01–2.00 μg g^{-1}), the method also showed recoveries of between 96.13 and 116.25% with acceptable inter-day and intra-day precision (RSD $\leq$ 13.60%).

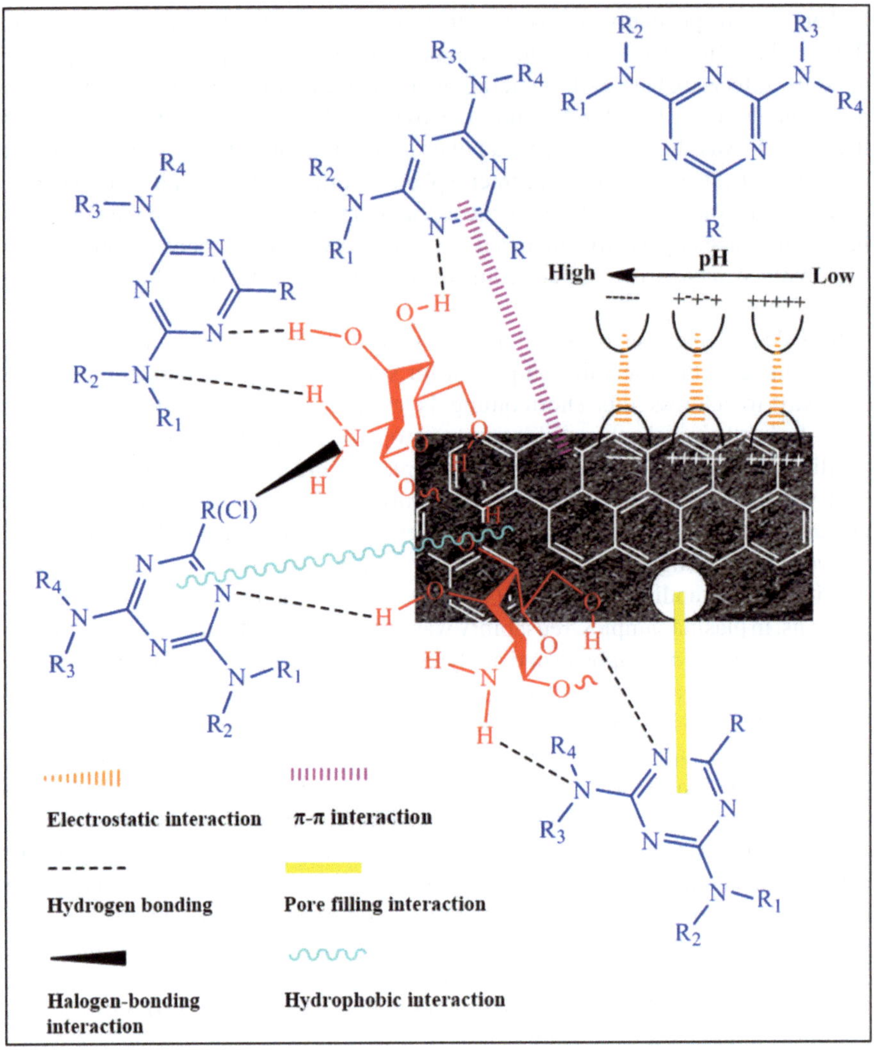

Fig. 3 Possible interaction mechanisms for extracting triazine herbicides by chitosan-modified phosphoric acid activated biochar (Modified from ref. [140] with permission from Elsevier)

Finally, microspheres of an MOF, MIL-88B(Fe) in combination with cellulose were prepared by an in situ method [141]. The composite was then used for the automated DPX of trace sulfonamides in milk samples prior to UHPLC-MS/MS analysis [141]. The sorbent exhibited reasonable reusability (eight extraction cycles). The method was considered sensitive, environmentally friendly and sorbent-saving. The method provided a LOD of 0.00660–0.0136 $\mu g\ kg^{-1}$ and satisfactory recoveries of between 69.8% and 100.9%.

In this section, some of the more interesting applications of μ-SPE in its various formats have been presented and discussed. Exemplifying the robust research activities in this sample preparation field in recent years, there are many more studies reported in the literature that could not be described in detail. Table 6 represents more examples of using chitosan- and cellulose-based materials in μ-SPE, together with relevant analytical details.

4 Conclusions and Future Trends

After its commercial introduction (ca. 1976), solid-phase extraction (SPE) became extensively used for the pre-concentration and extraction of many different analytes from a wide variety of liquid (primarily aqueous) samples. However, it became apparent over the succeeding years that the technique could be improved in terms of reducing the moderately large sample volumes, quantities of sorbents, volumes of elution solvents, the number of steps and types of operations (filtration, centrifugation, etc.) needed and so on. This realization provided the impetus to develop METs progressively over the following 40–45 years. Up to the present day, in addressing the limitations of SPE, there have emerged sorbent-based METs in a great diversity of different formats. μ-SPE is one such example of an MET. The main goal here is the use of low quantities of sorbents and solvents, small-scale devices or apparatus, meso- to nano-dimension materials, and invention of new tools and gadgets, etc. GAC has been based on, from the birth to now, the sustainability of the analytical laboratories in terms of the costs, and energy. In addition, GAC has been employed efforts for reducing the analytical operations, instrumentation, and apparatus, in terms of their scales and dimensions. Most importantly, dealing with the consumption of large amounts of solvents, reagents, and sorbents, is another goal of the GAC that has been emerge for. Considering these attributes and descriptions, it can be deemed that μ-SPE (in its various forms such as PT-SPE, DSPE, DPX, NDT, MEPS, SC-SPE and MB-μ-SPE, etc.,) conforms to GAC principles. Several these μ-SPE-based techniques can also be automated. Some of these techniques like MEPS and DPX are available on commercial autosampler platforms. The most dynamic activity in solid-based extraction research, including μ-SPE currently is possibly in the design, synthesis/preparation and application of novel materials as sorbents. These materials have largely been developed to address the shortcomings of conventional sorbents like unsatisfactorily low stability and surface area, inadequate adsorption capacity, lacking in selectivity, etc. Although many, if not all, authors normally claim that their sorbents are environmentally friendly and their synthesis procedures are simple, easy and fast, it is often difficult to verify at least some of the latter assertions without actually repeating the experiments. Be that as it may, based on the analytical results presented, most methods making use of the new sorbents generally show at least some improvements over existing procedures. The key question is whether the new materials are likely to be adopted by the community, as replacements for whatever are presently available commercially, and which performance has been proven over

Table 6 Applications of chitosan- and cellulose-based materials in various μ-SPE techniques

Material	Extraction mode	Analyte(s)	Real sample	Linear range	LOD	LOQ	Detection	Ref.
Chitosan cryogel	PT-SPE	Heavy metal ions	Environmental water	0.5–1000 ng L^{-1}	0.080–0.21 ng L^{-1}	–	GFAAS	[142]
MGCP[a]	MEPS	Carvedilol	Biological samples	1.5–300 ng mL^{-1}	0.5 ng mL^{-1}	–	HPLC–UV	[143]
PSMC[b]	NTD	PAHs	Water	5–1000 ng L^{-1}	0.75–1 ng L^{-1}	3–5 ng L^{-1}	–	[144]
EPN-MOF74Ns[c]	SC-SPE	Atenolol	Biological fluids and wastewater	0.5–500 ng mL^{-1}	0.15 ng mL^{-1}	–	HPLC–DAD	[145]
		Captopril		0.3–500 ng mL^{-1}	0.13 ng mL^{-1}			
PMMMC[d]	SC-SPE	β-blockers	Urine	–	1.4–40 μg L^{-1}	–	HPLC–UV	[146]

[a]Montmorillonite grafted on a cellulosic paper
[b]Polybutylene succinate/modified cellulose
[c]Electrospun polyacrylonitrile/nickel-MOF-74 nanofibres
[d]Polymeric monolith modified with 1-allyl-3-methylimidazolium chloride

many years of use. This chapter was focused on μ-SPE techniques, some of their fundamental characteristics and their applications in handling analytes in different sample types. The advantages and disadvantages of each method were discussed. Also, the applications of several novel material classes as sorbents in μ-SPE were assessed. Our survey has shown that despite the already broad applicability of these sample preparation techniques, it is clear that whenever new sorbents are designed and synthesized, these materials are likely to continue to spur lively research in the μ-SPE field in the foreseeable future. Thus far, there have been only a few reports on automated μ-SPE. Therefore, it is believed that this is a potentially fertile research area that should receive greater attention. Another main area is in the more rigorous and thorough exploration of the interactions between the sorbents and the analytes they are designed to extract. This vital area has not been investigated comprehensively in much of the literature as yet. Fuller understanding of possible interaction mechanisms will certainly provide useful guidance for chemists to design better sorbents, with desirable properties of greater stability, selectivity or universality, recyclability, resistance to interferences (matrix effects), stronger invulnerability to very complex (e.g., biological) matrices, and amenability to automated extraction workflows, amongst others. Some reports described in this chapter considered the application of new sorbents for only single analytes which in our view is wasteful of resources, efforts and time [21]. This is contrary to one GAC principle that says analysis should take into consideration multiple compounds [147]. So, research on the versatility and universality of new materials for multiple analyte extractions is urged. With the advent of non-conventional solvents such as DESs, ionic liquids, supramolecular solvents, etc., μ-SPE studies making use of these liquids can be anticipated. For example, DESs are considered green solvents which are easy to design and prepare (without need of purification) that have desirable physicochemical features. They have been used for different applications as modifiers of sorbents, parts of sorbents, solvents in their own right, and elution solvents [148]. Nemati et al. used dichloroacetic acid DES as an elution solvent for different pesticides (dimethoate, imidacloprid, pirimicarb, carbaryl, fenitrothion, hexythiazox, and phosalone) from polystyrene [149]. In another work, a novel technique namely homogenous DSPE was applied for the extraction of four polycyclic aromatic hydrocarbons hydroxylated metabolites from urine samples [150]. The technique was based on the dissolution of a water-miscible organic polymer as a sorbent in urine to form a homogenous phase. By the addition of NaCl, the sorbent was re-precipitated in the bulk solution as particles. The analytes were then adsorbed onto the sorbent particles. After separating, the sorbent was eluted by choline chloride:butyric acid DES. Analysis was by HPLC-FLD. Only 87 μL of the choline chloride:butyric acid was enough to achieve the maximum extraction. Besides these, we can also expect more innovations in the fabrication of devices, gadgets and apparatus specially designed for conducting μ-SPE.

Acknowledgements The authors are grateful for the financial support from their respective universities.

The authors have declared no conflict of interest.

References

1. Bagheri AR, Aramesh N, Gong Z, Cerda V, Lee HK (2022) Two-dimensional materials as a platform in extraction methods: a review. Trends Anal Chem 152:116606. https://doi.org/10.1016/j.trac.2022.116606
2. Nasiri M, Ahmadzadeh H, Amiri A (2020) Sample preparation and extraction methods for pesticides in aquatic environments: a review. Trends Anal Chem 123:115772. https://doi.org/10.1016/j.trac.2019.115772
3. Arabi M, Ostovan A, Bagheri AR, Guo X, Wang L, Li J, Wang X, Li B, Chen L (2020) Strategies of molecular imprinting-based solid-phase extraction prior to chromatographic analysis. Trends Anal Chem 128:115923. https://doi.org/10.1016/j.trac.2020.115923
4. Bagheri AR, Aramesh N, Haddad PR (2021) Applications of covalent organic frameworks and their composites in the extraction of pesticides from different samples. J Chromatogr A 1661:462612. https://doi.org/10.1016/j.chroma.2021.462612
5. Jon CS, Meng LY, Li D (2019) Recent review on carbon nanomaterials functionalized with ionic liquids in sample pretreatment application. Trends Anal Chem 120:115641. https://doi.org/10.1016/j.trac.2019.115641
6. Bagheri AR, Aramesh N, Liu Z, Chen C, Shen W, Tang S (2022) Recent advances in the application of covalent organic frameworks in extraction: a review. Crit Rev Anal Chem 1–34. https://doi.org/10.1080/10408347.2022.2089838
7. Bagheri AR, Aramesh N, Chen J, Liu W, Shen W, Tang S, Lee HK (2022) Polyoxometalate-based materials in extraction, and electrochemical and optical detection methods: a review. Anal Chim Acta 1209:339509. https://doi.org/10.1016/j.aca.2022.339509
8. Bagheri AR, Aramesh N, Lee HK (2022) Chitosan- and/or cellulose-based materials in analytical extraction processes: a review. Trends Anal Chem 157:116770. https://doi.org/10.1016/j.trac.2022.116770
9. Zhang J, Wang Y, Yu J, Wang Q, Khattak KN, Yang X (2022) Determination of pyrethroids in water samples by dispersive solid-phase extraction coupled with high-performance liquid chromatography. Water Environ Res 94:e10813. https://doi.org/10.1002/wer.10813
10. Gentili A (2020) Cyclodextrin-based sorbents for solid phase extraction. J Chromatogr A 1609:460654. https://doi.org/10.1016/j.chroma.2019.460654
11. Pena-Pereira F, Romero V, De La Calle I, Lavilla I, Bendicho C (2021) Graphene-based nanocomposites in analytical extraction processes. Trends Anal Chem 142:116303. https://doi.org/10.1016/j.trac.2021.116303
12. Daniels KD, Park M, Huang Z, Jia A, Flores GS, Lee HK, Snyder SA (2020) A review of extraction methods for the analysis of pharmaceuticals in environmental waters. Crit Rev Environ Sci Technol 50:2271–2299. https://doi.org/10.1080/10643389.2019.1705723
13. Feng J, Feng J, Ji X, Li C, Han S, Sun H, Sun M (2021) Recent advances of covalent organic frameworks for solid-phase microextraction. Trends Anal Chem 137:116208. https://doi.org/10.1016/j.trac.2021.116208
14. Jing W, Wang J, Kuipers B, Bi W, Chen DY (2021) Recent applications of graphene and graphene-based materials as sorbents in trace analysis. Trends Anal Chem 137:116212. https://doi.org/10.1016/j.trac.2021.116212
15. Li Y, Lan S, Zhu T (2021) Recent advances of graphene-based sorptive materials in extraction: a review. Trends Anal Chem 142:116319. https://doi.org/10.1016/j.trac.2021.116319
16. Naing NN, Lee HK (2020) Microextraction and analysis of contaminants adsorbed on atmospheric fine particulate matter: a review. J Chromatogr A 1627:461433. https://doi.org/10.1016/j.chroma.2020.461433
17. Jiménez-Skrzypek G, Ortega-Zamora C, González-Sálamo J, Hernández-Borges J (2022) Miniaturized green sample preparation approaches for pharmaceutical analysis. J Pharm Biomed Anal 207:114405. https://doi.org/10.1016/j.jpba.2021.114405
18. Kotova AA, Thiebaut D, Vial J, Tissot A, Serre C (2022) Metal-organic frameworks as stationary phases for chromatography and solid phase extraction: a review. Coord Chem Rev 455:214364. https://doi.org/10.1016/j.ccr.2021.214364

19. Naing NN, Tan SC, Lee HK (2020) Micro-solid-phase extraction. in Solid-Phase Extraction. Elsevier, 443–471
20. Sajid M, Nazal MK, Ihsanullah I (2020) Novel materials for dispersive (micro) solid-phase extraction of polycyclic aromatic hydrocarbons in environmental water samples: a review. Anal Chim Acta 1141:246–262. https://doi.org/10.1016/j.aca.2020.07.064
21. Yan H, Yang C, Sun Y, Row KH (2014) Ionic liquid molecularly imprinted polymers for application in pipette-tip solid-phase extraction coupled with gas chromatography for rapid screening of dicofol in celery. J Chromatogr A 1361:53–59. https://doi.org/10.1016/j.chroma.2014.07.102
22. Yan H, Sun N, Liu S, Row KH, Song Y (2014) Miniaturized graphene-based pipette tip extraction coupled with liquid chromatography for the determination of sulfonamide residues in bovine milk. Food Chem 158:239–244. https://doi.org/10.1016/j.foodchem.2014.02.089
23. Yan Z, Wu M, Hu B, Yao M, Zhang L, Lu Q, Pang J (2018) Electrospun UiO-66/polyacrylonitrile nanofibers as efficient sorbent for pipette tip solid phase extraction of phytohormones in vegetable samples. J Chromatogr A 1542:19–27. https://doi.org/10.1016/j.chroma.2018.02.030
24. Sun N, Han Y, Yan H, Song Y (2014) A self-assembly pipette tip graphene solid-phase extraction coupled with liquid chromatography for the determination of three sulfonamides in environmental water. Anal Chim Acta 810:25–31. https://doi.org/10.1016/j.aca.2013.12.013
25. Turazzi FC, Mores L, Carasek E, Merib J, De Oliveira Barra GM (2019) A rapid and environmentally friendly analytical method based on conductive polymer as extraction phase for disposable pipette extraction for the determination of hormones and polycyclic aromatic hydrocarbons in river water samples using high-performance liquid chromatography/diode array detection. J Environ Chem Eng 7:103156. https://doi.org/10.1016/j.jece.2019.103156
26. Carasek E, Morés L, Huelsmann RD (2021) Disposable pipette extraction: A critical review of concepts, applications, and directions. Anal Chim Acta 1192:339383. https://doi.org/10.1016/j.aca.2021.339383
27. Gomes IC, Martins RO, Machado LS, Cardoso AT, De Souza PS, Coltro WKT, De Tarso GP, Chaves AR (2022) Molecularly imprinted polymer as sorbent phase for disposable pipette extraction: A potential approach for creatinine analysis in human urine samples. J Pharm Biomed Anal 211:114625. https://doi.org/10.1016/j.jpba.2022.114625
28. Bordin DCM, Alves MNR, De Campos EG, De Martinis BS (2016) Disposable pipette tips extraction: fundamentals, applications and state of the art. J Sep Sci 39:1168–1172. https://doi.org/10.1002/jssc.201500932
29. Abdel-Rehim M (2004) New trend in sample preparation: on-line microextraction in packed syringe for liquid and gas chromatography applications: I. Determination of local anaesthetics in human plasma samples using gas chromatography–mass spectrometry. J Chromatogr B 801:317–321. https://doi.org/10.1016/j.jchromb.2003.11.042
30. Pereira JA, Gonçalves J, Porto-Figueira P, Figueira JA, Alves V, Perestrelo R, Medina S, Câmara JS (2019) Current trends on microextraction by packed sorbent-fundamentals, application fields, innovative improvements and future applications. Analyst 144:5048–5074. https://doi.org/10.1039/C8AN02464B
31. Abdel-Rehim M (2011) Microextraction by packed sorbent (MEPS): a tutorial. Anal Chim Acta 701:119–128. https://doi.org/10.1016/j.aca.2011.05.037
32. Moein MM, Abdel-Rehim A, Abdel-Rehim M (2015) Microextraction by packed sorbent (MEPS). Trends Anal Chem 67:34–44. https://doi.org/10.1016/j.aca.2011.05.037
33. Dalvand K, Ghiasvand A (2019) Simultaneous analysis of PAHs and BTEX in soil by a needle trap device coupled with GC-FID and using response surface methodology involving Box-Behnken design. Anal Chim Acta 1083:119–129. https://doi.org/10.1016/j.aca.2019.07.063
34. Zeinali S, Khalilzadeh M, Pawliszyn J (2022) The evolution of needle-trap devices with focus on aerosol investigations. Trends Anal Chem 153:116643. https://doi.org/10.1016/j.trac.2022.116643
35. Zeinali S, Ghosh C, Pawliszyn J (2022) Simultaneous determination of exhaled breath vapor and exhaled breath aerosol using filter-incorporated needle-trap devices: a comparison of

gas-phase and droplet-bound components. Anal Chim Acta 1203:339671. https://doi.org/10.1016/j.aca.2022.339671
36. Poormohammadi A, Bahrami A, Farhadian M, Shahna FG, Ghiasvand A (2017) Development of Carbotrap B-packed needle trap device for determination of volatile organic compounds in air. J Chromatogr A 1527:33–42. https://doi.org/10.1016/j.chroma.2017.10.062
37. Lord HL, Zhan W, Pawliszyn J (2010) Fundamentals and applications of needle trap devices: a critical review. Anal Chim Acta 677:3–18. https://doi.org/10.1016/j.aca.2010.06.020
38. Kaykhaii M, Hashemi SH, Miniaturized solid phase extraction. in Emerging Freshwater Pollutants. Elsevier, 49–61
39. Esrafili A, Ghambarian M, Tajik M, Baharfar M (2020) Spin-column micro-solid phase extraction of chlorophenols using MFU-4l metal-organic framework. Microchim Acta 187:1–9. https://doi.org/10.1007/s00604-019-4023-3
40. Namera A, Nakamoto A, Nishida M, Saito T, Kishiyama I, Miyazaki S, Yahata M, Yashiki M, Nagao M (2008) Extraction of amphetamines and methylenedioxyamphetamines from urine using a monolithic silica disk-packed spin column and high-performance liquid chromatography-diode array detection. J Chromatogr A 1208:71–75. https://doi.org/10.1016/j.chroma.2008.08.091
41. Crowell AM, Maclellan DL, Doucette AA (2015) A two-stage spin cartridge for integrated protein precipitation, digestion and SDS removal in a comparative bottom-up proteomics workflow. J Proteome Res 118:140–150. https://doi.org/10.1016/j.jprot.2014.09.030
42. Giyahban F, Amini S, Ebrahimzadeh H, Kandeh SH (2021) Spin-column micro-solid phase extraction of phthalate esters using electrospun polyacrylonitrile/iron (III)/Mg-based metal-organic framework 88B followed by GC analysis. Microchem J 170:106634. https://doi.org/10.1016/j.microc.2021.106634
43. Seidi S, Tajik M, Baharfar M, Rezazadeh M (2019) Micro solid-phase extraction (pipette tip and spin column) and thin film solid-phase microextraction: miniaturized concepts for chromatographic analysis. Trends Anal Chem 118:810–827. https://doi.org/10.1016/j.trac.2019.06.036
44. Basheer C, Alnedhary AA, Rao BM, Valliyaveettil S, Lee HK (2006) Development and application of porous membrane-protected carbon nanotube micro-solid-phase extraction combined with gas chromatography/mass spectrometry. Anal Chem 78:2853–2858. https://doi.org/10.1021/ac060240i
45. Lim TH, Hu L, Yang C, He C, Lee HK (2013) Membrane assisted micro-solid phase extraction of pharmaceuticals with amino and urea-grafted silica gel. J Chromatogr A 1316:8–14. https://doi.org/10.1016/j.chroma.2013.09.034
46. Martín-Esteban A (2021) Membrane-protected molecularly imprinted polymers: towards selectivity improvement of liquid-phase microextraction. Trends Anal Chem 138:116236. https://doi.org/10.1016/j.trac.2021.116236
47. Sajid M (2017) Porous membrane protected micro-solid-phase extraction: a review of features, advancements and applications. Anal Chim Acta 965:36–53. https://doi.org/10.1016/j.aca.2017.02.023
48. Chisvert A, Cárdenas S, Lucena R (2019) Dispersive micro-solid phase extraction. Trends Anal Chem 112:226–233. https://doi.org/10.1016/j.trac.2018.12.005
49. Ghorbani M, Aghamohammadhassan M, Ghorbani H, Zabihi A (2020) Trends in sorbent development for dispersive micro-solid phase extraction. Microchem J 158:105250. https://doi.org/10.1016/j.microc.2020.105250
50. Criado-García L, Arce L (2016) Extraction of toxic compounds from saliva by magnetic-stirring-assisted micro-solid-phase extraction step followed by headspace-gas chromatography-ion mobility spectrometry. Anal Bioanal Chem 408:6813–6822. https://doi.org/10.1007/s00216-016-9808-1
51. Khezeli T, Daneshfar A (2017) Development of dispersive micro-solid phase extraction based on micro and nano sorbents. Trends Anal Chem 89:99–118. https://doi.org/10.1016/j.trac.2017.01.004

52. Hakme E, Poulsen ME (2021) Evaluation of the automated micro-solid phase extraction clean-up system for the analysis of pesticide residues in cereals by gas chromatography-Orbitrap mass spectrometry. J Chromatogr A 1652:462384. https://doi.org/10.1016/j.chroma.2021.462384
53. Fernandes SR, Barreiros L, Sá P, Miró M, Segundo MA (2022) Automatic and renewable micro-solid-phase extraction based on bead injection lab-on-valve system for determination of tranexamic acid in urine by UHPLC coupled with tandem mass spectrometry. Anal Bioanal Chem 414:649–659. https://doi.org/10.1007/s00216-021-03606-y
54. Zhang Y, Zhang Y, Akakuru OU, Xu X, Wu A (2021) Research progress and mechanism of nanomaterials-mediated in-situ remediation of cadmium-contaminated soil: a critical review. J Environ Sci 104:351–364. https://doi.org/10.1016/j.jes.2020.12.021
55. Acauan LH, Kaiser AL, Wardle BL (2021) Direct synthesis of carbon nanomaterials via surface activation of bulk copper. Carbon 177:1–10. https://doi.org/10.1016/j.carbon.2021.02.045
56. Wang G, Liu L, Zhang Z (2021) Interface mechanics in carbon nanomaterials-based nanocomposites. Compos Part A Appl Sci 141:106212. https://doi.org/10.1016/j.compositesa.2020.106212
57. Sotolongo AC, Messina MM, Ibañez FJ, Wuilloud RG (2020) Hybrid ionic liquid-3D graphene-Ni foam for on-line preconcentration and separation of Hg species in water with atomic fluorescence spectrometry detection. Talanta 210:120614. https://doi.org/10.1016/j.talanta.2019.120614
58. Yuan Y, Wang M, Jia N, Zhai C, Han Y, Yan H (2019) Graphene/multi-walled carbon nanotubes as an adsorbent for pipette-tip solid-phase extraction for the determination of 17β-estradiol in milk products. J Chromatogr A 1600:73–79. https://doi.org/10.1016/j.chroma.2019.04.055
59. Tsai PC, Pundi A, Brindhadevi K, Ponnusamy VK (2021) Novel semi-automated graphene nanosheets based pipette-tip assisted micro-solid phase extraction as eco-friendly technique for the rapid detection of emerging environmental pollutant in waters. Chemosphere 276:130031. https://doi.org/10.1016/j.chemosphere.2021.130031
60. Seidi S, Sadat Karimi E, Rouhollahi A, Baharfar M, Shanehsaz M, Tajik M (2019) Synthesis and characterization of polyamide-graphene oxide-polypyrrole electrospun nanofibers for spin-column micro solid phase extraction of parabens in milk samples. J Chromatogr A 1599:25–34. https://doi.org/10.1016/j.chroma.2019.04.014
61. Tan SC, Lee HK (2021) Fully automated graphitic carbon nitride-based disposable pipette extraction-gas chromatography-mass spectrometric analysis of six polychlorinated biphenyls in environmental waters. J Chromatogr A 1637:461824. https://doi.org/10.1016/j.chroma.2020.461824
62. Arabsorkhi B, Sereshti H (2018) Determination of tetracycline and cefotaxime residues in honey by micro-solid phase extraction based on electrospun nanofibers coupled with HPLC. Microchem J 140:241–247. https://doi.org/10.1016/j.microc.2018.04.030
63. Golzari Aqda T, Behkami S, Raoofi M, Bagheri H (2019) Graphene oxide-starch-based micro-solid phase extraction of antibiotic residues from milk samples. J Chromatogr A 1591:7–14. https://doi.org/10.1016/j.chroma.2018.11.069
64. Yuan Y, Han Y, Yang C, Han D, Yan H (2020) Deep eutectic solvent functionalized graphene oxide composite adsorbent for miniaturized pipette-tip solid-phase extraction of toluene and xylene exposure biomarkers in urine prior to their determination with HPLC-UV. Microchim Acta 187:1–9. https://doi.org/10.1007/s00604-020-04370-z
65. Liu L, Tang W, Tang B, Han D, Row KH, Zhu T (2017) Pipette-tip solid-phase extraction based on deep eutectic solvent modified graphene for the determination of sulfamerazine in river water. J Sep Sci 40:1887–1895. https://doi.org/10.1002/jssc.201601436
66. Lv Y, Zhu T (2020) Polyethyleneimine-modified porous aromatic framework and silane coupling agent grafted graphene oxide composite materials for determination of phenolic acids in Chinese Wolfberry drink by HPLC. J Sep Sci 43:774–781. https://doi.org/10.1002/jssc.201900766

67. Qi M, Tu C, Li Z, Wang W, Chen J, Wang AJ (2018) Determination of sulfonamide residues in honey and milk by HPLC coupled with novel graphene oxide/polypyrrole foam material-pipette tip solid phase extraction. Food Anal Methods 11:2885–2896. https://doi.org/10.1007/s12161-018-1271-5
68. Zhang Y, Zhao YG, Chen WS, Cheng HL, Zeng XQ, Zhu Y (2018) Three-dimensional ionic liquid-ferrite functionalized graphene oxide nanocomposite for pipette-tip solid phase extraction of 16 polycyclic aromatic hydrocarbons in human blood sample. J Chromatogr A 1552:1–9. https://doi.org/10.1016/j.chroma.2018.03.039
69. Zhang H, Wu X, Yuan Y, Han D, Qiao F, Yan H (2018) An ionic liquid functionalized graphene adsorbent with multiple adsorption mechanisms for pipette-tip solid-phase extraction of auxins in soybean sprouts. Food Chem 265:290–297. https://doi.org/10.1016/j.foodchem.2018.05.090
70. Hashemi SH, Kaykhaii M, Keikha AJ, Parkaz A (2018) Application of Box-Behnken design in the optimization of a simple graphene oxide/zinc oxide nanocomposite-based pipette tip micro-solid phase extraction for the determination of Rhodamine B and Malachite green in seawater samples by spectrophotometry. Anal Method 10:5707–5714. https://doi.org/10.1039/C8AY02309C
71. Oliveira TC, Lanças FM (2023) Determination of selected herbicides in sugarcane-derived foods by graphene-oxide based disposable pipette extraction followed by liquid chromatography-tandem mass spectrometry. J Chromatogr A 1687:463690. https://doi.org/10.1016/j.chroma.2022.463690
72. Tan SC, Lee HK (2020) Graphitic carbon nitride as sorbent for the emulsification-enhanced disposable pipette extraction of eight organochlorine pesticides prior to GC-MS analysis. Microchim Acta 187:1–10. https://doi.org/10.1007/s00604-019-4107-0
73. Jordan-Sinisterra M, Lanças FM (2022) Microextraction by packed sorbent of selected pesticides in coffee samples employing ionic liquids supported on graphene nanosheets as extraction phase. Anal Bioanal Chem 414:413–423. https://doi.org/10.1007/s00216-021-03245-3
74. Ayazi Z, Shekari Esfahlan F, Matin P (2018) Graphene oxide reinforced polyamide nanocomposite coated on paper as a novel layered sorbent for microextraction by packed sorbent. Int. J. Environ.l Anal. Chem 98:1118–1134. https://doi.org/10.1080/03067319.2018.1535061
75. Ahmadi M, Moein MM, Madrakian T, Afkhami A, Bahar S, Abdel-Rehim M (2018) Reduced graphene oxide as an efficient sorbent in microextraction by packed sorbent: determination of local anesthetics in human plasma and saliva samples utilizing liquid chromatography-tandem mass spectrometry. J Chromatogr B 1095:177–182. https://doi.org/10.1016/j.jchromb.2018.07.036
76. Karimiyan H, Uheida A, Hadjmohammadi M, Moein MM, Abdel-Rehim M (2019) Polyacrylonitrile/graphene oxide nanofibers for packed sorbent microextraction of drugs and their metabolites from human plasma samples. Talanta 201:474–479. https://doi.org/10.1016/j.talanta.2019.04.027
77. Zare FD, Allahdadlalouni M, Baktash MY, Bagheri H (2020) Reduced graphene oxide-melamine formaldehyde as a highly efficient platform for needle trap microextraction of volatile organic compounds. Microchim Acta 157:104932. https://doi.org/10.1016/j.microc.2020.104932
78. Dincer BY, Balcı S, Tomul F (2020) In-situ mesoporous silica pillared clay synthesis and effect of titanium and iron incorporation to structural properties. Micropor Mesopor Mater 305:110342. https://doi.org/10.1016/j.micromeso.2020.110342
79. Zhang M, Yang J, Geng X, Li Y, Zha Z, Cui S, Yang J (2019) Magnetic adsorbent based on mesoporous silica nanoparticles for magnetic solid phase extraction of pyrethroid pesticides in water samples. J Chromatogr A 1598:20–29
80. Teo HL, Wong L, Liu Q, Teo TL, Lee TK, Lee HK (2016) Simple and accurate measurement of carbamazepine in surface water by use of porous membrane-protected micro-solid-phase extraction coupled with isotope dilution mass spectrometry. Anal Chim Acta 912:49–57. https://doi.org/10.1016/j.aca.2016.01.028

81. Augusto Santos Aguiar Júnior C, Dos Santos ALR, De Faria AM (2020) Disposable pipette extraction using a selective sorbent for carbendazim residues in orange juice. Food Chem 309:125756. https://doi.org/10.1016/j.foodchem.2019.125756
82. Khesina ZB, Iartsev SD, Revelsky AI, Buryak AK (2021) Microextraction by packed sorbent optimized by statistical design of experiment as an approach to increase the sensitivity and selectivity of HPLC-UV determination of parabens in cosmetics. J Pharm Biomed Anal 195:113843. https://doi.org/10.1016/j.jpba.2020.113843
83. Shen Q, Yang H, Li Y, Li S, Chen K, Wang H, Wang H, Ma J (2022) Rapid determination of antiviral drugs in yellow catfish (Pelteobagrus fulvidraco) using graphene/silica nanospheres (G/KCC-1) based pipette tip solid-phase extraction with ultra-performance liquid chromatography-tandem mass spectrometry. J Chromatogr B 1189:123097. https://doi.org/10.1016/j.jchromb.2022.123097
84. Lin H, Chen X, Ma J, Zhang X, Li T, Zhang Y, Wang H (2021) Determination of propofol in human plasma with C18 pipette-tip based solid-phase extraction followed by liquid chromatography atmospheric-pressure chemical ionization tandem mass spectrometry analysis. J Pharm Biomed Anal 193:113714. https://doi.org/10.1016/j.jpba.2020.113714
85. Cruz JC, Miranda LFC, Queiroz MEC (2021) Pipette tip micro-solid phase extraction (octyl-functionalized hybrid silica monolith) and ultra-high-performance liquid chromatography-tandem mass spectrometry to determine cannabidiol and tetrahydrocannabinol in plasma samples. J Sep Sci 44:1621–1632. https://doi.org/10.1002/jssc.202000906
86. Jordan-Sinisterra M, Vargas Medina DA, Lanças FM (2022) Microextraction by packed sorbent of polycyclic aromatic hydrocarbons in brewed coffee samples with a new zwitterionic ionic liquid-modified silica sorbent. J Food Compos Anal 114:104832. https://doi.org/10.1016/j.jfca.2022.104832
87. Ghiasvand A, Heidari N, Abdolhosseini S, Hamdi A, Haddad P (2018) Evaluation of a cooling/heating-assisted microextraction instrument using a needle trap device packed with aminosilica/graphene oxide nanocomposites, covalently attached to cotton. Analyst 143:2632–2640. https://doi.org/10.1039/C8AN00063H
88. Poormohammadi A, Bahrami A, Ghiasvand A, Shahna FG, Farhadian M (2019) Preparation of Carbotrap/silica composite for needle trap field sampling of halogenated volatile organic compounds followed by gas chromatography/mass spectrometry determination. J Environ Health Sci Eng 17:1045–1053. https://doi.org/10.1007/s40201-019-00418-2
89. He T, Li Q, Lin T, Li J, Bai S, An S, Kong X, Song Y-F (2023) Recent progress on highly efficient removal of heavy metals by layered double hydroxides. Chem Eng J 462:142041. https://doi.org/10.1016/j.cej.2023.142041
90. Wang MM, Chen Q, Zhang DD, Chen XW, Chen ML (2017) Tetra-nickel substituted polyoxotungsate as an efficient sorbent for the isolation of His6-tagged proteins from cell lysate. Talanta 171:173–178. https://doi.org/10.1016/j.talanta.2017.04.079
91. Li T, Miras HN, Song YF (2017) Polyoxometalate (POM)-layered double hydroxides (LDH) composite materials: design and catalytic applications. Catalysts 7:260. https://doi.org/10.3390/catal7090260
92. Tang S, Lee HK (2013) Application of dissolvable layered double hydroxides as sorbent in dispersive solid-phase extraction and extraction by co-precipitation for the determination of aromatic acid anions. Anal Chem 85:7426–7433. https://doi.org/10.1021/ac4013573
93. Tang S, Chia GH, Chang Y, Lee HK (2014) Automated dispersive solid-phase extraction using dissolvable Fe_3O_4-layered double hydroxide core-shell microspheres as sorbent. Anal Chem 86:11070–11076. https://doi.org/10.1021/ac503323e
94. Manouchehri M, Seidi S, Rouhollahi A, Noormohammadi H, Shanehsaz M (2020) Micro solid phase extraction of parabens from breast milk samples using Mg-Al layered double hydroxide functionalized partially reduced graphene oxide nanocomposite. Food Chem 314:126223. https://doi.org/10.1016/j.foodchem.2020.126223
95. Mirzaee MT, Seidi S, Alizadeh R (2021) Pipette-tip SPE based on Graphene/ZnCr LDH for Pb (II) analysis in hair samples followed by GFAAS. Anal Biochem 612:113949. https://doi.org/10.1016/j.ab.2020.113949

96. Huang Q, Chen Y, Yu H, Yan L, Zhang J, Wang B, Du B, Xing L (2018) Magnetic graphene oxide/MgAl-layered double hydroxide nanocomposite: one-pot solvothermal synthesis, adsorption performance and mechanisms for Pb^{2+}, Cd^{2+}, and Cu^{2+}. Chem Eng J 341:1–9. https://doi.org/10.1016/j.cej.2018.01.156
97. Liu Y, Wu H, Min L, Song S, Yang L, Ren Y, Wu Y, Zhao R, Wang H, Jiang Z (2020) 2D layered double hydroxide membranes with intrinsic breathing effect toward CO_2 for efficient carbon capture. J Membr Sci 598:117663. https://doi.org/10.1016/j.memsci.2019.117663
98. Xie Y, Yuan X, Wu Z, Zeng G, Jiang L, Peng X, Li H (2019) Adsorption behavior and mechanism of Mg/Fe layered double hydroxide with Fe_3O_4-carbon spheres on the removal of Pb (II) and Cu (II). J Colloid Interface Sci 536:440–455. https://doi.org/10.1016/j.jcis.2018.10.066
99. Seidi S, Sanàti SE (219) Nickel-iron layered double hydroxide nanostructures for micro solid phase extraction of nonsteroidal anti-inflammatory drugs, followed by quantitation by HPLC-UV. Microchim. Acta 186:297. https://doi.org/10.1007/s00604-019-3419-4.
100. Guo X, Ren T, Ji J, Yang Y, Di X (2022) An alternative analytical strategy based on QuEChERS and dissolvable layered double hydroxide dispersive micro-solid phase extraction for trace determination of sulfonylurea herbicides in wolfberry by LC-MS/MS. Food Chem 396:133652. https://doi.org/10.1016/j.foodchem.2022.133652
101. Wang X, Zhou W, Wang C, Chen Z (2018) In situ immobilization of layered double hydroxides onto cotton fiber for solid phase extraction of fluoroquinolone drugs. Talanta 186:545–553. https://doi.org/10.1016/j.talanta.2018.04.100
102. Seidi S, Doroudian M (2020) Electrospun NiFe layered double hydroxide/Nylon 6 composite nanofibers as a sorbent for micro solid phase extraction by packed sorbent of non-steroidal anti-inflammatory drugs in human blood. J Chromatogr A 1614:460718. https://doi.org/10.1016/j.chroma.2019.460718
103. Bagheri AR, Aramesh N, Bilal M (2020) New frontiers and prospects of metal-organic frameworks for removal, determination, and sensing of pesticides. Environ Res 194:110654. https://doi.org/10.1016/j.envres.2020.110654
104. Bagheri AR, Ghaedi M (2020) Application of Cu-based metal-organic framework (Cu-BDC) as a sorbent for dispersive solid-phase extraction of gallic acid from orange juice samples using HPLC-UV method. Arab J Chem 13:5218–5228. https://doi.org/10.1016/j.arabjc.2020.02.020
105. Liang W, Wied P, Carraro F, Sumby CJ, Nidetzky B, Tsung CK, Falcaro P, Doonan CJ (2021) Metal-organic framework-based enzyme biocomposites. Chem Rev 121:1077–1129. https://doi.org/10.1021/acs.chemrev.0c01029
106. Thorarinsdottir AE, Harris TD (2020) Metal-organic framework magnets. Chem Rev 120:8716–8789. https://doi.org/10.1021/acs.chemrev.9b00666
107. Amini S, Ebrahimzadeh H, Seidi S, Jalilian N (2021) Application of electrospun polyacrylonitrile/Zn-MOF-74@ GO nanocomposite as the sorbent for online micro solid-phase extraction of chlorobenzenes in water, soil, and food samples prior to liquid chromatography analysis. Food Chem 363:130330. https://doi.org/10.1016/j.foodchem.2021.130330
108. Gao G, Xing Y, Liu T, Wang J, Hou X (2019) UiO-66 (Zr) as sorbent for porous membrane protected micro-solid-phase extraction androgens and progestogens in environmental water samples coupled with LC-MS/MS analysis: the application of experimental and molecular simulation method. Microchem J 146:126–133. https://doi.org/10.1016/j.microc.2018.12.050
109. Soury S, Bahrami A, Alizadeh S, Shahna FG, Nematollahi D (2019) Development of a needle trap device packed with zinc based metal-organic framework sorbent for the sampling and analysis of polycyclic aromatic hydrocarbons in the air. Microchem J 148:346–354. https://doi.org/10.1016/j.microc.2019.05.019
110. Zhou P, Zhang W, Wang X (2021) Development of a syringe membrane-based microextraction method based on metal-organic framework mixed-matrix membranes for preconcentration/extraction of polycyclic aromatic hydrocarbons in tea infusion. Food Chem 361:130105. https://doi.org/10.1016/j.foodchem.2021.130105

111. Kaykhaii M, Hashemi SH, Andarz F, Piri A, Sargazi G (2021) Chromium-based metal organic framework for pipette tip micro-solid phase extraction: an effective approach for determination of methyl and propyl parabens in wastewater and shampoo samples. BMC Chem 15:60. https://doi.org/10.1186/s13065-021-00786-7
112. Zhang J, Yu C, Chen Z, Luo X, Zhao H, Wu F (2021) Zeolitic imidazolate framework-8/fluorinated graphene coated SiO_2 composites for pipette tip solid-phase extraction of chlorophenols in environmental and food samples. Talanta 228:122229. https://doi.org/10.1016/j.talanta.2021.122229
113. Sun M, Sun H, Feng J, Feng J, Fan J, Sun M, Feng Y (2022) Carbonized metal-organic framework-74/carbon aerogel composites for the efficient extraction of triazole fungicides from fruits and vegetables. J Chromatogr A 1683:463552. https://doi.org/10.1016/j.chroma.2022.463552
114. Feng S, Zhang A, Wu F, Luo X, Zhang J (2022) Boronic acid grafted metal-organic framework for selective enrichment of cis-diol-containing compounds. J Chromatogr A 1677:463281. https://doi.org/10.1016/j.chroma.2022.463281
115. Khodayari P, Jalilian N, Ebrahimzadeh H, Amini S (2022) Electrospun cellulose acetate /polyacrylonitrile /thymol /Mg-metal organic framework nanofibers as efficient sorbent for pipette-tip micro-solid phase extraction of anti-cancer drugs. React Funct Polym 173:105217. https://doi.org/10.1016/j.reactfunctpolym.2022.105217
116. Fornari F, Bianchi F, Riboni N, Casoli F, Bacchi A, Mazzeo PP, Pelagatti P, Careri M (2022) Metal-organic framework-based magnetic dispersive micro-solid-phase extraction for the gas chromatography-mass spectrometry determination of polycyclic aromatic compounds in water samples. J Chromatogr A 1671:463010. https://doi.org/10.1016/j.chroma.2022.463010
117. Li L, Li Y, Zhang S, Wang T, Hou X (2023) Monolithic and compressible MIL-101 (Cr)/cellulose aerogel/melamine sponge based microextraction in packed syringe towards trace nitroimidazoles in water samples prior to UPLC-MS/MS analysis. Talanta 253:123935. https://doi.org/10.1016/j.talanta.2022.123935
118. Soury S, Firoozichahak A, Nematollahi D, Alizadeh S, Kakaei H, Abbasi A (2021) Needle-trap device packed with the MIL-100(Fe) metal-organic framework for the extraction of the airborne organochlorine pesticides. Microchemi. J 171:106866. https://doi.org/10.1016/j.microc.2021.106866
119. Rahimpoor R, Firoozichahak A, Alizadeh S, Nematollahi D (2022) Urinary bio-monitoring of amphetamine derivatives by needle trap device packed with the zirconium-based metal-organic framework. Sci Rep 12:1–10. https://doi.org/10.1038/s41598-022-17861-1
120. Rahimpoor R, Firoozichahak A, Nematollahi D, Alizadeh S, Mohammad Alizadeh P, Alinaghi Langari AA (2021) Bio-monitoring of non-metabolized BTEX compounds in urine by dynamic headspace-needle trap device packed with 3D Ni/Co-BTC bimetallic metal-organic framework as an efficient absorbent. Microchem J 166:106229. https://doi.org/10.1016/j.microc.2021.106229
121. Giesbers M, Carrasco-Correa EJ, Simó-Alfonso EF, Herrero-Martínez JM (2019) Hybrid monoliths with metal-organic frameworks in spin columns for extraction of non-steroidal drugs prior to their quantitation by reversed-phase HPLC. Microchim Acta 186:1–9. https://doi.org/10.1007/s00604-019-3923-6
122. Zhang Y, Yuan ZL, Deng XY, Wei HD, Wang WL, Xu Z, Feng Y, Shi X (2022) Metal-organic framework mixed-matrix membrane-based extraction combined HPLC for determination of bisphenol A in milk and milk packaging. Food Chem 386:132753. https://doi.org/10.1016/j.foodchem.2022.132753
123. Mirzajani R, Kardani F, Ramezani Z (2020) Fabrication of UMCM-1 based monolithic and hollow fiber-Metal-organic framework deep eutectic solvents/molecularly imprinted polymers and their use in solid phase microextraction of phthalate esters in yogurt, water and edible oil by GC-FID. Food Chem 314:126179. https://doi.org/10.1016/j.foodchem.2020.126179
124. Sharma RK, Yadav P, Yadav M, Gupta R, Rana P, Srivastava A, Zbořil R, Varma RS, Antonietti M, Gawande MB (2020) Recent development of covalent organic frameworks (COFs):

synthesis and catalytic (organic-electro-photo) applications. Mater Horiz 7:411–454. https://doi.org/10.1039/C9MH00856J

125. Bagheri AR, Aramesh N, Sher F, Bilal M (2021) Covalent organic frameworks as robust materials for mitigation of environmental pollutants. Chemosphere 270:129523. https://doi.org/10.1016/j.chemosphere.2020.129523
126. Xu H, Gao J, Jiang D (2015) Stable, crystalline, porous, covalent organic frameworks as a platform for chiral organocatalysts. Nature Chem 7:905. https://doi.org/10.1038/nchem.2352
127. Kandambeth S, Shinde DB, Panda MK, Lukose B, Heine T, Banerjee R (2013) Enhancement of chemical stability and crystallinity in porphyrin-containing covalent organic frameworks by intramolecular hydrogen bonds. Angew Chem Int Ed 52:13052–13056. https://doi.org/10.1002/anie.201306775
128. Chen X, Addicoat M, Jin E, Zhai L, Xu H, Huang N, Guo Z, Liu L, Irle S, Jiang D (2015) Locking covalent organic frameworks with hydrogen bonds: General and remarkable effects on crystalline structure, physical properties, and photochemical activity. J Am Chem Soc 137:3241–3247. https://doi.org/10.1021/ja509602c
129. Wang R, Li C, Li Q, Zhang S, Lv F, Yan Z (2020) Electrospinning fabrication of covalent organic framework composite nanofibers for pipette tip solid phase extraction of tetracycline antibiotics in grass carp and duck. J Chromatogr A 1622:461098. https://doi.org/10.1016/j.chroma.2020.461098
130. Kurd N, Bahrami A, Afkhami A, Shahna FG, Assari MJ, Farhadian M (2022) Application of Fe_3O_4@TbBd nanobeads in microextraction by packed sorbent (MEPS) for determination of BTEXs biomarkers by HPLC-UV in urine samples. J Chromatogr B 1197:123197. https://doi.org/10.1016/j.jchromb.2022.123197
131. Safarpour Khotbesara N, Bahrami A, Habibi Mohraz M, Afkhami A, Farhadian M (2022) Development of a needle trap device packed with the Schiff base network-1/single-walled carbon nanotube for sampling phenolic compounds in air. Microchem J 172:106984. https://doi.org/10.1016/j.microc.2021.106984
132. Kang B, Liu H, Chen G, Hetong L, Chen S, Chen T (2022) Novel covalent organic frameworks based electrospun composite nanofiber membranes as pipette-tip strong anion exchange sorbent for determination of inorganic arsenic in rice. Food Chem 408:135192. https://doi.org/10.1016/j.foodchem.2022.135192
133. Tian X, Dai Y, Cheng Y, Zhang L, Kong R-M, Xia L, Kong C, Li G (2022) Combination of pipette tip solid phase extraction and high performance liquid chromatography for determination of plant growth regulators in food samples based on the electrospun covalent organic framework/polyacrylonitrile nanofiber as highly efficient sorbent. J Chromatogr A 1661:462692. https://doi.org/10.1016/j.chroma.2021.462692
134. Chen Z, Yu C, Xi J, Tang S, Bao T, Zhang J (2019) A hybrid material prepared by controlled growth of a covalent organic framework on amino-modified MIL-68 for pipette tip solid-phase extraction of sulfonamides prior to their determination by HPLC. Microchim Acta 186:393. https://doi.org/10.1007/s00604-019-3513-7
135. Zhang Y, Liao W, Dai Y, Wang W, Wang A (2020) Covalent organic framework Schiff base network-1-based pipette tip solid phase extraction of sulfonamides from milk and honey. J Chromatogr A 1634:461665. https://doi.org/10.1016/j.chroma.2020.461665
136. Yan Z, Hu B, Li Q, Zhang S, Pang J, Wu C (2019) Facile synthesis of covalent organic framework incorporated electrospun nanofiber and application to pipette tip solid phase extraction of sulfonamides in meat samples. J Chromatogr A 1584:33–41. https://doi.org/10.1016/j.chroma.2018.11.039
137. Militao IM, Roddick FA, Bergamasco R, Fan L (2021) Removing PFAS from aquatic systems using natural and renewable material-based adsorbents: a review. J Environ Chem Eng 9:105271. https://doi.org/10.1016/j.jece.2021.105271
138. Li G, Row KH (2021) Deep eutectic solvents cross-linked molecularly imprinted chitosan microsphere for the micro-solid phase extraction of p-hydroxybenzoic acid from pear rind. J Sep Sc 44:549–556. https://doi.org/10.1002/jssc.202000984

139. Hejabri Kandeh S, Amini S, Ebrahimzadeh H (2021) Simultaneous trace-level monitoring of seven opioid analgesic drugs in biological samples by pipette-tip micro solid phase extraction based on PVA-PAA/CNT-CNC composite nanofibers followed by HPLC-UV analysis. Microchim Acta 188:1–10. https://doi.org/10.1007/s00604-021-04931-w
140. Wang Q, Wang T, Zhang Y, Ma J, Tuo Y (2022) Preparation and evaluation of a chitosan modified biochar as an efficient adsorbent for pipette tip-solid phase extraction of triazine herbicides from rice. Food Chem 396:133716. https://doi.org/10.1016/j.foodchem.2022.133716
141. Li L, Zhang H, Zhang M, Wang T, Hou X (2022) MIL-88B(Fe)/cellulose microspheres as sorbent for the fully automated dispersive pipette extraction towards trace sulfonamides in milk samples prior to UPLC-MS/MS analysis. Anal Chim Acta 1232:340420. https://doi.org/10.1016/j.aca.2022.340420
142. Wang Z, Liao Y, Liu J, Huang X (2022) On-site separation and enrichment of heavy metal ions in environmental waters with multichannel in-tip microextraction device based on chitosan cryogel. Microchem J 175:107107. https://doi.org/10.1016/j.microc.2021.107107
143. Ayazi Z, Matin P (2021) Montmorillonite grafted on a cellulosic paper as a novel layered sorbent for microextraction by packed sorbent in combination with HPLC for determination of carvedilol in biological samples. Microchem J 171:106795. https://doi.org/10.1016/j.microc.2021.106795
144. Roostaie A, Abdolrasouli MH, Mohammadiazar S, Hossenipour A (2022) Polybutylene succinate/modified cellulose bionanocomposites as sorbent for needle trap microextraction. J Chromatogr A 1689:463715. https://doi.org/10.1016/j.chroma.2022.463715
145. Amini S, Ebrahimzdeh H, Seidi S, Jalilian N (2020) Preparation of electrospun polyacrylonitrile/Ni-MOF-74 nanofibers for extraction of atenolol and captopril prior to HPLC-DAD. Microchim Acta 187:1–12. https://doi.org/10.1007/s00604-020-04483-5
146. Mompó-Roselló O, Ribera-Castelló A, Simó-Alfonso EF, Ruiz-Angel MJ, García-Alvarez-Coque MC, Herrero-Martínez JM (2020) Extraction of β-blockers from urine with a polymeric monolith modified with 1-allyl-3-methylimidazolium chloride in spin column format. Talanta 214:120860. https://doi.org/10.1016/j.talanta.2020.120860
147. Sajid M, Płotka-Wasylka J (2022) Green analytical chemistry metrics: a review. Talanta 238:123046. https://doi.org/10.1016/j.talanta.2021.123046
148. Werner J, Zgoła-Grześkowiak A, Płatkiewicz J, Płotka-Wasylka J, Jatkowska N, Kalyniukova A, Zaruba S, Andruch V (2023) Deep eutectic solvents in analytical sample preconcentration Part B: Solid-phase (micro)extraction. Microchem J 191:108898. https://doi.org/10.1016/j.microc.2023.108898
149. Nemati M, Tuzen M, Farazajdeh MA, Kaya S, Afshar Mogaddam MR (2022) Development of dispersive solid-liquid extraction method based on organic polymers followed by deep eutectic solvents elution; application in extraction of some pesticides from milk samples prior to their determination by HPLC-MS/MS. Anal Chim Acta 1199:339570. https://doi.org/10.1016/j.aca.2022.339570
150. Jouyban A, Nemati M, Farazajdeh MA, Yazdani A, Afshar Mogaddam MR (2022) Salt-induced homogenous solid phase extraction of hydroxylated metabolites of polycyclic aromatic hydrocarbons from urine samples using a deep eutectic solvent as an elution solvent prior to HPLC-FLD analysis. Microchem J 172:106932. https://doi.org/10.1016/j.microc.2021.106932

Dispersive-Micro-Solid Phase Extraction

Giovanni D'Orazio

Abstract Dispersive extraction techniques are valuable for their ability to maximize the interaction between the sample and the extractant, leading to increased extraction efficiency. Dispersive solid phase extraction (dSPE) was initially developed to enhance the selectivity of the analytical process by using a solid phase to retain potential interferents from the sample matrix. While dSPE efficiently cleans up the sample, its lack of preconcentration limits sensitivity. Recently, a new miniaturized extraction technique called dispersive micro solid-phase extraction (d-μSPE) is a novel and appealing type of solid phase extraction (SPE) method, offering numerous benefits for analytical applications such as pre-concentration, clean-up, and extraction and especially gain of method sensitivity. The chapter provides an introduction to this extraction technique, starting with dSPE as the initial approach. It then describes the main contributions in the d-μSPE field based on the solid sorbent's nature and the strategies for sorbent dispersion. Additionally, it explored the future trends in this technique, with particular emphasis on the combination of d-μSPE with dispersive liquid phase microextraction (DLLME) and automation of sample preparation procedure coupling with instrumental analytical techniques. d-μSPE remains promising as a powerful sample preparation method with the potential for further advancements and applications in various analytical fields.

Keywords Dispersion modes · Dispersive-micro solid phase extraction · Green Chemistry · Miniaturisation · Sorbents

Abbreviations

AIBN	Azobisisobutyronitrile
APDC	ammonium pyrrolidine dithiocarbamate
AuNPs	gold nanoparticles
CE	capillary electrophoresis

G. D'Orazio (✉)
Consiglio Nazionale delle Ricerche (CNR), Istituto per i Sistemi Biologici (ISB), Roma, Italy
e-mail: giovanni.dorazio@cnr.it

M. Á. Rodríguez-Delgado et al. (eds.), *Microextraction Techniques*, Integrated Analytical Systems, https://doi.org/10.1007/978-3-031-50527-0_3

CNTs	carbon nanotubes
CTAB	cetyltrimethylammonium bromide
DDT	dichlorodiphenyltrichloroethane
DMIPs	dummy molecularly imprinted polymers
dLLME	dispersive liquid-phase microextraction
d-μSPE	dispersive-micro solid phase extraction
dSPE	dispersive solid phase extraction
dSPME	dispersive solid-phase microextraction method
EA-D-μSPE	Effervescence-assisted dispersive micro-solid phase extraction
G	graphene
GC	gas chromatography
GCB	graphitized black carbon
GC-MS	Gas chromatography-mass spectrometry
GDY	graphdiyne
GNRs	graphene nanoribbons
GO	Graphene oxide
GQDs	graphene quantum dots
HPLC	high pressure liquid chromatography
LC	liquid Chromatography
LLE	liquid-liquid extraction
mag-MIMs	dummy molecularly imprinted microspheres
MET	microextraction techniques
MIPSs	molecularly imprinted Polymers
MNPs	magnetic nanoparticles
MOFs	Metal-organic frameworks
MSPD	Matrix Solid Phase Dispersion
MSPME	magnetic solid phase microextraction
MWCNT	multi-walled carbon nanotubes
NP	nanoparticle
OPPs	organophosphorus pesticides
o-SWNHs	oxidized SWNHs
PAX	polymer anion exchange
PAHs	polycyclic Aromatic Hydrocarbons
PCX	cation exchange materials
PIX	ion exchange materials
PSA	primary-secondary amines
QuEChERS	Quick, Easy, Cheap, Effective, Rugged, and Safe
SAdSPE	solvent assisted dispersive solid-phase extraction
SAX	strong anion exchange
SBSE	stir bar sorptive extraction
SPE	solid phase extraction
SPME	solid-phase microextraction
SPNE	solid-phase nanoextraction
SWCNT	single-walled carbon nanotubes
SWNHs	single-walled carbon nanohorns

UA-d-μSPE	Ultrasonic-assisted-d-μSPE
VA-d-μSPE	Vortex-assisted d-μSPE

1 Introduction

In accordance with the trend in chemistry toward a more sustainable approach, Analytical Chemistry often struggles to meet the principle of "Real-time analysis for pollution prevention". Although direct analytical procedures are simple and fast, they rarely achieve the sensitivity required by legislation or our expectations. In addition, the compatibility between the sample matrix and the instrumental analytical technique can be an insurmountable difficulty in real-time analysis.

To address these limitations, the concept of sample preparation procedure within the analytical methodology has developed as a response to these shortcomings.

Sample preparation is a crucial phase in analytical methodology, often associated with potential environmental pollution due to the frequent use of organic solvents. Additionally, multiple manipulations, such as extraction, mineralization, filtration, distillation, etc., are required to concentrate and isolate target analytes.

In pursuit of the twelve principles outlined by Anastas and Warner [1], Green Analytical Chemistry is defined as an approach grounded in sustainable chemistry. This emerging trend aims to minimize chemical analysis's environmental and human health impact by eliminating or reducing the utilization of solvents, reagents, and other hazardous substances. To achieve this objective, scientific research focuses on developing more efficient, portable, and multi-parameter analytical instruments and adopting environmentally friendly solvents, reagents, and devices for sample handling. Moreover, the miniaturization of analytical systems is pivotal in reducing energy consumption and usage during analysis.

In an endeavor to adhere to the Green Chemistry principles [1], Analytical Chemistry has become "Green" inscribing in Sustainable Chemistry philosophy. This emerging paradigm seeks to minimize chemical analysis's environmental and human health impacts by eliminating or reducing reliance on solvents, reagents, and other hazardous substances. To achieve this objective, scientific research is focused on advancing the development of more efficient, portable, and multi-parameter analytical instruments and enhancing sample handling techniques through environmentally friendly solvents, reagents, and devices. Additionally, the miniaturization of analytical systems is crucial in reducing resource consumption and energy usage during the analysis process. Concerning sample preparation, analytical research has introduced the concept of Sustainable Analytical Procedures (SAP) [2], combining environmentally friendly solvents and utilizing miniaturized devices/procedures for extractions. In this context, microextraction techniques (METs) have been defined [3] as a means to address these two requirements of low environmental impact solvents and miniaturized devices/procedures for extraction.

It is widely acknowledged that sample preparation remains a critical bottleneck in many analytical methods, impacting the analytical process's sensitivity, selectivity, and overall speed. To overcome these challenges, the evolution of sample preparation has followed six general trends: miniaturization, simplification, automation, speed, cost-effectiveness, and safety. METs have emerged as a promising approach to enhance the classical sample treatment procedures used in chemical analysis.

Generally, METs involve the miniaturization of traditional extraction modes, such as solid-phase extraction (SPE) and liquid-liquid extraction (LLE).

2 Fundamentals

2.1 Dispersive Solid Phase Extraction

Although liquid-liquid extraction (LLE) is still utilized in various standards and reference methods, it is considered a secondary option compared to SPE due to its complexities, high consumption of organic solvents, and difficulties in automation. In addition, SPE is widely recognized as the most versatile sample treatment technique available today. This is mainly attributed to its extensive range of sorbent materials, including porous carbon, silica-based solids, and polymeric substances, as well as specialized materials like immunosorbents and polymers with molecular imprinting. Using such diverse sorbents allows for tailoring the SPE method to match the unique characteristics of analytes and matrices, thereby enhancing the analytical process' efficiency and reliability. Additionally, SPE can be easily automated, minimizing human involvement while ensuring greater reproducibility and precision in the analytical results [4, 5].

The latest trend in SPE is twofold: (i) using sorbent materials based on nanostructured carbon particles, such as fullerenes, nanotubes, nanodiamonds, nanocones and nanohorns, and (ii) miniaturizing the extraction device, leading to the development of μSPE. To this end, conventional configurations with a cartridge have been replaced by coated fibers, stirrers, disks, membranes and miniaturized devices, which have shown to be highly effective in extracting and purifying desired analytes [6]. These solutions meet, of course, numerous advantages characteristic of METs.

In general, conventional SPE is a procedure carried out using a device or cartridge made of a polymeric material that houses a solid sorbent through which the sample/extract is percolated. However, this method, for instance, limits the flow rate, and it becomes particularly critical when dealing with nanoscale particles, leading to increased back pressure. Moreover, the passage through the sorbent material reduces the analyte-sorbent interaction due to the compact geometry that hinders broader diffusion and the competitiveness of matrix interferences with the active sites of the solid sorbent [7].

Solid-phase dispersive extraction (dSPE) is an alternative approach to conventional SPE, where the sorbent is dispersed in the sample matrix (or its extract). The dispersive phenomenon promotes homogeneous mixing and increases the contact surface between the sorbent particles and the analyte, positively influencing the absorption kinetics and enhancing the overall efficiency of the extraction process.

In 2003, Anastassiades et al. were the first to report the use of dispersed sorbents to improve the efficiency of the extraction method, creating a protocol now known as QuEChERS (Quick, Easy, Cheap, Effective, Rugged, and Safe), which is still widely recognized for its main features [8].

Unlike SPE, dSPE or QuEChERS method aims to clean up the sample's organic extract, removing interferents and co-extracts from the matrix, thus leaving the analytes clean in the liquid phase. In this methodology, a small quantity (approximately 50 mg) of an appropriate sorbent (a mixture of salts and absorbents) not previously conditioned is added to the real sample extract and subsequently dispersed by manual agitation or vortex.

In this approach, the extraction of analytes occurs in a negative mode, meaning they remain in a cleaner liquid phase, improving the determination selectivity. This clean-up step is of fundamental importance as it allows for removing interferences that may be present in the sample matrix, focusing instead on the analytes of interest [6].

dSPE procedure has sometimes been identified with the Matrix Solid Phase Dispersion (MSPD) extraction for sample treatment. However, there are substantial differences between the two dispersive approaches. MSPD was introduced by Barker et al. [9] involves dispersing the sample on the surface of a solid support or sorbent in a glass or ceramic mortar to achieve complete disaggregation and dispersion of the real sample. The mixture is then transferred to an empty column or SPE cartridge without further treatment. The procedure is similar to SPE, where analytes are eluted with an appropriate solvent and collected for analysis. To enhance the selectivity of the MSPD technique, it could be coupled offline or online with SPE.

Generally, MSPD is commonly applied to solid, semi-solid, and viscous samples, while its application to liquid samples is quite rare. On the other hand, dSPE is different because a small amount of sorbent is added to an organic extract rather than the original sample.

So if dSPE presents several advantages, including enhanced interaction between the sample and the extracting phase, leading to improved selectivity for effective sample clean-up, however, there are two significant drawbacks closely associated with this technique: firstly, the lack of pre-concentration capability results in limited control over the analytical sensitivity of the process. Without pre-concentration, the detection limit may not be as low as desired, potentially affecting the accuracy of the analysis; secondly, the use of substantial amounts of organic solvent, often measured in milliliters, goes against the principles of microextraction techniques and then the greener and more sustainable practices in analytical chemistry.

2.2 Dispersive Micro Solid Phase Extraction

The Dispersive micro solid phase extraction (d-μSPE) or dispersive solid-phase microextraction method (dSPME) originates from the miniaturized extraction techniques such as solid-phase microextraction (SPME) and stir bar sorptive extraction (SBSE), which use sorbents to extract and concentrate solutes from a liquid sample [10].

Introduced by Tsai and co-workers to preconcentration and extraction of four tetracyclines in the water and milk samples in 2009, the D-μSPE is a cutting-edge hybrid technique that incorporates the principles of miniaturized extraction procedures [11].

Unlike DLLME, where a liquid acceptor phase is used, d-μSPE employs a solid acceptor phase, a sorbent material similar to conventional dispersive solid-phase extraction (dSPE) and solid-phase extraction (SPE).

The dispersion phenomenon enhances the interaction between the analytes and the sorbent, allowing compensation for the low amount of sorbent used, typically at milligram levels. This extraction phase by dispersion is often assisted by the external energy, such as vortex, ultrasound, or microwave, or by using specific solvents or chemical reagents akin to DLLME.

Subsequently, the solid phase that has trapped the target analytes is extracted from the solution or sample extract to proceed with the desorption step. For this purpose, compatible procedures for instrumental analysis have been developed, such as thermal desorption, more suitable for gas chromatography (GC), or chemical extraction with reduced volume of appropriate solvent followed by drying and reconstitution for liquid-phase separation techniques (LC and CE). A schematic illustration of the d-μSPE procedure is showed in Fig. 1

In this respect, in d-μSPE, two critical aspects should be considered [13, 14]: (i) the selection of the most suitable sorbent based on the target analyte class and the sample matrix; (ii) the application of different strategies to promote the dispersion and disintegration of the sorbent itself, optimizing the extraction efficiency; (iii) sorbent separation procedure; (iv) type and properties of desorption solvent.

3 Sorbents in Dispersive Micro-solid Phase Extraction

The sorbents used in d-μSPE play a dual role in the overall extraction procedure: (i) clean-up and (ii) enrichment/pre-concentration of target analytes. Their selection is dictated, on the one hand, by their affinity towards the target analytes, i.e., how easily they can trap them, and on the other hand, by the desorption phase, which is equally important.

An important requirement of any extraction methodology is that the analyte-sorbent interaction be driven by a mechanism of adsorption/absorption characterized by interactions such as van der Waals forces, electrostatic interactions, hydrogen

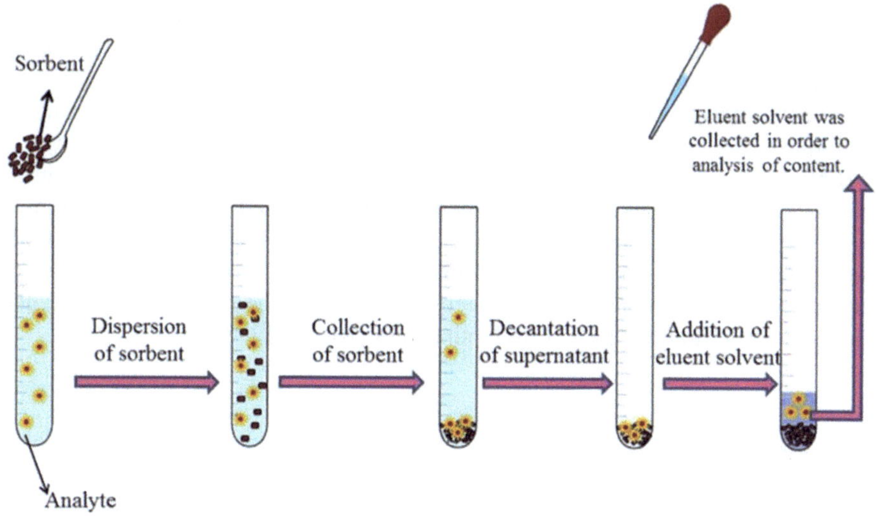

Fig. 1 Schematic illustration of the main steps in d-μSPE. Reprinted from [12] with permission from Elsevier

bonding, dispersion forces, π–π interactions, dipole–dipole interactions, which are reversible under different conditions to allow prompt desorption.

The efficiency of both the adsorption and desorption processes depends on the contact surface between the sample and the sorbent and, subsequently between the sorbent and the desorbing solvent. In this regard, the choice of the solid phase has encouraged the development of new materials with different sorbents, not only to achieve different analytical/chemical selectivity and chemical/physical stability but also with structures possessing a high surface-to-volume ratio.

Over the years, numerous sorbents have been employed in d-μSPE, which could be classified according to various criteria via chemical composition or main structure.

In this context, three main groups of sorbents could be identified [7]:

(i) Micro materials: this group includes classical sorbents originally used in SPE, such as the well-known octadecyl silica (C_{18}), which were initially proposed at the beginning of d-μSPE and more selective polymer-based materials like molecularly imprinted Polymers (MIPs);
(ii) Nanostructured sorbent materials offer unique advantages in d-μSPE due to their small size and high surface area. This group includes nanoparticles (NPs) metal and nonmetal oxides and their derivatives, carbon-based nanomaterials or composites (carbon nanotubes, graphene and its derivatives, etc.) Within the NPs group are included the magnetic nanoparticles (MNPs) that characterize the magnetic solid phase microextraction (MSPME);
(iii) Hybrid Materials (Composites): The third group consists of hybrid materials or composites that combine properties of micromaterials and nanostructured adsorbents.

Over the years, numerous sorbent materials have been synthesized specifically for the extraction of different classes of compounds from various matrices, including environmental, agro-food, and biological samples. Therefore, in this chapter, a careful selection was carried out to provide a comprehensive overview of this microextraction technique.

Bibliographic research was achieved by using the most important scientific databases such as Scopus (http://www.scopus.com/), PubMed (http://www.ncbi.nlm.nih.gov/pubmed) and ISI Web of Knowledge (https://webofknowledge.com/) and Google scholar (https://scholar.google.it/) allowing for a selection of a series of research papers and reviews. This selection is listed in the reference section, and readers can refer to it for further in-depth exploration [7, 12–21].

3.1 Micromaterials

During the early development of the technique, the research initially focused on using sorbent materials commonly employed in dSPE, which consist of easily dispersible micrometric particles suitable for the extraction process. Given their crucial role in d-μSPE, these sorbent materials act as traps for analytes of various natures.

3.1.1 Silica-Based Sorbents

Sorbent materials based on silica functionalized with primary-secondary amines (PSA), RP-C18, aminopropyl ($-NH_2$), cyanopropyl (–CN), strong anion exchange (SAX), neutral alumina, graphitized black carbon (GCB) and alumina were extensively investigated.

Taking advantage of the knowledge acquired from the SPE technique, RP-C18 silica-based particles have proven to be among the most versatile sorbents thanks to their excellent adsorption capacity, high mechanical stability, and long-lasting performance. Therefore, they were proposed as initial sorbents in the dispersive μ-SPE technique.

This sorbent has been used to extract organic contaminants such as Polycyclic Aromatic Hydrocarbons (PAHs). When combined with GC-MS, the sorbent material was vigorously dispersed in water using a specialized glass device and subsequently subjected to desorption and filtration through a microextraction unit before direct injection into the GC-MS system [22].

A remarkable improvement in d-μSPE, when coupled with GC-MS, was proposed by Galán-Cano et al., where the sorbent was previously dispersed in acetone, and the elution phase was replaced with thermal desorption in the injection port using an external heater and an airflow [23].

In a broader context, Tsai et al. conducted a pioneering study evaluating the utility of different silica-based sorbents for the preconcentration of tetracyclines from surface waters and milk samples. The authors presented a comprehensive investigation that also included polymer-based sorbents commonly used in conventional SPE, such as the Oasis® series from Waters (Milford, MA, USA), the Strata series from Phenomenex® (Torrance, CA, USA), and the polyamide resin (DPA6S) from Supelco® (Bellefonte, PA, USA).

The study examined the recovery factors under various extraction conditions in water and acetonitrile, revealing that silica-based sorbents significantly outperformed the polymer-based ones [11].

In recent studies, silica-based particles such as hemimicelles were coated with ionic liquids [24] and core–shell particles coated with polymers, such as polyaniline, have shown promising extraction capabilities [25]. Similar silica material such as zeolites, which are aluminosilicate minerals known for their high porosity, has been modified with surfactants (e.g., cetyltrimethylammonium bromide), or it has been coated with polyaniline [26].

3.1.2 Polymeric Sorbents

As for silica-based sorbents, polymeric sorbents are a widely applied category traditionally used in SPE, and they have found a technological boost in their miniaturized application, partly due to their reduced cost. Their notoriety is dictated by their key features, including good stability, high porosity, simple synthesis, modifiable structure, and suitable adsorption capacity. The vast combinatorial possibilities of monomers to synthesize polymers with diverse chemical interactions, such as polar, non-polar, and ion-exchange properties, allow for the extraction of organic and inorganic analytes from real samples of any chemical properties.

In 2011, OASIS®-HLB (a copolymer of polyvinylpyrrolidone-divinylbenzene) was employed as the first non-polar polymeric adsorbent for the extraction and determination of nitroaromatic compounds in water samples, serving as a model analytical problem. The success of the extraction was attributed not only to the hydrophobic interaction but also to the π-π interaction between the benzene ring in the sorbent and the analytes. The authors present an alternative method for dispersing the sorbent material in this experimental study. Instead of using typical methods like vortexing and/or combining with an appropriate organic solvent to ensure maximum dispersibility of the sorbent particles, the authors propose a green alternative: the effervescence-assisted μ-SPE. This procedure was achieved using a suitable tablet containing effervescence precursors (sodium carbonate as a carbon dioxide source and phosphate salt as a proton donor). In this case, when the mixture of sorbent/tablet was into the sample, the effervescence phenomena began releasing CO_2, causing the homogeneous dispersion of the sorbent [27].

Widely used sorbents are polymer ion exchange materials (PIX) where the extraction mechanism relies on strong electrostatic interaction analyte/sorbent that can be changed by the pH of the sample solution, influencing the protonation or deprotonation of their functional groups, both analyte and polymer. In this respect, polymer cation exchange materials (PCX) and polymer anion exchange (PAX) has been applied to extract various analytes such as triazine herbicides in drinking water sample [28], cyromazine and melamine in milk samples [29], adamantane drugs in chicken muscle [30], glycopyrrolate stereoisomers in rat plasma [31], extracted bisphenols in edible oils [32].

Although d-μSPE can be included in the green trend Analytical Chemistry due to its miniaturization, sometimes, the same cannot be said for using less sustainable sorbent materials. As in this case, synthetic polymers are a concern for environmental pollution due to their high stability and persistence in the environment. A greener alternative approach is using natural polymers that offer ready availability, non-toxicity, excellent ability to decompose and additional advantages, including ease of functionalization, and good stability. Pourmand et al. describe the possibility to use a biopolymer, Agarose, that was functionalized by co-polymerization with polymethylmethacrylate. Improving its reactivity, stability, and surface area, the polymeric sorbent Agarose based proved successful in extracting four metal ions, including Cd(II), Ni(II), Cu(II), and Zn(II), from water and vegetable samples [33].

When the magnetic sorbents are employed, magnetic solid phase extraction is used. The first magnetic polymeric sorbent was proposed in 2013 by Gang Zhao et al., who developed a tetraethylenepentamine-functionalized magnetic polymer. Several stages for the synthesis of this sorbent were involved: (i) Preparation of Fe_3O_4 using a co-precipitation technique, (ii) Coating of Fe_3O_4 with oleic acid, (iii) Co-polymerizing Fe_3O_4 coated with oleic acid along with glycidyl-methacrylate, divinyl-benzene, and methyl methacrylate using the suspension polymerization strategy, and (iv) functionalizing the final product with tetraethylenepentamine [34].

Thanks to the hydrogen bonds formation between the amine groups on the sorbent surface and the hydroxyl groups of the analytes, this novel material achieved a high extraction efficiency of phenolic environmental estrogens from human plasma samples.

Magnetic sorbent involved in polymeric structure can be synthesized using a Sol-gel technique as a Sol-gel technique proposed by Behbahani et al. After several steps of the synthesis process, both magnetic particles by Fe_3O_4, SiO_2, tetraethoxysilane and triethoxyvinylsilane and its coating of polymeric material using 3-vinylthiopene as a monomer and AIBN as a polymerization initiator was applied to the extraction of Cd(II) from biological and water samples [35].

3.1.3 Molecularly Imprinted Polymers

Molecular imprinting polymers (MIPs) are synthetic polymers created from monomers, cross-linkers, and templates. Briefly, during the polymerization process, the template is enclosed within the 3D polymer network. The choice of monomer depends on the type of template and its functional groups. Suitable interactions between the monomer and the analyte (template) lead to the development of specific MIPs for target analyte extraction. Hydrogen bonding or ionic interactions between templates and monomers are preferred over covalent bonding to facilitate template removal. Chemical initiators like peroxide and azo compounds, photon initiators (ultraviolet light), and thermal initiators are used to initiate the polymerization procedure based on the type of monomer and cross-linker model. The solvent used as a progeny agent plays a crucial role in determining the morphology of the MIP structure, particularly the size of the pores and surface area. MIPs with high surface areas and low pore distributions are achieved by increasing the solubility of monomers and templates in the solvent or reducing the solvent volume during the MIPs synthesis. MIPs used as sorbents in d-μSPE offer several advantages, including simple preparation, high thermal and chemical stability, reusability, excellent selectivity, and cost-effectiveness [36, 37].

In 2012, Chen et al. developed the first MIP as a sorbent for d-μSPE to determine sulfamethazine. The sorbent was prepared using sulfamethazine as the template, methacrylic acid as a monomer, and ethylene glycol dimethacrylate as a cross-linker, with acetonitrile as the solvent and AIBN as the initiator. The MIP exhibited high selectivity and sensitivity, with a short equilibrium time of 5 min and excellent enrichment capability for sulfamethazine extraction [38].

Various modifications have been attempted to improve molecularly imprinted polymers (MIPs), including, the synthesis of dummy molecularly imprinted polymers (DMIPs).

Although MIPs generally exhibit excellent selectivity for analyte extraction, their selectivity can be compromised due to template retention within the MIPs or residual template leakage. In contrast, DMIPs offer a different approach by using a dummy template instead of the actual analyte. This solution proves beneficial in scenarios involving costly or hazardous analytes during polymerization, low solubility in the polymerization solvent, or analyte degradation during the process [39, 40].

Different technologies and synthesis approaches are combined to make the microextraction process more effective, stable, and user-friendly. Using a strong magnetic field for sorbent separation is a convenient and faster alternative to centrifugation. The synthesis procedure of MIPs can be integrated with magnetic nanoparticle formation technology.

Cheng et al. successfully prepared magnetic dummy molecularly imprinted microspheres (mag-MIMs) using dichlorodiphenyltrichloroethane (DDT) as the dummy template for extracting dicofol from tea products [41].

The preparation of magnetic molecularly imprinted polymers (MIPs) involves combining the excellent selectivity of MIPs with the ease of magnetic separation provided by magnetic nanoparticles (Fe_3O_4 NPs). Using a two-step process, bazmandegan-Shamili et al. developed a magnetic MIP (MMIPs) for diazinon extraction. First, magnetic $Fe_3O_4@SiO_2$ was prepared via the sol-gel technique. In the second step, an MIP was synthesized on the surface of the magnetic $Fe_3O_4@SiO_2$ using the precipitation polymerization technique. This novel sorbent was proved effective in the extraction of diazinon.

$Fe_3O_4@SiO_2@MIP$ proved effective in the extraction of fenitrothion and quercetin. Furthermore, studies showed that this sorbent exhibited excellent stability and could be reused for several rounds of analyte extraction without significantly declining extraction efficiency [12, 42–44].

The extraction efficiency of MIPs can be enhanced by reducing the extraction time, improving the mass transfer of the analyte, and increasing the surface area of the sorbent, for example, through the MOF@MIP or Carbon-Based Material @MIP preparation strategy.

3.2 *Nanostructured Sorbents*

3.2.1 Nanoparticles

Nanoscience and nanotechnology have had a significant impact on recent Analytical Chemistry research. The introduction of nanoparticles (NPs), both inorganic and organic, represents a revolutionary milestone in analytical sciences. NPs are particularly attractive due to their inherent chemical, electrical, optical, thermal, and magnetic properties, which have been successfully utilized in sample treatments, especially microextraction techniques [12, 44].

The researchers demonstrated that reducing the particle size leads to an exponential increase in the area-to-volume ratio, resulting in an increasing nanomaterial sorption capacity [45].

An essential aspect of this extraction procedure, crucial for fully exploiting the sorption capacity of a given nanomaterial, is to achieve efficient dispersion of the sorbent within the sample.

Early experimental works on this topic concern the use of metal nanoparticles as sorbents, they have not received much attention, possibly due to their limited ability to interact effectively with the analyte. In this respect, the gold nanoparticles (AuNPs) were among the earliest nanoparticles used for solid-phase nanoextraction (SPNE). In a study focused on extracting PAHs, commercial colloids of AuNPs were utilized [46]. An intriguing finding in this study was the correlation between extraction efficiency and the diameter of AuNPs. This led to the conclusion that smaller particles, such as the 20 nm AuNPs, were more efficient for extraction, emphasizing the importance of nanometric size in miniaturized extraction techniques. Another application for determining Hg(II) ions was achieved by using silver nanoparticles (Ag NPs) as a sorbent for the extraction. The authors reported the rapid formation of an amalgam resulting from the interaction between inorganic mercury and Ag-NPs during the extraction process [47].

Metal oxide nanoparticles, including TiO_2, ZnO, and CeO_2, have been developed as sorbents for extracting specific analytes such as mercury species, germanium, and beryllium ions [48, 49]. These studies have revealed that the sorbent cavities' size, shape, and surface properties play a crucial role in the analyte absorption process, primarily through physisorption. In this respect, experimental conditions, such as sample solution pH, do not directly impact the size and shape of the sorbent cavities but significantly influence the extraction process by altering the surface charge of the sorbent.

Metal oxide nanoparticles have primarily been employed for the extraction of inorganic cations and anions, with limited reported applications for organic compounds [50–53]. In general, the limited application can be attributed to their high dispersibility and stability in suspension, which poses challenges in their recovery from the bulk sample solution [7].

Nonetheless, these nanoparticles are frequently amalgamated with other compounds to create nanocomposites or functionalized nanoparticles, augmenting their extraction and improving the selectivity as sorbents. Silica, a non-metal oxide, finds extensive use in d-μa SPE owing to its distinctive attributes such as straightforward synthesis, elevated specific surface area, controllable morphology, environmental friendliness, and exceptional stability. As mentioned, silica-based sorbents are subject to modification with an ionic liquid to enhance their extraction capabilities for organophosphate pesticides from water samples. [24].

Magnetic iron oxide nanoparticles (Fe_3O_4@NPs) are extensively used as sorbents in d-μSPE due to their easy and rapid separation using a strong magnet, eliminating the need for centrifugation or filtration step (Fig. 2). They offer benefits such as low toxicity, simple synthesis, easy modification, large surface area, and high adsorption capacity [54, 55].

However, using pure Fe_3O_4 as sorbents faces challenges due to poor oxidative stability and the tendency to form large aggregates, making them non-selective. However, these issues can be addressed using cobalt-ferrite ($CoFe_2O_4$) or growing a silica shell over Fe_3O_4 MNPs to improve their performance as sorbents [56, 57].

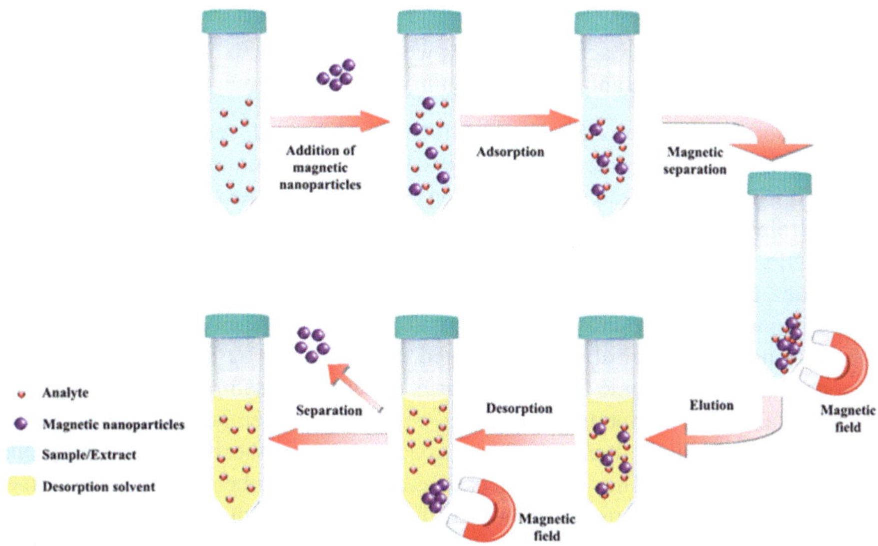

Fig. 2 Schematic illustration of the main steps in magnetic dispersive micro-solid phase extraction. Reprinted from [18] with permission from Taylor & Francis

To address these issues, Fe_3O_4 NPs are often functionalized with appropriate compounds to enhance their selectivity and reduce aggregation, with various procedures applied for surface modification. Researchers usually prefer using MNPs coated with different sorbents, creating core-shell structures or forming hemimicelles and/or admicelles with surfactants, ionic liquids, or polymeric chains. Additionally, MNPs can be embedded in polymers or other more selective materials to confer on them the magnetism needed for easy recovery, resulting in hybrid materials. This versatility of available coatings in the analytical literature allows for a wide range of applications and improved selectivity [58].

Since the pioneering application in d-μSPE by Shen et al. in 2007 [59], where C18-coated Fe_3O_4 MNPs were utilized for the extraction of organophosphorus pesticides, the field has seen exponential growth with nearly half a thousand published articles exploring the use of magnetic materials in d-μSPE.

Several procedures were applied to modify the Fe_3O_4 NPs surface.

(i) Direct formation of metal oxide on Fe_3O_4 NPs.

In this approach, metal oxide nanoparticles are deposited directly onto the surface of Fe_3O_4 NPs using a chemical precipitation method. Fe_3O_4 NPs are initially dispersed in a solvent, typically water, and an ionic compound of the metal is dissolved in the Fe_3O_4 NPs suspension. As the pH of the solution is increased to above 9, metal hydroxide nanoparticles precipitate on the Fe_3O_4 NPs. Finally, a calcination step is carried out to form the metal oxide nanoparticles on the Fe_3O_4 NPs surface. For example, Fe_3O_4@Al_2O_3 NPs were prepared using this method for the extraction of Mo(VI), Sb(V), and V(V) ions, [60];

(ii) In the second approach, Fe_3O_4 NPs were modified with conducting polymers like polydopamine and polypyrrole through chemical reactions [61, 62].

The modification process is straightforward and rapid, requiring no complex equipment. During this process, Fe_3O_4 NPs and a suitable monomer are dispersed in an appropriate solvent, and polymerization occurs to form the sorbent. For instance, polypyrrole/Fe_3O_4 NPs were synthesized by polymerizing pyrrole on the surface of Fe_3O_4 NPs and used as a sorbent for extracting antidepressant drugs [62], Socas-Rodríguez et al. [61] utilized classical polydopamine-coated Fe_3O_4 nanoparticles to determine 21 compounds with estrogenic activity in various milk samples. On the other hand, Vakh et al. [63] employed carbon-coated ZrO_2@Fe_3O_4 magnetic nanoparticles to determine fluoroquinolones in baby food samples.

This class novel compounds has gained popularity as alternative extraction media in analytical chemistry due to their low toxicity, easy synthesis, low volatility, and high thermal stability. They are often used as coatings on sorbents to enhance extraction selectivity and sensitivity. However, their use as solvents can pose challenges related to phase separation. ILs can be immobilized covalently on sorbents to address this issue, ensuring easy separation. For instance, C16mimBr-coated Fe_3O_4 MNPs were utilized for extracting PAHs from water, where the IL and MNPs were mixed in the sample solution, enabling simultaneous coating and extraction processes. This approach offers a solution to overcome the limitations of IL elution in the final extract, making it compatible with separation systems like GC [64].

Different approaches are achieved through self-assembly adsorption. This simple and efficient procedure involves dispersing the IL and Fe_3O_4 NPs into a polar aprotic solvent. Following this procedure, Peng et al. synthesized [C_8 MIM][PF_6]@Fe_3O_4 NPs based on the self-assembly adsorption with dispersing [C_8MIM][PF_6] and Fe_3O_4 for performed extraction of clofentezine and chlorfenapyr from water sample [65]. Fe_3O_4 NPs can be functionalized with a ligand. In a notable study, Fe_3O_4@decanoic acid nanoparticles (NPs) and CTAB (cetyltrimethylammonium bromide) as a ligand were used for extracting acidic (diclofenac, DIC) and basic (diphenhydramine, DPH) drugs from the Biological fluid sample [66].

The sorbent (CTAB-Fe_3O_4@decanoic acid NPs) was formed in situ through interactions between Fe_3O_4@decanoic acid NPs, CTAB, and the analytes, involving hydrophobic interactions, ion-pair formation, and π-cation interactions as shown in the Fig. 3.

In another study, CTAB-Fe_3O_4@caprylic acid NPs were used for extracting 17β-estradiol, estrone, and diethylstilbestrol. Increasing CTAB enhanced the dispersion of the sorbent in the sample solution, leading to improved extraction efficiency and reduced extraction time due to increased contact between the sorbent and analytes [66].

The authors found that increasing the amount of CTAB enhances the dispersion of the sorbent in the sample solution. This improvement in dispersion is beneficial in DμSPE, as it leads to increased extraction efficiency and reduced extraction time.

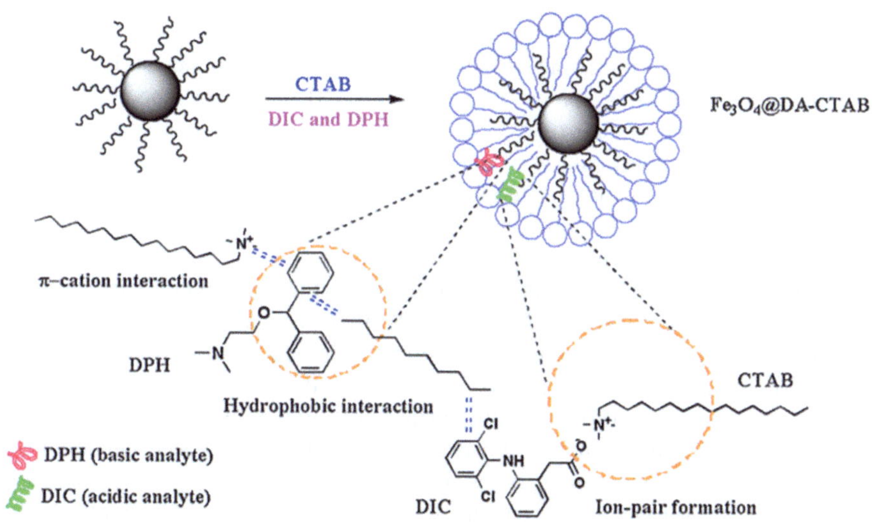

Fig. 3 Schematic diagram of synthesis procedure and probable interactions between the analytes and CTAB-coated Fe_3O_4@decanoic acid nanoparticles. Reprinted from [66] with permission from Elsevier

Enhancing the sorbent dispersion efficiency in the sample solution increases the contact surface between the sorbent and the analyte, allowing for more efficient extraction of the analyte [67].

Another approach is the sol-gel technique for functionalized Fe_3O_4. The sol-gel method is a highly suitable technique for coating sorbents onto Fe_3O_4 due to its numerous advantages, such as high surface area, strong adhesion, homogeneity, and porosity. This process involves two steps under mild thermal conditions: (1) hydrolysis of a precursor and (2) condensation of water or alcohol. Alkyl alkoxysilane is commonly used as a precursor, and an acid or base catalyst is applied to control the reaction rate and morphology of the resulting sol-gel material. The size of the material pores produced in the sol-gel process differs depending on the catalyst used, with acidic catalysts producing smaller pores than basic catalysts. Several methods have been developed to functionalize $Fe_3O_4@SiO_2$. For instance, Ghazaghi et al. heated $Fe_3O_4@SiO_2$ NPs with Ni(II) solution and ammonia to prepare Fe_3O_4@NixSiOy sorbents. To extract tyrosine kinase inhibitors (TKIs) from serum and cerebrospinal fluid samples [68].

The polymeric sorbents, formed by dispersing $Fe_3O_4@SiO_2$ in an organic solvent along with a monomer, cross-linking monomer, and initiator, have proven to be highly stable in various solvents. This stability is attributed to the strong chemical bonding between the polymeric coating and the magnetic core of $Fe_3O_4@SiO_2$. Additionally, the presence of the magnetic core enables these sorbents to be easily separated from sample solutions using a strong magnet [69].

3.2.2 Carbon-Based Nanomaterials

Carbon nanomaterials, including carbon nanotubes (CNTs), graphene (G), graphene quantum dots (GQDs), graphene nanoribbons (GNRs), graphdiyne (GDY), and fullerenes, are versatile allotropic forms that have found extensive applications in analytical chemistry, particularly in sample preparation, separation techniques, and detection systems. These nanomaterials possess a hexagonal lattice with sp2 carbon atoms, except for GDY, which contains both sp and sp2 carbon atoms. Due to a π-conjugated structure, these nanomaterials can engage in non-covalent interactions with other substances, making them suitable for interactions with aromatic compounds through π-stacking. Consequently, they serve as effective sorbents for extracting aromatic compounds rather than ionic compounds. Moreover, their advantages include high extraction capacity, chemical inertness, mechanical and thermal stability, good dispensability, and large surface area, making them valuable tools in analytical chemistry applications. However, it is essential to address the challenge of nanoparticle aggregation, which arises from the limited solubility of carbon nanomaterials in both aqueous and organic media. This issue is particularly significant for carbon nanotubes, a well-known allotropic form of carbon. The aggregation hinders their practical use in conventional extraction devices like SPE cartridges or miniaturized units, where the sample flow is critical. The presence of nanoparticle bundles also reduces the extraction process's efficiency and leads to system overpressure.

To fully harness the sorbent capabilities of carbon nanoparticles in microextraction, immobilizing them on an inert or active surface within miniaturized extraction devices proves to be a promising approach. This strategy prevents nanoparticle aggregation and allows for efficient utilization of their sorption properties. Available comprehensive reviews are available for additional insights on this topic [70, 71].

Considering that as-grown carbon nanotubes often exhibit bundle formation, surface modification becomes crucial to prevent aggregation and enhance the selectivity when utilizing carbon nanoparticles as extracting media. In the context of carbon nanoparticles, oxidation is commonly employed to introduce hydroxyl, carboxyl, or carbonyl groups, which improves their solubility in aqueous media and reduces aggregation. Additionally, less frequently, functionalization with various active groups has been proposed further to facilitate the use of carbon nanoparticles in dispersive μ-SPE, enhancing their applicability and selectivity in sample preparation processes.

Carbon nanotubes (CNTs) come in two main types: single-walled carbon nanotubes (SWCNT) and multi-walled carbon nanotubes (MWCNT) [72].

SWCNTs have a higher absorption capacity due to their higher surface-to-volume ratio than MWCNTs, consisting of several cylinders with interlayer spacing. As sorbent in dispersive μ-SPE, it can be oxidised to overcome the aggregation and reduce the surface area for analyte interaction. For instance, the oxidation of single-walled carbon nanohorns (SWNHs, the same family of SWCNTs) into oxidized SWNHs (o-SWNHs) using a microwave device increased the number of polar functional groups on the sorbent surface, enhancing the dispersion efficiency in the aqueous/polar organic sample solution. The ionic strength and pH of the sample solution also play a role in the sorbent dispersion efficiency.

The study investigated the dispersion efficiency of o-SWNHs for of extracting polycyclic aromatic hydrocarbons (PAHs) from environmental water samples [73]. The o-SWNHs have been successfully used for the extraction of triazines from water samples, mainly through π-stacking interactions [74].

To improve the dispersion of CNTs in solution, surfactants have been employed. Cationic surfactants have shown to be particularly effective in enhancing the stable and homogenous dispersion of CNTs. This is achieved by the adsorption of cationic surfactants on the CNTs' surface, creating a positive charge that induces electrostatic repulsion between CNT particles, preventing aggregation, and resulting in a stable dispersion of the sorbent particles in the solution [75].

CNTs are not suitable sorbents for the extraction of ionic analytes due to their non-polar nature. However, a straightforward solution to enhance their extraction efficiency is applying a ligand that can form a hydrophobic complex with cations in the sample solution. This leads to an increased adsorption of the analyte on the CNTs surface through hydrophobic interactions and van der Waals forces. The pH and ionic strength of the sample solution play a crucial role in the complex formation by affecting the structure of the functional groups at the ligand surface and the analyte species. By controlling these factors, the extraction recovery can be improved by increasing the efficiency of complex formation. An example of this approach is demonstrated with MWCNTs and graphene by Skorek and co-workers, who utilized ammonium pyrrolidine dithiocarbamate (APDC) as a chelating agent to form a hydrophobic complex with Se(IV) [76, 76].

This hydrophobic complex was successfully extracted using MWCNTs or oxidized MWCNTs (O-MWCNTs) as sorbents from water and biological samples, showcasing the potential of this method for efficient microextraction in analytical chemistry [78].

The functionalization of CNTs is a commonly employed method to enhance the sorbent's selectivity by altering its interaction with the analyte. A simple and direct approach involves the physical adsorption of a suitable compound into the pores of CNTs. For example, a green sorbent for the extraction of flavonoids was developed by dispersing MWCNTs into a chitosan/acetic acid suspension through sonication treatment [79].

A straightforward same procedure was also applied to prepare the magnetic CNTs as magnetic sorbent of Fe_3O_4@CNTs [80, 81].

In addition to physical adsorption, chemical bonding between the components of the sorbent can enhance its stability. Two steps are typically involved in synthesising functionalized CNTs as sorbents. The first step is CNT oxidation using various strategies such as microwave devices, chemical procedures, electrochemical oxidation, and sonochemical methods. The second step includes substituting the OH group in the carboxyl functional group on CNTs with appropriate functional groups or compounds through substitute-on reactions [82].

The extraction procedure involves preparing a suspension of graphene with ammonium pyrrolidine dithiocarbamate (APDC) as a ligand and Triton-X-100 as a disperser agent. This results in the in-situ formation of a non-ionic complex between metal ions and APDC, which adsorb on the graphene surface through van der Waals forces. The use Ofof Triton-X-100 as a surfactant helps create a homogeneous graphene suspension and enhances the repeatability of the extraction method [77, 83].

Graphene oxide (GO) is a significant derivative of graphene that finds extensive use as a sorbent in various applications. The presence of functional groups, such as carboxyl and hydroxyl functional groups on the edges of GO sheets and ether functional groups on the GO sheets, gives it unique properties. These functional groups enable GO to interact with analytes through hydrogen bonding and polar-polar interactions, facilitating the absorption and extraction of analytes. Moreover, the functional groups on GO can be easily modified with different functional groups, enhancing the sorbent's selectivity and allowing analytes to interact selectively with various analytes.

In analytical chemistry, GO as a sorbent in d-μSPE has been used in numerous studies for the extraction of various compounds, including theophylline, theobromine, caffeine, nicotine, phenolic compounds, malachite green, crystal violet, metal ions, and organophosphorus pesticides (OPPs) [84–90].

The interaction between GO and the analytes in many of these studies was found to be pH-dependent, where the pH of the solution affects the protonation or deprotonation of the functional groups on GO, resulting in a positive or negative surface charge on the GO, respectively.

GO offers various advantages as a sorbent due to its easy functionalization, presence of functional groups for analyte interactions, and aromaticity, making it particularly effective for extracting aromatic compounds. It can be easily dispersed in aqueous samples, making it convenient for sample preparation in analytical chemistry.

Hybrid materials based on clays and graphene have shown promise for enhancing specificity in extraction processes. Clay minerals possess 2D structures with high intercalation capacity and porosity, making them suitable support materials for other adsorbents. They can also interact with analytes through polar and hydrogen bonding interactions [91].

Magnetic composites of graphene, such as Fe_3O_4/rGO and magnetic GO, have been extensively used to extract polycyclic aromatic hydrocarbons (PAHs) from water samples. These nanocomposites offer the advantage of easy separation using magnets and can be tuned for specific extraction needs by adjusting their composition [92].

In summary, GO and its derivatives, along with hybrid materials and magnetic composites of graphene, are valuable sorbents with various advantages for dispersive micro-solid phase extraction (d-μSPE) applications in analytical chemistry, enabling efficient and specific extraction of target compounds from complex samples.

3.2.3 Metal-organic Frameworks

Metal-organic frameworks (MOFs) are a versatile class of mesoporous materials consisting of metal cations or clusters linked by organic ligands. These materials offer several advantages over traditional porous materials like zeolites, primarily due to their structural flexibility during preparation. MOFs can be tailored to have specific cavity sizes, surface areas, and functionalization by selecting different building blocks, synthesis strategies, and conditions. Combining these main features, MOFs have emerged as reliable and innovative solid sorbents in various sample preparation procedures [93–95].

Overall, MOFs offer exceptional possibilities for sorbent applications due to their tunable properties and the ability to control their structure and size through different synthesis methods, and in particular, MOF's pore environment, pore size, pore aperture widths, presence of unsaturated metal sites, and the type of metal used significantly influence their performance d-μSPE [96].

4 Dispersion Techniques in Dispersive Micro-Solid Phase Extraction

4.1 External Energy Source in the Dispersion Step

In d-μSPE, the dispersion of the sorbent plays a crucial role, especially when using sorbents that tend to aggregate easily, such as carbon nanotubes. The best extraction performances are ensured when the dispensability phenomenon occurs throughout the entire extraction process, including analyte isolation and elution steps. The critical factor affecting the efficiency of analyte extraction in d-μSPE is the uniform and complete dispersion of sorbent particles in the sample solution. In this respect, increasing the number of sorbent particles and reducing their size in the sample solution enhances the contact surface between the sorbent and analyte, resulting in two main advantages: reduced sorbent consumption and shorter extraction or desorption time.

However, due to the polarity difference between the sorbent and the sample matrix, the initial phase dispersion step is often more difficult for various reasons. To overcome these challenges, the dispersion process may require assistance from external energy sources, such as mechanical stirring etc. or the use of specific chemicals.

When an external energy source aids the dispersion of sorbent, the d-μSPE method can be categorized into three main types:

(i) Traditional d-μSPE; (ii) Vortex-assisted d-μSPE (VA-d-μSPE), (iii) Ultrasonic-assisted-d-μSPE. (UA-d-μSPE).

Each of these methods can further be classified based on the type of sorbent used, falling into two groups: neat materials and composites. They can also be divided into two subgroups: magnetic sorbents and non-magnetic sorbents. Generally, a review of published papers reveals that the application of the d-μSPE method has significantly expanded over the years. Among the three methods, ultrasonic-assisted d-μSPE is the most frequently applied, followed closely by vortex-assisted d-μSPE, with traditional d-μSPE having a lower frequency of application (including chemical dispersion mode).

In the traditional d-μSPE method, as previously described, the sample and the sorbent is placed in an extraction vessel where an external energy source is applied for stirring or dispersion for a specified duration. Once the extraction process is complete, the sorbent is typically retrieved through centrifugation or filtration. In the case of magnetic sorbents, an external magnetic field can be used for retrieval. Subsequently, the elution process, which may involve re-dispersion, is carried out, and the sorbent and eluate are separated before the instrumental analysis takes place.

Studies conducted on mild dispersion, such as the stirring mode, involve sorbent materials that exhibit a strong tendency for dispersion due to their chemical properties. Generally, sorbent materials based on silica and polymers those that also possess hydrophilic properties, can be easily dispersed in both aqueous and organic phases.

Traditional d-μSPE has been successfully applied to environmental, agri-food and biological matrix samples [14].

The use of external energies in the extraction process accelerates the dispersion of the sorbent and enhances the diffusion of analytes. However, if not appropriately controlled, the application of energy can increase the vessel's temperature, negatively impacting the extraction thermodynamics and potentially degrading the analytes.

The vortex agitation mode can be regarded as a mild dispersion technique and applied in sample preparation as in Vortex assisted d-μSPE procedure. Although its simplicity, wide availability, cost-effectiveness, and capacity to extract multiple samples simultaneously make it an attractive choice for improving mass transfer in microextraction techniques [24, 97].

Although the extraction kinetics may be slower compared to more energetic methods, they still fall within a reasonable time range of minutes. Additionally, the mechanical dispersion caused by the vortex is less likely to lead to analyte degradation during the extraction procedure, which is considered a significant advantage. VA-d-μSPE, a technique introduced to extract triazines from water samples, utilizes oxidized single-walled carbon nanohorns as the sorbent [74]. In this approach, microwave energy is used during sorbent preparation to achieve a homogeneous and stable dispersion of the sorbent in water, modifying pristine material. In this case, the VA-d-μSPE procedure demonstrated significantly performed for this study. VA-d-μSPE was effective for IL-modified silica composite sorbents for extracting organophosphate pesticides from water samples [98]. Vortex-assisted was also introduced in the d-μSPE by using MOFs sorbent. Several studies have demonstrated the effectiveness of MOFs sorbents in VA-d-μSPE for extracting different analytes, such as parabens and heavy metal ions, contaminants from various environmental and biological samples [41, 99].

Ultrasonic waves find numerous applications in chemistry, such as electrochemistry, organic compound analysis, material synthesis, and sample preparation. When the ultrasonic bath (or probe) is considered in laboratory procedures, it is a potent energy valuable source in analytical chemistry procedures. The bubbles formation that form inside as a result of waves passing through the liquid collapse to produce the cavitation phenomenon. This formincludes very short-life micro-reactors of extreme conditions that are usefuladvantageous in the sample preparation step. The main effects, especially on d-μSPE, are:

(i) Effective method to disperse nanoparticles and eliminate agglomeration in aqueous suspensions due to increased interparticle collisions for a mechanical effect of collapsing cavitation and shockwave; (ii) reduces the viscosity and size distribution of nanoparticles in the sample solution and then the time required to reach equilibrium and enhancing extraction efficiency; (iii) add the extraction (or desorption) procedure and reducing the amount of sorbent needed.

In this regard, using ultrasonic assistance to disperse the sorbent material into the sample solution is known as ultrasonic-assisted dispersive micro solid-phase microextraction (UA-d-μSPE). Researchers have successfully applied UA-d-μSPE in various studies.

Leading to reduced sorbent and organic solvent consumption, the miniaturized technique was suitable for expensive precious materials and complicated synthesis as MOF or GO for extracting hormones from cosmetics and extracting theophylline, theobromine, and caffeine from tea samples, respectively [84, 100].

The use of UA-d-μSPE can lead to two main challenges: the potential for analyte or sorbent degradation and the formation of free radicals due to high temperatures and pressures during cavitation and the difficulty of separating the sorbent from the sample solution when a stable suspension of the sorbent is formed. To address these issues, Liu et al. implemented a combination of ultrasonication and vortex techniques to disperse the sorbent into the sample solution to extract food colorants in beverages. Although ultrasonication promotes a homogeneous sorbent dispersion and enhances the interaction between the analyte and the sorbent, prolonged ultrasonication can increase the solution temperature and lead to analyte desorption from the sorbent [101].

This non-negligible aspect of ultrasound raises the question of an accurate study of its time in the extraction procedure as a compromise between the degradation of the analytes and reduction of the matrix effect. This issue was faced by of developing a new method for extracting naproxen and ibuprofen from real water and milk samples [102].

Not least, of ultrasonic waves could give rise to a potential issue of sorbent degradation due to the prolonged dispersion stage into the solvent. To address this concern, researchers introduced a solution by incorporating a small amount of surfactant to aid the sorbent dispersion process, reducing the ultrasonication time [103].

4.2 Chemical Dispersion

Chemical dispersion is a technique used to enhance the dispersibility of a sorbent in a sample. This approach was coined 2013 as solvent-assisted dispersive solid-phase extraction (SAdSPE), which involves using a non-conventional sorbent, such as benzophenone, dispersed into an aqueous sample using a water-miscible organic solvent. The process is similar to classical DLLME, but the sorbent becomes solid in the aqueous environment and can be recovered through centrifugation or filtration. However, of using water-miscible solvents as dispersers may negatively impact analyte partitioning due to increased solubility in the donor solution [104, 105].

As previously reported in the Sect. 1.2, Lasarte-Aragones et al. proposed the Effervescence-assisted dispersive micro-solid phase extraction (EA-d-μSPE) as an alternative solution for the dispersion step. Briefly, this method relies on an effervescent reaction to disperse the sorbent. A tablet containing a proton donor, a carbon dioxide source (e.g., sodium carbonate), and the sorbent are prepared and directly added to the sample. The tablet's dissolution and reaction generate carbon dioxide bubbles that efficiently disperse the sorbent. The EA-d-μSPE allows for easy on-site extraction, and the disperser does not significantly affect analyte partition equilibrium as only the ionic strength changes during the extraction [106].

These proposed methods offer efficient and convenient ways to disperse sorbents for solid-phase extraction, enabling effective sample preparation for various analytical applications.

5 Trends in Dispersive Micro-Solid Phase Extraction

An interesting attractivetrend that deserves special attention in developing of d-μSPEconcerns its combination with other miniaturized extraction techniques. To overcome the limitations of individual techniques, this approach holds great promise for enhancing sample preparation and analysis methods.

Recently, Shi and Lee proposed an innovative combination of DLLME with d-μSPE to determine PAHs in water samples [107].

The key feature of this approach is the reversal of the techniques' order and the introduction of magnetic nanoparticles to eliminate the need for final centrifugation. The simplified analytical procedure involves 1-octanol (extractant) intoaqueous phase and hydrophobic magnetic nanoparticles. Due to their hydrophobic nature, the 1-octanol impregnates the nanoparticles, leading to their separation from the aqueous phase in the same step. The donor phase was discarded by decanting sorbent keeping by magnet device. Subsequently, a washing step with water is performed, and the 1-octanol and PAHs are eluted from the nanoparticles using acetonitrile and ultrasound. The two-step microextraction technique strategy shows exceptional performance in this case, achieving low picogram per milliliter level detection limits.

Another aspect of the d-μSPE combination is the involvement in an automated extraction system. Automatic systems offer numerous positive aspects, including reducing errors associated with operator handling and minimizing their exposure to chemical substances. Additionally, automation increases sample throughput, enabling the analysis of a more significant number of samples in a shorter time, and it contributes to the reduction of required resources such as chemicals and energy.

Since the efficient analyte/sorbent interaction and easier handling of the sorbent throughout the extraction workflow are crucial, the magnetic sorbents can be an optimal solution. Yamaguchi et al. developed a specialized device for automatically handling magnetic silica-coated Fe_3O_4 beads, specifically extracting nucleic acids [108].

A similar strategy, Vakh et al. proposed an automated magnetic dispersive micro solid-phase extraction procedure in a fluidized reactor for the determination of fluoroquinolone antimicrobial drugs in meat-based baby food samples [63].

The developed automated procedure involves injecting the sample solution into the fluidized reactor, where online separation of the analytes from the sample matrix is achieved using a fluidized beds strategy employing magnetic nanoparticles. Subsequently, elution of the analytes occurs, and their determination is carried out through the HPLC system with fluorescence detection. The scheme of automation of magnetic dispersive micro-solid phase extraction is shown in Fig. 4.

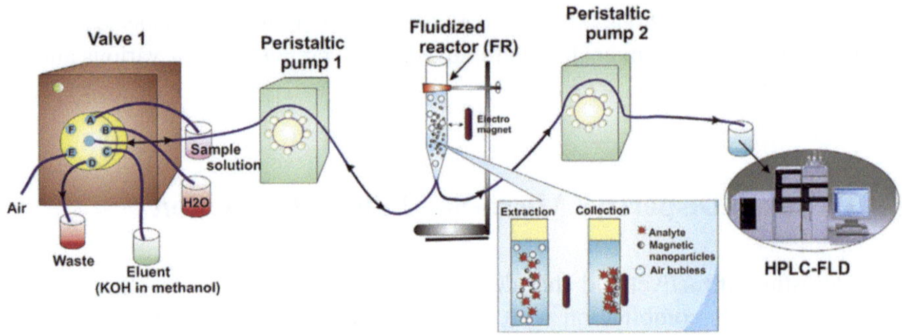

Fig. 4 Scheme of automation of magnetic dispersive micro-solid phase extraction. Reprinted from [63] with permission from Elsevier

6 Conclusions and Future Trends

Dispersive solid phase extraction (dSPE) has become a well-established technique in food analysis, offering efficient selectivity improvement and easy on sample pre-treatment. The synthesis of novel sorbents has enabled the miniaturization of the technique, shifting the focus to the analyte rather than interferents. This approach aligns with Green Chemistry principles, as it reduces sample and reagent consumption. In this trend, the miniaturized extraction techniques such as D-μSPE are perfectly included.

Its simplicity, speed, efficiency due to the dispersion of the sorbent and highly versatile applicability and selectivity due to the wide range of available solids sorbent with different chemical functionalities and manifold options are some of its main features.

In the future, the applicability of dispersive micro solid phase extraction is expected to increase along two research lines. First, the significance of nanostructured solids is evident, as they offer exceptional sorbent properties and high extraction recoveries using minimal amounts of solids. The second automation solution will be crucial to enable routine use in laboratories. Additionally, direct integration with instrumental techniques will facilitate the development of rapid analysis platforms, making them valuable in various fields, such as environmental, forensic, and safety control applications.

Acknowledgements This work was supported by the National Recovery and Resilience Plan (NRRP), Mission 4 Component 2 Investment 1.3—Call for proposals No. 341 of 15 March 2022 of Italian Ministry of University and Research funded by the European Union—NextGenerationEU; Award Number: Project code PE00000003, Concession Decree No. 1550 of 11 October 2022 adopted by the Italian Ministry of University and Research, CUP D93C22000890001, Project title "ON Foods - Research and innovation network on food and nutrition Sustainability, Safety and Security- Working ON Foods".

Conflict of Interest
The authors have declared no conflict of interest.

References

1. Gałuszka A, Migaszewski Z, Namieśnik J (2013) The 12 principles of green analytical chemistry and the SIGNIFICANCE mnemonic of green analytical practices. TrAC—Trends Anal Chem 50:78–84. https://doi.org/10.1016/j.trac.2013.04.010
2. Pena-Pereira F, Lavilla I, Bendicho C (2010) Liquid-phase microextraction techniques within the framework of green chemistry. TrAC—Trends Anal Chem 29:617–628. https://doi.org/10.1016/j.trac.2010.02.016
3. Tobiszewski M, Mechlińska A, Zygmunt B, Namieśnik J (2009) Green analytical chemistry in sample preparation for determination of trace organic pollutants. TrAC—Trends Anal Chem 28:943–951. https://doi.org/10.1016/j.trac.2009.06.001
4. Płotka-Wasylka J, Szczepańska N, de la Guardia M, Namieśnik J (2016) Modern trends in solid phase extraction: new sorbent media. TrAC—Trends Anal Chem 77:23–43. https://doi.org/10.1016/j.trac.2015.10.010
5. Calderilla C, Maya F, Leal LO, Cerdà V (2018) Recent advances in flow-based automated solid-phase extraction. TrAC—Trends Anal Chem 108:370–380. https://doi.org/10.1016/j.trac.2018.09.011
6. Valcárcel M, Cárdenas S, Simonet BM et al (2008) Carbon nanostructures as sorbent materials in analytical processes. TrAC—Trends Anal Chem 27:34–43. https://doi.org/10.1016/j.trac.2007.10.012
7. Chisvert A, Cárdenas S, Lucena R (2019) Dispersive micro-solid phase extraction. TrAC—Trends Anal Chem 112:226–233. https://doi.org/10.1016/j.trac.2018.12.005
8. Anastassiades M, Lehotay SL, Stajnbaher D, Schenck FJ (2003) Fast and easy multiresidue method employing acetonitrile extraction/partitioning and "dispersive solid-phase extraction" for the determination of pesticide residues in produce. J AOAC Int 86:412–431. https://doi.org/10.1371/journal.pone.0029268.
9. Publishers ES, Barker A., Long R et al (1989) Veterinary April received 17th. 1989. 475:353–361
10. Pérez-Rodríguez M, Pellerano RG, Pezza L, Pezza HR (2018) An overview of the main foodstuff sample preparation technologies for tetracycline residue determination. Talanta 182:1–21. https://doi.org/10.1016/j.talanta.2018.01.058
11. Tsai WH, Huang TC, Huang JJ et al (2009) Dispersive solid-phase microextraction method for sample extraction in the analysis of four tetracyclines in water and milk samples by high-performance liquid chromatography with diode-array detection. J Chromatogr A 1216:2263–2269. https://doi.org/10.1016/j.chroma.2009.01.034
12. Khezeli T, Daneshfar A (2017) Development of dispersive micro-solid phase extraction based on micro and nano sorbents. TrAC—Trends Anal Chem 89:99–118. https://doi.org/10.1016/j.trac.2017.01.004
13. Ghorbani M, Aghamohammadhassan M, Ghorbani H, Zabihi A (2020) Trends in sorbent development for dispersive micro-solid phase extraction. Microchem J 158:105250. https://doi.org/10.1016/j.microc.2020.105250
14. Ghorbani M, Aghamohammadhassan M, Chamsaz M et al (2019) Dispersive solid phase microextraction. TrAC—Trends Anal Chem 118:793–809. https://doi.org/10.1016/j.trac.2019.07.012
15. Faraji M, Yamini Y, Gholami M (2019) Recent advances and trends in applications of solid-phase extraction techniques in food and environmental analysis
16. Sajid M, Nazal MK, Ihsanullah I (2021) Novel materials for dispersive (micro) solid-phase extraction of polycyclic aromatic hydrocarbons in environmental water samples: a review. Anal Chim Acta 1141:246–262. https://doi.org/10.1016/j.aca.2020.07.064
17. Cruz-Vera M, Lucena R, Cárdenas S, Valcárcel M (2011) Sample treatments based on dispersive (micro)extraction. Anal Methods 3:1719–1728. https://doi.org/10.1039/c1ay05201b

18. Khatibi SA, Hamidi S, Siahi-Shadbad MR (2021) Current trends in sample preparation by solid-phase extraction techniques for the determination of antibiotic residues in foodstuffs: a review. Crit Rev Food Sci Nutr 61:3361–3382. https://doi.org/10.1080/10408398.2020.1798349
19. Gu YX, Yan TC, Yue ZX, et al. (2022) Recent developments and applications in the microextraction and separation technology of harmful substances in a complex matrix. Microchem J 176:107241. https://doi.org/10.1016/j.microc.2022.107241
20. da Silva Soares, Burato J, Vargas Medina DA, de Toffoli AL et al (2020) Recent advances and trends in miniaturized sample preparation techniques. J Sep Sci 43:202–225. https://doi.org/10.1002/jssc.201900776
21. Cárdenas Aranzana MS (2010) Dispersive Solid-Phase (Micro)Extraction. Encycl Anal Chem. https://doi.org/10.1002/9780470027318.a9167
22. Nascimento MM, Olímpio da Rocha G, Bittencourt de Andrade J (2019) Simple and effective dispersive micro-solid phase extraction procedure for simultaneous determination of polycyclic aromatic compounds in fresh and marine waters. Talanta 204:776–791. https://doi.org/10.1016/j.talanta.2019.06.061
23. Galán-Cano F, Lucena R, Cárdenas S, Valcárcel M (2011) Direct coupling of dispersive micro-solid phase extraction and thermal desorption for sensitive gas chromatographic analysis. Anal Methods 3:991–995. https://doi.org/10.1039/c1ay05033h
24. Galán-Cano F, Lucena R, Cárdenas S, Valcárcel M (2013) Dispersive micro-solid phase extraction with ionic liquid-modified silica for the determination of organophosphate pesticides in water by ultra performance liquid chromatography. Microchem J 106:311–317. https://doi.org/10.1016/j.microc.2012.08.016
25. Sowa I, Wójciak-Kosior M, Strzemski M, et al. (2018) Silica modified with polyaniline as a potential sorbent for matrix solid phase dispersion (MSPD) and dispersive solid phase extraction (d-SPE) of plant samples. Materials (Basel) 11. https://doi.org/10.3390/ma11040467
26. Arnnok P, Patdhanagul N, Burakham R (2017) Dispersive solid-phase extraction using polyaniline-modified zeolite NaY as a new sorbent for multiresidue analysis of pesticides in food and environmental samples. Talanta 164:651–661. https://doi.org/10.1016/j.talanta.2016.11.003
27. Lasarte-Aragonés G, Lucena R, Cárdenas S, Valcárcel M (2011) Effervescence-assisted dispersive micro-solid phase extraction. J Chromatogr A 1218:9128–9134. https://doi.org/10.1016/j.chroma.2011.10.042
28. Chen D, Zhang Y, Miao H et al (2015) Determination of triazine herbicides in drinking water by dispersive micro solid phase extraction with ultrahigh-performance liquid chromatography-high-resolution mass spectrometric detection. J Agric Food Chem 63:9855–9862. https://doi.org/10.1021/acs.jafc.5b03973
29. Chen D, Zhao Y, Miao H, Wu Y (2015) A novel dispersive micro solid phase extraction using PCX as the sorbent for the determination of melamine and cyromazine in milk and milk powder by UHPLC-HRMS/MS. Talanta 134:144–152. https://doi.org/10.1016/j.talanta.2014.10.036
30. Chen D, Miao H, Zhao Y, Wu Y (2017) Dispersive micro solid phase extraction of amantadine, rimantadine and memantine in chicken muscle with magnetic cation exchange polymer. J Chromatogr B Anal Technol Biomed Life Sci 1051:92–96. https://doi.org/10.1016/j.jchromb.2017.03.005
31. Liu Y, Yu L, Zhang H, Chen D (2018) Dispersive micro-solid-phase extraction combined with online preconcentration by capillary electrophoresis for the determination of glycopyrrolate stereoisomers in rat plasma. J Sep Sci 41:1395–1404. https://doi.org/10.1002/jssc.201700753
32. Xian Y, Wu Y, Dong H et al (2017) Dispersive micro solid phase extraction (DMSPE) using polymer anion exchange (PAX) as the sorbent followed by UPLC–MS/MS for the rapid determination of four bisphenols in commercial edible oils. J Chromatogr A 1517:35–43. https://doi.org/10.1016/j.chroma.2017.08.067

33. Pourmand N, Sanagi MM, Naim AA et al (2015) Dispersive micro-solid phase extraction method using newly prepared poly(methyl methacrylate) grafted agarose combined with ICP-MS for the simultaneous determination of Cd, Ni, Cu and Zn in vegetable and natural water samples. Anal Methods 7:3215–3223. https://doi.org/10.1039/c4ay02889a
34. Zhao YG, Chen XH, Pan SD et al (2013) Simultaneous analysis of eight phenolic environmental estrogens in blood using dispersive micro-solid-phase extraction combined with ultra fast liquid chromatography-tandem mass spectrometry. Talanta 115:787–797. https://doi.org/10.1016/j.talanta.2013.05.053
35. Behbahani M, Veisi A, Omidi F et al (2018) Application of a dispersive micro-solid-phase extraction method for pre-concentration and ultra-trace determination of cadmium ions in water and biological samples. Appl Organomet Chem 32:1–10. https://doi.org/10.1002/aoc.4134
36. Li G, Row KH (2018) Recent applications of molecularly imprinted polymers (MIPs) on micro-extraction techniques. Sep Purif Rev 47:1–18. https://doi.org/10.1080/15422119.2017.1315823
37. Turiel E, Martín-Esteban A (2019) Molecularly imprinted polymers-based microextraction techniques. TrAC—Trends Anal Chem 118:574–586. https://doi.org/10.1016/j.trac.2019.06.016
38. Chen C, Zhang X, Long Z et al (2012) Molecularly imprinted dispersive solid-phase microextraction for determination of sulfamethazine by capillary electrophoresis. Microchim Acta 178:293–299. https://doi.org/10.1007/s00604-012-0833-2
39. Xu S, Lu H, Li J et al (2013) Dummy molecularly imprinted polymers-capped CdTe quantum dots for the fluorescent sensing of 2,4,6-trinitrotoluene. ACS Appl Mater Interfaces 5:8146–8154. https://doi.org/10.1021/am4022076
40. Wu X, Li Y, Zhu X et al (2017) Dummy molecularly imprinted magnetic nanoparticles for dispersive solid-phase extraction and determination of bisphenol A in water samples and orange juice. Talanta 162:57–64. https://doi.org/10.1016/j.talanta.2016.10.007
41. Cheng X, Yan H, Wang X et al (2014) Vortex-assisted magnetic dispersive solid-phase microextraction for rapid screening and recognition of dicofol residues in tea products. Food Chem 162:104–109. https://doi.org/10.1016/j.foodchem.2014.04.023
42. Bazmandegan-Shamili A, Dadfarnia S, Haji Shabani AM et al (2018) MultiSimplex optimization of the dispersive solid-phase microextraction and determination of fenitrothion by magnetic molecularly imprinted polymer and high-performance liquid chromatography. J Iran Chem Soc 15:1181–1189. https://doi.org/10.1007/s13738-018-1316-0
43. Asfaram A, Arabi M, Ostovan A et al (2018) Simple and selective detection of quercetin in extracts of plants and food samples by dispersive-micro-solid phase extraction based on core-shell magnetic molecularly imprinted polymers. New J Chem 42:16144–16153. https://doi.org/10.1039/C8NJ03349H
44. Castillo-García ML, Aguilar-Caballos MP, Gómez-Hens A (2016) Nanomaterials as tools in chromatographic methods. TrAC—Trends Anal Chem 82:385–393. https://doi.org/10.1016/j.trac.2016.06.019
45. Witham CS, Oppenheimer C, Horwell CJ (2005) Volcanic ash-leachates: a review and recommendations for sampling methods. J Volcanol Geotherm Res 141:299–326. https://doi.org/10.1016/j.jvolgeores.2004.11.010
46. Wang H, Campiglia AD (2008) Determination of polycyclic aromatic hydrocarbons in drinking water samples by solid-phase nanoextraction and high-performance liquid chromatography. Anal Chem 80:8202–8209. https://doi.org/10.1021/ac8014824
47. Krawczyk M, Stanisz E (2015) Silver nanoparticles as a solid sorbent in ultrasound-assisted dispersive micro solid-phase extraction for the atomic absorption spectrometric determination of mercury in water samples. J Anal Atomic Spectrom 30:2353–2358. https://doi.org/10.1039/c5ja00344j
48. Krawczyk M, Stanisz E (2016) Ultrasound-assisted dispersive micro solid-phase extraction with nano-TiO_2 as adsorbent for the determination of mercury species. Talanta 161:384–391. https://doi.org/10.1016/j.talanta.2016.08.071

49. Murauer A, Bakry R, Schottenberger H et al (2017) An innovative monolithic zwitterionic stationary phase for the separation of phenolic acids in coffee bean extracts by capillary electrochromatography. Anal Chim Acta 963:136–142. https://doi.org/10.1016/j.aca.2017.01.048
50. Nyaba L, Matong JM, Dimpe KM, Nomngongo PN (2016) Speciation of inorganic selenium in environmental samples after suspended dispersive solid phase microextraction combined with inductively coupled plasma spectrometric determination. Talanta 159:174–180. https://doi.org/10.1016/j.talanta.2016.06.023
51. Shirkhanloo H, Khaligh A, Mousavi HZ, Rashidi A (2016) Ultrasound assisted-dispersive-micro-solid phase extraction based on bulky amino bimodal mesoporous silica nanoparticles for speciation of trace manganese (II)/(VII) ions in water samples. Microchem J 124:637–645. https://doi.org/10.1016/j.microc.2015.10.008
52. Zhang Y, Yang X, Wang J et al (2017) In situ solvent formation microextraction combined with magnetic dispersive micro-solid-phase extraction for the determination of benzoylurea insecticides in water samples. J Sep Sci 40:442–448. https://doi.org/10.1002/jssc.201600787
53. Dastkhoon M, Ghaedi M, Asfaram A et al (2017) Cu@SnS/SnO_2 nanoparticles as novel sorbent for dispersive micro solid phase extraction of atorvastatin in human plasma and urine samples by high-performance liquid chromatography with UV detection: application of central composite design (CCD). Ultrason Sonochem 36:42–49. https://doi.org/10.1016/j.ultsonch.2016.10.030
54. Chen S, Yan J, Li J, Lu, D (2019) Magnetic $ZnFe_2O_4$ nanotubes for dispersive micro solid-phase extraction of trace rare earth elements prior to their determination by ICP-MS. Microchim Acta 186:1–7. https://doi.org/10.1007/s00604-019-3342-8
55. Chen S, Yan J, Li J, Lu D (2019) Dispersive micro-solid phase extraction using magnetic $ZnFe_2O_4$ nanotubes as adsorbent for preconcentration of Co(II), Ni(II), Mn(II) and Cd(II) followed by ICP-MS determination. Microchem J 147:232–238. https://doi.org/10.1016/j.microc.2019.02.066
56. Khajeh M, Laurent S, Dastafkan K (2013) Nanoadsorbents: Classification, preparation, and applications (with emphasis on aqueous media). Chem Rev 113:7728–7768. https://doi.org/10.1021/cr400086v
57. Román IP, Chisvert Alberto A, Canals A (2011) Dispersive solid-phase extraction based on oleic acid-coated magnetic nanoparticles followed by gas chromatography-mass spectrometry for UV-filter determination in water samples. J Chromatogr A 1218:2467–2475. https://doi.org/10.1016/j.chroma.2011.02.047
58. Ríos Á, Zougagh M (2016) Recent advances in magnetic nanomaterials for improving analytical processes. TrAC—Trends Anal Chem 84:72–83. https://doi.org/10.1016/j.trac.2016.03.001
59. Shen HY, Zhu Y, Wen XE, Zhuang YM (2007) Preparation of Fe_3O_4-C18 nano-magnetic composite materials and their cleanup properties for organophosphorous pesticides. Anal Bioanal Chem 387:2227–2237. https://doi.org/10.1007/s00216-006-1082-1
60. Nyaba L, Matong JM, Nomngongo PN (2016) Nanoparticles consisting of magnetite and Al_2O_3 for ligandless ultrasound-assisted dispersive solid phase microextraction of Sb, Mo and V prior to their determination by ICP-OES. Microchim Acta 183:1289–1297. https://doi.org/10.1007/s00604-016-1766-y
61. Socas-Rodríguez B, Hernández-Borges J, Herrera-Herrera AV, Rodríguez-Delgado MÁ (2018) Multiresidue analysis of oestrogenic compounds in cow, goat, sheep and human milk using core-shell polydopamine coated magnetic nanoparticles as extraction sorbent in micro-dispersive solid-phase extraction followed by ultra-high-performance liquid chroma. Anal Bioanal Chem 410:2031–2042. https://doi.org/10.1007/s00216-018-0882-4
62. Asgharinezhad AA, Karami S, Ebrahimzadeh H et al (2015) Polypyrrole/magnetic nanoparticles composite as an efficient sorbent for dispersive micro-solid-phase extraction of antidepressant drugs from biological fluids. Int J Pharm 494:102–112. https://doi.org/10.1016/j.ijpharm.2015.08.001

63. Vakh C, Alaboud M, Lebedinets S et al (2018) An automated magnetic dispersive micro-solid phase extraction in a fluidized reactor for the determination of fluoroquinolones in baby food samples. Anal Chim Acta 1001:59–69. https://doi.org/10.1016/j.aca.2017.11.065
64. Zhang Q, Yang F, Tang F et al (2010) Ionic liquid-coated Fe_3O_4 magnetic nanoparticles as an adsorbent of mixed hemimicelles solid-phase extraction for preconcentration of polycyclic aromatic hydrocarbons in environmental samples. Analyst 135:2426–2433. https://doi.org/10.1039/c0an00245c
65. Peng B, Zhang J, Lu R et al (2013) Dispersive micro-solid phase extraction based on self-assembling, ionic liquid-coated magnetic particles for the determination of clofentezine and chlorfenapyr in environmental water samples. Analyst 138:6834–6843. https://doi.org/10.1039/c3an00814b
66. Asgharinezhad AA, Mollazadeh N, Ebrahimzadeh H et al (2014) Magnetic nanoparticles based dispersive micro-solid-phase extraction as a novel technique for coextraction of acidic and basic drugs from biological fluids and waste water. J Chromatogr A 1338:1–8. https://doi.org/10.1016/j.chroma.2014.02.027
67. Wang J, Chen Z, Li Z, Yang Y (2016) Magnetic nanoparticles based dispersive micro-solid-phase extraction as a novel technique for the determination of estrogens in pork samples. Food Chem 204:135–140. https://doi.org/10.1016/j.foodchem.2016.02.016
68. Ghazaghi M, Mousavi HZ, Shirkhanloo H, Rashidi A (2016) Ultrasound assisted dispersive micro solid-phase extraction of four tyrosine kinase inhibitors from serum and cerebrospinal fluid by using magnetic nanoparticles coated with nickel-doped silica as an adsorbent. Microchim Acta 183:2779–2789. https://doi.org/10.1007/s00604-016-1927-z
69. Fernandes VC, Freitas M, Pacheco JPG et al (2018) Magnetic dispersive micro solid-phase extraction and gas chromatography determination of organophosphorus pesticides in strawberries. J Chromatogr A 1566:1–12. https://doi.org/10.1016/j.chroma.2018.06.045
70. Trojanowicz M (2006) Analytical applications of carbon nanotubes: a review. TrAC—Trends Anal Chem 25:480–489. https://doi.org/10.1016/j.trac.2005.11.008
71. Socas-Rodríguez B, Herrera-Herrera AV, Asensio-Ramos M, Hernández-Borges J (2014) Recent applications of carbon nanotube sorbents in analytical chemistry. J Chromatogr A 1357:110–146. https://doi.org/10.1016/j.chroma.2014.05.035
72. Liu WW, Chai SP, Mohamed AR, Hashim U (2014) Synthesis and characterization of graphene and carbon nanotubes: a review on the past and recent developments. J Ind Eng Chem 20:1171–1185. https://doi.org/10.1016/j.jiec.2013.08.028
73. Jiménez-Soto JM, Cárdenas S, Valcárcel M (2012) Evaluation of single-walled carbon nanohorns as sorbent in dispersive micro solid-phase extraction. Anal Chim Acta 714:76–81. https://doi.org/10.1016/j.aca.2011.11.055
74. Jiménez-Soto JM, Cárdenas S, Valcárcel M (2012) Dispersive micro solid-phase extraction of triazines from waters using oxidized single-walled carbon nanohorns as sorbent. J Chromatogr A 1245:17–23. https://doi.org/10.1016/j.chroma.2012.05.016
75. Grijalba AC, Escudero LB, Wuilloud RG (2015) Ionic liquid-assisted multiwalled carbon nanotube-dispersive micro-solid phase extraction for sensitive determination of inorganic As species in garlic samples by electrothermal atomic absorption spectrometry. Spectrochim Acta—Part B Atomic Spectrosc 110:118–123. https://doi.org/10.1016/j.sab.2015.06.005
76. Skorek R, Zawisza B, Marguí E et al (2013) Dispersive micro solid-phase extraction using multiwalled carbon nanotubes for simultaneous determination of trace metal ions by energy-dispersive x-ray fluorescence spectrometry. Appl Spectrosc 67:204–209. https://doi.org/10.1366/12-06738
77. Kocot K, Sitko R (2014) Trace and ultratrace determination of heavy metal ions by energy-dispersive X-ray fluorescence spectrometry using graphene as solid sorbent in dispersive micro solid-phase extraction. Spectrochim Acta—Part B Atomic Spectrosc 94–95:7–13. https://doi.org/10.1016/j.sab.2014.02.003
78. Skorek R, Turek E, Zawisza B et al (2012) Determination of selenium by X-ray fluorescence spectrometry using dispersive solid-phase microextraction with multiwalled carbon nanotubes as solid sorbent. J Anal Atomic Spectrom 27:1688–1693. https://doi.org/10.1039/c2ja30179b

79. Cao W, Hu SS, Ye LH et al (2015) Trace-chitosan-wrapped multi-walled carbon nanotubes as a new sorbent in dispersive micro solid-phase extraction to determine phenolic compounds. J Chromatogr A 1390:13–21. https://doi.org/10.1016/j.chroma.2015.02.060
80. Jalilian N, Ebrahimzadeh H, Asgharinezhad AA (2018) Determination of acidic, basic and amphoteric drugs in biological fluids and wastewater after their simultaneous dispersive micro-solid phase extraction using multiwalled carbon nanotubes/magnetite nanoparticles@poly(2-aminopyrimidine) composite. Microchem J 143:337–349. https://doi.org/10.1016/j.microc.2018.08.037
81. Hooshmand S, Es'haghi Z (2018) Hydrophilic modified magnetic multi-walled carbon nanotube for dispersive solid/liquid phase microextraction of sunitinib in human samples. Anal Biochem 542:76–83. https://doi.org/10.1016/j.ab.2017.11.019
82. Ghorbani M, Chamsaz M, Rounaghi GH (2016) Glycine functionalized multiwall carbon nanotubes as a novel hollow fiber solid-phase microextraction sorbent for pre-concentration of venlafaxine and o-desmethylvenlafaxine in biological and water samples prior to determination by high-performance liquid. Anal Bioanal Chem 408:4247–4256. https://doi.org/10.1007/s00216-016-9518-8
83. Ghazaghi M, Shirkhanloo H, Mousavi HZ, Rashidi AM (2015) Ultrasound-assisted dispersive solid phase extraction of cadmium(II) and lead(II) using a hybrid nanoadsorbent composed of graphene and the zeolite clinoptilolite. Microchim Acta 182:1263–1272. https://doi.org/10.1007/s00604-015-1446-3
84. Sereshti H, Khosraviani M, Samadi S, Amini-Fazl MS (2014) Simultaneous determination of theophylline, theobromine and caffeine in different tea beverages by graphene-oxide based ultrasonic-assisted dispersive micro solid-phase extraction combined with HPLC-UV. RSC Adv 4:47114–47120. https://doi.org/10.1039/c4ra06412g
85. Mahpishanian S, Sereshti H (2014) Graphene oxide-based dispersive micro-solid phase extraction for separation and preconcentration of nicotine from biological and environmental water samples followed by gas chromatography-flame ionization detection. Talanta 130:71–77. https://doi.org/10.1016/j.talanta.2014.06.004
86. Hu SS, Cao W, Da JH et al (2015) Dispersive micro solid-phase extraction with graphene oxide for the determination of phenolic compounds in dietary supplements by ultra high performance liquid chromatography coupled with quadrupole time-of-flight tandem mass spectrometry. Food Anal Methods 8:833–840. https://doi.org/10.1007/s12161-014-9959-7
87. Li G, Zhang X, Zhang L et al (2015) Salt-assisted graphene oxide dispersive solid phase microextraction for sensitive detection of malachite green and crystal violet by HPLC. Chromatographia 78:979–985. https://doi.org/10.1007/s10337-015-2913-z
88. Ghazaghi M, Mousavi HZ, Rashidi AM et al (2016) Innovative separation and preconcentration technique of coagulating homogenous dispersive micro solid phase extraction exploiting graphene oxide nanosheets. Anal Chim Acta 902:33–42. https://doi.org/10.1016/j.aca.2015.11.011
89. Pytlakowska K (2016) Dispersive micro solid-phase extraction of heavy metals as their complexes with 2-(5-bromo-2-pyridylazo)-5-diethylaminophenol using graphene oxide nanoparticles. Microchim Acta 183:91–99. https://doi.org/10.1007/s00604-015-1596-3
90. Akbarzade S, Chamsaz M, Rounaghi GH, Ghorbani M (2018) Zero valent Fe-reduced graphene oxide quantum dots as a novel magnetic dispersive solid phase microextraction sorbent for extraction of organophosphorus pesticides in real water and fruit juice samples prior to analysis by gas chromatography-mass spectrom. Anal Bioanal Chem 410:429–439. https://doi.org/10.1007/s00216-017-0732-9
91. Mateos R, Vera-López S, Saz M et al (2019) Graphene/sepiolite mixtures as dispersive solid-phase extraction sorbents for the analysis of polycyclic aromatic hydrocarbons in wastewater using surfactant aqueous solutions for desorption. J Chromatogr A 1596:30–40. https://doi.org/10.1016/j.chroma.2019.03.004
92. Pang L, Zhang W, Zhang W et al (2017) Magnetic graphene solid-phase extraction in the determination of polycyclic aromatic hydrocarbons in water. RSC Adv 7:53720–53727. https://doi.org/10.1039/c7ra10551g

93. Hashemi B, Zohrabi P, Raza N, Kim KH (2017) Metal-organic frameworks as advanced sorbents for the extraction and determination of pollutants from environmental, biological, and food media. TrAC—Trends Anal Chem 97:65–82. https://doi.org/10.1016/j.trac.2017.08.015
94. Rocío-Bautista P, González-Hernández P, Pino V et al (2017) Metal-organic frameworks as novel sorbents in dispersive-based microextraction approaches. TrAC—Trends Anal Chem 90:114–134. https://doi.org/10.1016/j.trac.2017.03.002
95. Ma W, Li X, Bai Y, Liu H (2018) Applications of metal-organic frameworks as advanced sorbents in biomacromolecules sample preparation. TrAC—Trends Anal Chem 109:154–162. https://doi.org/10.1016/j.trac.2018.10.003
96. Rocío-Bautista P, Pino V, Pasán J et al (2018) Insights in the analytical performance of neat metal-organic frameworks in the determination of pollutants of different nature from waters using dispersive miniaturized solid-phase extraction and liquid chromatography. Talanta 179:775–783. https://doi.org/10.1016/j.talanta.2017.12.012
97. Cai Q, Zhang L, Zhao P et al (2017) A joint experimental-computational investigation: metal organic framework as a vortex assisted dispersive micro-solid-phase extraction sorbent coupled with UPLC-MS/MS for the simultaneous determination of amphenicols and their metabolite in aquaculture wa. Microchem J 130:263–270. https://doi.org/10.1016/j.microc.2016.09.014
98. History A (2010) Manuscript guidelines. Landsc J 29:106–106. https://doi.org/10.3368/lj.29.1.106
99. Rocío-Bautista P, Martínez-Benito C, Pino V et al (2015) The metal-organic framework HKUST-1 as efficient sorbent in a vortex-assisted dispersive micro solid-phase extraction of parabens from environmental waters, cosmetic creams, and human urine. Talanta 139:13–20. https://doi.org/10.1016/j.talanta.2015.02.032
100. Zhai Y, Li N, Lei L et al (2014) Dispersive micro-solid-phase extraction of hormones in liquid cosmetics with metal-organic framework. Anal Methods 6:9435–9445. https://doi.org/10.1039/c4ay01763c
101. Liu FJ, Liu CT, Li W, Tang AN (2015) Dispersive solid-phase microextraction and capillary electrophoresis separation of food colorants in beverages using diamino moiety functionalized silica nanoparticles as both extractant and pseudostationary phase. Talanta 132:366–372. https://doi.org/10.1016/j.talanta.2014.09.014
102. Ghorbani M, Chamsaz M, Rounaghi GH (2016) Ultrasound-assisted magnetic dispersive solid-phase microextraction: a novel approach for the rapid and efficient microextraction of naproxen and ibuprofen employing experimental design with high-performance liquid chromatography. J Sep Sci 39:1082–1089. https://doi.org/10.1002/jssc.201501246
103. Ghorbani M, Chamsaz M, Rounaghi GH et al (2016) Development of a novel ultrasonic-assisted magnetic dispersive solid-phase microextraction method coupled with high performance liquid chromatography for determination of mirtazapine and its metabolites in human urine and water samples employing experimen. Anal Bioanal Chem 408:7719–7729. https://doi.org/10.1007/s00216-016-9869-1
104. Jamali MR, Firouzjah A, Rahnama R (2013) Solvent-assisted dispersive solid phase extraction. Talanta 116:454–459. https://doi.org/10.1016/j.talanta.2013.07.023
105. Nakhaei JM, Jamali MR, Sohrabnezhad S, Rahnama R (2019) In-syringe solvent-assisted dispersive solid phase extraction followed by flame atomic absorption spectrometry for determination of nickel in water and food samples. Microchem J 144:88–92. https://doi.org/10.1016/j.microc.2018.08.063

106. Lasarte-Aragonés G, Lucena R, Cárdenas S, Valcárcel M (2013) Effervescence-assisted carbon nanotubes dispersion for the micro-solid-phase extraction of triazine herbicides from environmental waters. Anal Bioanal Chem 405:3269–3277. https://doi.org/10.1007/s00216-013-6718-3
107. Shi ZG, Lee HK (2010) Dispersive liquid-liquid microextraction coupled with dispersive μ-solid-phase extraction for the fast determination of polycyclic aromatic hydrocarbons in environmental water samples. Anal Chem 82:1540–1545. https://doi.org/10.1021/ac9023632
108. Yamaguchi A, Matsuda K, Uehara M et al (2016) A novel automated device for rapid nucleic acid extraction utilizing a zigzag motion of magnetic silica beads. Anal Chim Acta 906:1–6. https://doi.org/10.1016/j.aca.2015.10.011

Solid-Phase Microextraction

Khaled Murtada and Janusz Pawliszyn

Abstract Sample preparation represents a pivotal stage within the analytical workflow. This chapter delves into the latest advancements in solid-phase microextraction (SPME), a technology renowned for its ability to facilitate uncomplicated, highly sensitive, swift, and solvent-free extraction of analytes from gaseous, liquid, and solid samples. This versatile approach extends its utility to trace-level analysis of compounds even within intricate matrices. Consequently, SPME has emerged as a preeminent sample preparation technique in the past decade, frequently employed in the form of an automated fiber-injection system in conjunction with chromatographic separation modules. Its primary application pertains to the extraction of volatile and semi-volatile organic compounds.

Keywords Solid-phase microextraction · Fiber · Membrane · Headspace · Sample preparation · Solvent free

Abbreviations

SPME	Solid-phase microextraction
GC	Gas chromatography
LC	Liquid chromatography
MS	Mass spectrometry
FID	Flame ionization detector
HS	Headspace
DI	Direct immersion
LOD	Limit of detection
MOF	Metal organic framework

K. Murtada (✉) · J. Pawliszyn (✉)
Department of Chemistry, University of Waterloo, Waterloo, ON, Canada
e-mail: kmurtada@uwaterloo.ca

J. Pawliszyn
e-mail: Janusz@uwaterloo.ca

M. Á. Rodríguez-Delgado et al. (eds.), *Microextraction Techniques*, Integrated Analytical Systems, https://doi.org/10.1007/978-3-031-50527-0_4

COF	Covalent organic framework
PDMS	Polydimethylsiloxane
PAN	Polyacrylonitrile
DVB	divinylbenzene
MWCNTs	Multiwalled carbon nanotubes
Car	Carboxen
GO	Graphene oxide
HLB	Hydrophilic-lipophilic balance
NPs	Nanoparticles
OPPs	Organophosphorous pesticides
PAHs	Polycyclic aromatic hydrocarbons
PAEs	Phthalate esters
PCBs	Polychlorinated biphenyls
PFASs	Per- and polyfluorinated alkyl substances
PPY	Polypyrrole
PANI	Polyaniline
PAN	Polyacrylonitrile
PEG	Polyethylene glycol
PA	Polyacrylate
TF-SPME	Thin-film solid-phase microextraction
VOCs	Volatile organic compounds

1 Introduction

In recent decades, researchers in chemistry and technology communities have achieved significant advances in separation science and sample-preparation technologies. However, an efficient universal sample-pretreatment method capable of isolating target compounds from a sample matrix for instrumental analysis, irrespective of sample type and complexity or the chromatographic technique used for quantitative and qualitative analysis, remains elusive. Sample preparation is a critical component in all analytical workflows, as the clean extracts produced via such methods enable effective separation and seamless analysis, and help ensure the analytical instrument is operating under optimum working conditions [1–3].

Liquid-liquid extraction (LLE) and solid-phase extraction (SPE) are classical exhaustive sample-preparation techniques that have been successfully applied for the analysis of various samples [4–7]. Unfortunately, LLE techniques are characterized by numerous limitations, such as inadequate selectivity for target compounds, the need for large amounts of toxic organic solvents, unwanted emulsion formation, and long preparation times due to solvent evaporation and sample reconstitution. Conversely, SPE is a time-consuming, multi-step procedure that requires clean, particle-free samples, and often involves solvent evaporation and sample reconstitution in solvents, which can result in analyte loss. Moreover, the application of SPE for the isolation of polar compounds and metabolites, especially in biological samples, is limited by the availability of only a handful of suitable sorbents. Solid-phase microextraction (SPME) is an innovative sample-preparation technology that addresses many of the limitations of SPE and LLE, particularly the ability to offer high sensitivity without the use of solvent, which has led to its wide application in analytical chemistry [8–11]. SPME is an equilibrium-based extraction technique based on the migration of analytes from the sample to a sorbent material applied to a substrate via a free radical cross-linking reaction. The most widely employed sorbents in SPME devices include polydimethylsiloxane (PDMS), polyethylene glycol (PEG), polyacrylate (PA), carboxen/polydimethylsiloxane (Car/PDMS), and carboxen/divinylbenzene/polydimethylsiloxane (Car/DVB/PDMS). In SPME, sampling continues until the sorbent has reached its maximum capacity (equilibrium), at which point the device is removed and subjected to direct or indirect instrumental desorption.

The literature contains a large (and continuously growing) number of reports of novel SPME workflows developed for a wide variety of applications, including the analysis of environmental, biological, and pharmaceutical samples; the analysis of foods, beverages, flavors, and fragrances; forensic and toxicology studies; and product testing [12–19]. In recent years, several authors have published reviews/articles surveying the application of SPME in areas such as the analysis of wine volatiles, in vivo analysis of pollutants, on-site soil analysis, water sample analysis, food analysis, in vitro and in vivo metabolomics studies, and pharmaceutical and biomedical analysis [8, 9, 20–25].

This chapter provides an overview of recent, innovative work focusing on SPME. The remainder of this chapter can be divided into four primary sections. Firstly, we will introduce the core principles of SPME. Next, we will delve into recent research that has led to innovative advancements in SPME geometries and coating materials. Following that, we will provide a recap of noteworthy recent applications of SPME in the realms of food, environmental, and bioanalytical studies. Lastly, we will engage in a discourse on the prospective paths for future research and developments in the field of SPME.

2 Fundamentals

SPME operates by establishing equilibrium between the target compounds within the sample matrix and the extraction phase adhered to the SPME device's surface (as depicted in Fig. 1). In this regard, SPME shares foundational principles with electrochemical methods like potentiometry and amperometry, albeit with key distinctions, primarily in terms of capacity. This distinction is crucial because it permits SPME to be seamlessly integrated with various readout techniques such as GC or LC-MS, facilitating qualitative and quantitative analysis, particularly when employing coatings compatible with the sample matrix. Analogous to biosensors, matrix-compatible coatings adopt a membrane protection strategy, enabling their application in highly intricate samples. The comprehension and optimization of coating extraction kinetics

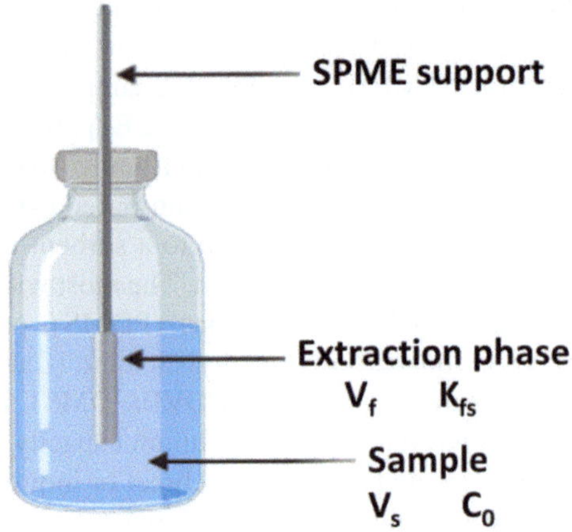

Fig. 1 V_f represents the volume of the fiber coating, K_{fs} denotes the distribution coefficient between the fiber coating and the sample, V_s represents the sample volume, and C_0 signifies the initial analyte concentration within the sample

and thermodynamics are of paramount importance, ensuring the swift accumulation of analytes and prompt pre-equilibrium determinations with the requisite sensitivity. The mass transfer of analytes from the matrix to the extraction phase can be elucidated employing Fick's second law, a second-order partial differential equation solvable via the initial and boundary conditions of the specific system under investigation [26–28].

In addition to its applicability in the analysis of organic compounds across various disciplines, SPME also enables researchers to compute the distribution coefficients of analytes between the coating material and the sample matrix. Several studies have made efforts to estimate the SPME distribution constant (K_{SPME}) and correlate it with the partition coefficient (log K_{ow}) [29–32]. However, predictions regarding the quantity of extracted compounds display variability, and as of now, there is no clear-cut relationship established between the partition coefficients and the characteristics of the analytes [33, 34].

In the context of SPME, the process is typically deemed finished when distribution equilibrium is attained between the sample matrix and the fiber coating, as described by Eq. 1. Adhering to the principles of mass conservation, particularly when only two phases are under consideration (for example, the sample matrix and the fiber coating), then

$$C_0 V_s = C_s^{\infty} V_s + C_f^{\infty} V_f \tag{1}$$

where C_0. represents the initial analyte concentration within the sample, V_s denotes the sample volume, C_s^{∞} stands for the equilibrium concentration within the sample, C_f^{∞} represents the equilibrium concentration on the coating, and V_f signifies the volume of the coating.

The distribution coefficient (K_{fs}). between the coating and the sample matrix is formally defined as:

$$K_{fs} = \frac{C_f^\infty V_f}{C_s^\infty V_s} \quad (2)$$

The quantity of analyte moles absorbed (n) by the coating when it reaches equilibrium can be succinctly expressed using Eq. (3), which results from the amalgamation of Eqs. (1) and (2):

$$n = C_f^\infty V_f = \frac{K_{fs} V_f V_s C_0}{K_{fs} V_f + V_s} \quad (3)$$

Here, K_{fs} represents the distribution coefficient governing the interaction between the coating and the sample matrix. Equation (4) proves useful in establishing the equilibrium state for a three-phase system, encompassing scenarios that include the headspace,

$$n = C_f^\infty V_f = \frac{K_{fs} V_f V_s C_0}{K_{fs} V_f + K_{hs} V_h + V_s}. \quad (4)$$

where K_{hs}. represents the distribution coefficient between the coating and the headspace. Equation (4) stipulates that the quantity of analyte extracted remains unaffected by the positioning of the fiber within the system. Therefore, the fiber can be positioned either in the headspace or directly within the sample, provided that the volumes of the fiber coating, headspace, and sample are maintained at a constant level.

The fiber constant serves as a useful metric for assessing the fiber's performance, particularly in situations involving coatings with solid particles. What makes it particularly valuable is that it doesn't necessitate data regarding the active surface area or adsorption distribution constant. For assessing mass transfer within the coating, it's advantageous to treat the entire extraction phase as a liquid phase, even when it contains particles. In this context, we use the extraction phase diffusion coefficient as the effective diffusion coefficient (D_{eff}) [35]. Equation (5), derived and adapted from theories related to mass transfer in porous media and chromatography [36], elucidates the concept of this effective diffusion coefficient.

$$D_{eff} = \frac{D_E}{1 + k} \quad (5)$$

where D_E represents the diffusion coefficient ($m^2 s^{-1}$) within a single binder material (e.g., PDMS, PAN, etc.) and k stands for the phase capacity, which denotes the extraction ratio between the composite mixed-phase sorbent (e.g., HLB/PDMS, HLB/PAN, etc.) and the extraction phase composed solely of PDMS or PAN. In the case of mixed-phase sorbents (e.g., HLB/PDMS, HLB/PAN, C_{18}/PAN, etc.), the majority of analytes tend to accumulate on the sorbent material (e.g., HLB, C_{18}, etc.), as evidenced by variations in their respective distribution constants (K).

3 Novel Developments

SPME has been studied extensively, producing a variety of different configurations. Currently, there are a variety of available SPME geometries, including: (A) fibers [14], (B) in-tube [37], (C) in-tip [38], (D) vessel wall, (E) arrow [39], (F) suspended particles [40], (G) stirrer, (H) disk, (I) thin-film [41], and (J) 96-blade configuration [42]. Figure 2 illustrates the different forms of SPME considered in this chapter.

Among the above-listed techniques, fibers can be sequenced prior to being introduced to the GC instrument, while in-tube SPME can be used for liquid chromatography. SPME arrows are an evolution of SPME fibers and can be applied for headspace analysis or direct immersion in liquid matrices [39, 43]. Thin-film solid-phase microextraction (TF-SPME) is a new geometry that has emerged as an attractive sample-preparation technique, as its high surface area-to-volume ratio—and thus, its greater volume of extraction phase—enables enhanced sensitivity without sacrificing sampling time [10, 11, 44, 45]. On the other hand, the use of 96-well plates has also received much attention due to their potential to provide high-throughput when performing multiple microextractions in parallel [46, 47]. As the above examples show, the variety of available SPME geometries allows researchers to select the

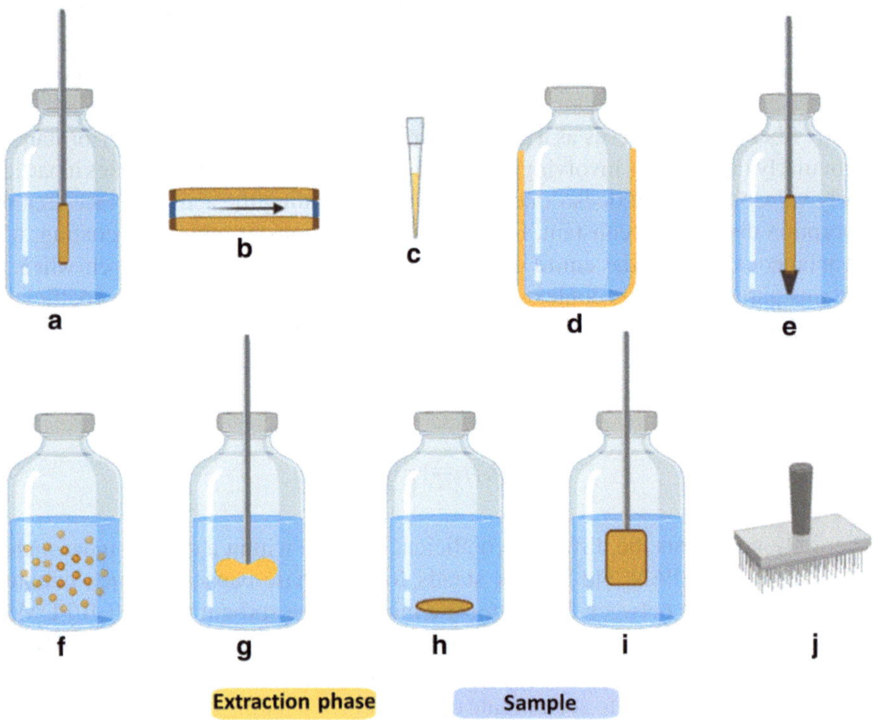

Fig. 2 Different SPME geometries

most optimal configuration for a given application. Currently, SPME fibers are the most widely used geometry due to their small size, high portability, and easy automation. Due to these advantages, the preparation of SPME fibers has been studied more extensively compared to other SPME geometries; however, these strategies can be adapted to prepare SPME devices with other configurations.

The coating's properties determine the method's analyte enrichment efficiency, and consequently, its sensitivity and reliability. Typically, extraction phases are constructed using either polymeric absorbents with liquid-like properties or solid adsorbents [48]. The extraction capabilities of liquid-like absorbents are determined by the distribution coefficient of the target analytes between the coating and the sample. In recent times, there has been a growing prevalence of solid-phase adsorbents characterized by their substantial surface areas, extensive porosities, and a profusion of interaction sites in research pertaining to SPME..

In the case of solid sorbents, analytes engage in interactions with the adsorbent surface through mechanisms such as hydrogen bonding, π-π interactions, dipole-dipole forces, electrostatic attractions, or hydrophobic/hydrophilic interactions. These various interaction types collectively exert a significant influence on both the quantity and the rate at which analyte adsorption occurs [49]. Two principles should be considered when designing an SPME coating. Firstly, there should be strong interaction between the coating materials and target analytes, as this will ensure excellent analyte enrichment. Secondly, it should be easy to firmly immobilize the coating material onto the supporting substrate. Table 1 presents an overview of recent developments in different SPME coating materials.

4 Main Applications

SPME has been applied successfully for the analysis of analytes in various samples, including organic analytes in environmental [8, 19, 112], food [113, 114], biological matrices [115–117], particularly whole blood, pharmaceuticals, and air. To date, researchers and separation experts have published several thorough reviews detailing the use of SPME for the analysis of different sample matrices and gaps that need to be addressed.

4.1 Environmental Applications

The development of effective SPME methods in the environmental field has been critical in enabling the extraction and analysis of several analytes. Various traditional SPME methods have been employed to analyze certain analytes in environmental samples, despite possessing notable limitations such as the need for a post-treatment step, higher costs, and limited efficiency [19, 112, 114]. However, despite these challenges, SPME remains the prevailing microextraction technique,

Table 1 Recent developments in SPME coating materials

Coating material	Analytes	LOD (ng L^{-1} or ng g^{-1})	Instrument	Sample matrix	Refs.
Silver	PAEs and PAHs	20–100 for PAHs; 20–50 for PAEs	GC-FID	Disposable paper cup and instant noodle barrel	[50]
Palladium	PAEs and PAHs	50–100 for PAHs; 30 for PAEs	GC-FID	Various aqueous samples	[51]
Etched stainless-steel wire	PAHs	240–630	GC-FID	River water and wastewater samples	[52]
PPY-Ag NP nanocomposite	Parabens	10	LC-UV	Water, fruit juice, and beer	[53]
Octanedithiol-functionalized Au nanoparticles	UV filters	25–56	LC-UV	River water, wastewater, and rainwater	[54]
Tungsten	Antimalarial drug	8 for pyrimethamine	High-resolution MS	Fish and paramecium	[55]
Silver nanoparticles	Monounsaturated fatty acid methyl esters	5.2	HPLC	Food	[56]
Silver nanoparticles	PAHs	60	GC-FID	Underground water	[57]
Gold nanoparticles	PAHs	10–200 (μg L^{-1})	GC-FID	Sea water	[58]
Gold nanoparticles	Aromatic hydrocarbons	8–37 for PAHs	LC-UV	Water	[59]
Anodized aluminum	Alcohols, benzene homologues, and alkanes	30–300 for alcohols	GC-FID	Gaseous samples	[60]
Anodized aluminum	Volatile compounds	–	GC-MS	Medicinal plants	[61]
Ordered nanoporous anodic alumina	VOCs	0.7–3.4	GC-MS	Human exhaled breath	[62]

(continued)

Table 1 (continued)

Coating material	Analytes	LOD (ng L^{-1} or ng g^{-1})	Instrument	Sample matrix	Refs.
Titanium dioxide nanowires	Phosphopeptides and phospholipids	≤25	High-resolution MS	Biological matrices	[63]
Zinc-zinc oxide	UV filters	52–84	LC-UV	Environmental water samples	[64]
Titanium dioxide@Carbon	PAHs	0.4–7.1	GC-MS	River water	[65]
Zirconium dioxide	Halophenols	201–300	GC-ECD	Wastewater	[66]
Lead (IV) oxide nanoparticles	Volatile organoselenium	11–16	GC-MS	Beverages, urine, and plasma	[67]
Nanostructured cobalt tetraoxide	Benzene homologues	1–11 (μg L^{-1})	GC-MS	Water and fruit juice	[68]
Oxidized MWCNTs	Amphetamine-type stimulants	200–1300	GC-MS	Human urine	[69]
MWCNTs/PANI-PPy@PDMS	Pesticide residues	0.39–2.49	GC-MS	Garlic	[70]
Tetraethylene-pentamine-functionalized MWCNTs	BPAs	450	UV-Vis	Water	[71]
Functionalized and carboxylated carbon nanotubes	Morphine	150	HPLC	Urine	[72]
MWCNTs/MnO_2 nanocomposite-based polythiophene	PAHs	0.1–0.8	GC	Soil	[73]
MIL-88(Fe)/GO composite	PAEs	0.5–2.0	GC	Vegetable oil	[74]
(PANI/PPy/GO) composite	VOCs	1.0–12	GC-MS	Water	[75]
Molybdenum disulfide/reduced GO	PCBs	50–90	GC-MS	Food	[76]
Hollow carbon nanobubbles	PCBs	0.0017–0.0042	GC-MS	Water	[77]
Nitrogen-doped porous biochar	Chlorobenzenes	0.007–0.079	GC	Water	[78]

(continued)

Table 1 (continued)

Coating material	Analytes	LOD (ng L^{-1} or ng g^{-1})	Instrument	Sample matrix	Refs.
Carboxylation modified mesoporous carbon aerogel	Tetracyclines	0.36–0.71	LC	Water	[79]
Large-pore ordered mesoporous carbon	PAHs	1.6–10	GC	Water	[80]
Polydopamine modified ordered mesoporous carbon	Phenols	0.08–0.38	GC-MS	Water	[81]
Nitrogen-doped porous carbon derived from g-C_3N_4 templated MOF	OPPs	0.23–7.5	GC-MS	Food	[82]
Porous carbon derived from MOF-74-C	Odorous organic contaminants	0.01–0.9	GC-MS	Water	[83]
Ni-Zn MOF	PAHs	0.1–3.0	GC-MS	Water and soil	[84]
Cd-MOF	BTEX	1–10	GC-FID	Seawater	[85]
HKUST-1	PAHs	0.12–0.99	GC-MS	Water	[86]
MOF	UV filters	0.6–2.1	GC-MS	Aqueous samples	[87]
ZIF-8, UiO-66, MIL-88, $Tb_2(BDC)_3$	PFOA	11(ZIF-8)	MS	Water	[88]
Zn-FMOF	VOCs	150–900	GC-FID	Wastewater	
	[89]				
MOF-5 and MIL-101 (Fe)	PAHs	0.0.2–0.30 for MIL-101@MON	GC-MS	Water, particulate, and food	[90]
MOFs/PANI	Chlorobenzenes	0.1–0.2	GC-MS	Water	[91]
ZIF-8	PAHs, NPAHs	0.3–27.0	GC-MS	Water	[92]
UiO-66	Polycyclic musks	0.015–0.010	GC-MS	Fortified river water	[93]

(continued)

Table 1 (continued)

Coating material	Analytes	LOD (ng L^{-1} or ng g^{-1})	Instrument	Sample matrix	Refs.
MOF-5	Volatile organic sulfur compounds	200–1700; 6000–23,800	GC-MS	Chinese chive and garlic sprouts	[94]
HKUST-1	PAHs	0.12–9.9	GC-MS	Water	[95]
(Cu-DAT) MOFs	PAHs	1.3–6.7	GC-MS	River water and living fish	[96]
TFPB-BD COF	PCBs	0.07–0.35	GC-MS/MS	Aquatic products	[97]
$TpPaBD_{50}$ COF	Tetrabromobisphenol A derivatives	0.5–12	CFDI-MS	Water	[98]
CF_3-COF	PFASs	0.0001–0.0008	UHPLC-MS/MS	Milk products	[99]
Dioxin-linked COF	PFASs	0.002–0.0045	UPLC-MS/MS	Water	[100]
$TpPaNO_2$-COF	Pesticides	0.04–0.25	GC-ECD	Fruit	[101]
TPB-DMTP-COF	Phenols	0.0048–0.015	GC-MS	Water	[102]
TpPa-1-COF	Synthetic musks	0.04–0.31	GC-MS/MS	Water	[103]
TpPa-1-COF	Polybrominated diphenyl ethers	0.0058–0.022	GC-MS	Water	[104]
TpBD-COF	PAHs	0.02–1.66	GC-MS	Grilled meat	[105]
SNW-1-COF	Phenols	0.06–0.2	GC-MS	Honey	[106]
TpBD-COF	Chlorophenols	0.3–0.7 and 0.8–1.8	GC-MS	Honey and canned yellow peaches	[107]
Hydrazine-COF	Pyrethroids	0.11–0.23	GC-ECD	Vegetable and fruit	[108]
GO/SNW-1	PAEs	10–500	GC-MS	Water and picked cucumber solution	[109]

(continued)

Table 1 (continued)

Coating material	Analytes	LOD (ng L^{-1} or ng g^{-1})	Instrument	Sample matrix	Refs.
TPT-COF	PAEs	5–95	GC-FID	Juice	[110]
COF-SCU1	Benzene homologues	0.03–0.15	GC-MS	Indoor air	[111]

and researchers are consistently exploring ways to address the mentioned limitations. The conventional commercially available SPME device comprises a fused silica or stainless-steel fiber, either coated or uncoated, with a thin sorbent layer. This fiber is typically affixed to a syringe-like device. In pursuit of enhanced extraction efficiency for environmental applications, researchers have delved into diverse strategies, encompassing the adoption of varied coating materials and alternative device configurations [118, 119]. As a result, SPME techniques have received great attention in the analytical and environmental fields due to their enhanced properties and high selectivity for certain target analytes. Commercial SPME fibers featuring non-polar or semipolar coatings fabricated from conventional and newly synthesized materials (e.g., PDMS, DVB, or Car) have been successfully employed to extract selected analytes from environmental samples. For instance, Wu et al. [120] developed and deployed two novel monolith-based electrodes for electric field assisted SPME (EFA-SPME) aimed at the simultaneous detection of phenylurea and sulfonylurea herbicides. In this work, the authors applied poly(vinylimidazole-co-ethylene dimethacrylate) and poly(methylacrylic acid co-ethylene dimethacrylate/divinylbenzene) monolith onto the surfaces of stainless steel wires, which were then respectively used as the anode and cathode for EFA-SPME. Figure 3 shows the preparation of the poly(vinylimidazole-co-ethylene dimethacrylate)-monolith-based anode and the poly(methylacrylic acid co-ethylene dimethacrylate/divinylbenzene)-monolith-based cathode, as well as the electric field assisted SPME protocol applied for the simultaneously extraction of sulfonylureas and phenylureas.

Grandy et al. [121] developed a drone-equipped TF-SPME sampler featuring HLB/PDMS membranes, enabling the remote assessment of environmental water pollutants (see Fig. 4). In order to enhance mobility, this drone-assisted sampling method was integrated with portable hand-held GC-MS instrumentation, thereby bolstering the method's suitability for on-site sampling, extraction, and analyte identification.

In 2020 [44], our research team introduced an innovative in-vial standard gas generation system that employed thin-film membranes supported by mixed-sorbent carbon mesh as carriers for analytes. These vials were designed with carbon mesh membranes loaded with various sorbents such as pure PDMS, DVB/PDMS, HLB/PDMS, and Car/PDMS, which were subsequently spiked with modified McReynolds standards. The results obtained indicated that the TF-SPME gas generation vial exhibited comparable, and in certain instances, superior performance when compared to the PS/DVB silicone-oil-based vial (as illustrated in Fig. 5). Additionally, the TF-SPME vial boasted a much cleaner, reusable, and user-friendly design. Moreover, the outcomes also confirmed the suitability of these novel TF-SPME-based standard gas generation vials for the consistent generation of gaseous standards essential for GC-MS analysis and quality control purposes.

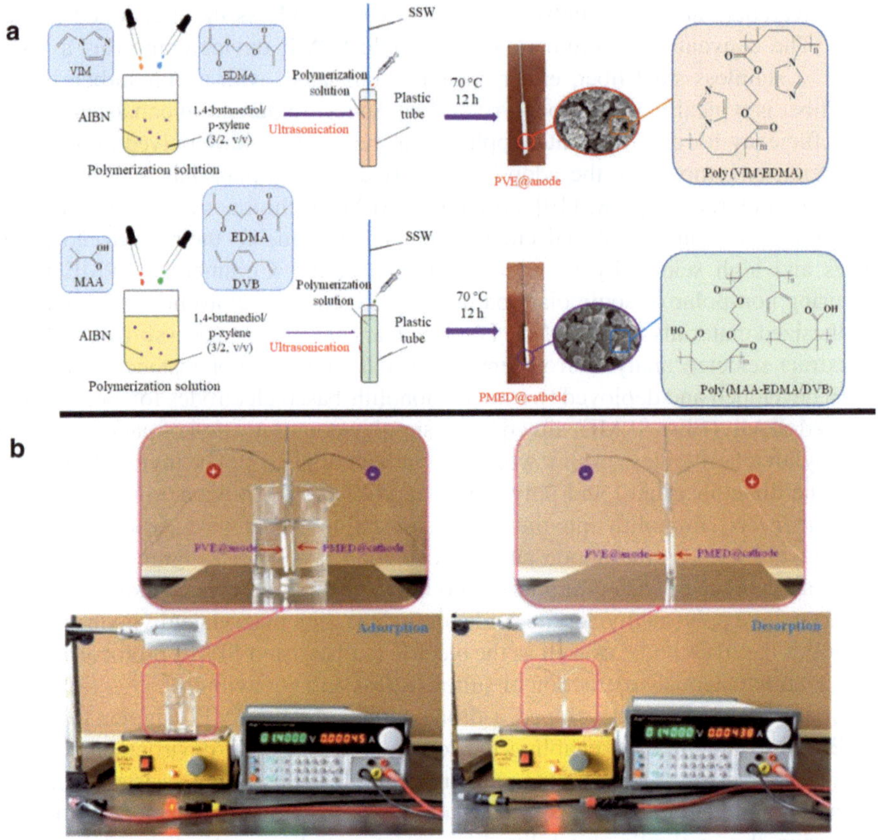

Fig. 3 **a** Synthesis of the monolithic anode using poly(vinylimidazole-co-ethylene dimethacrylate) and the cathode using poly(methylacrylic acid co-ethylene dimethacrylate/divinylbenzene). **b** Operation of the developed monolith-based electrodes/electric field assisted-SPME protocol in the adsorption and desorption steps. Reprinted with permission from [120] with permission from Elsevier

4.2 Food Applications

SPME has become one of the most popular methods for the pretreatment of food samples, having been applied for a range of matrices including liquids, such as milks, wines, and oils; semifluids, such as honey; and solids, such as meats, vegetables, and fruits. One reason for SPME's popularity with such samples is that it can be applied for targeted or untargeted analysis. Researchers have fabricated various SPME devices (fibers, thin films, in tube, and coated blades) using a range of functional materials and extraction models to satisfy the wide range of extraction requirements when using food samples [122–126]. Food matrices are inherently intricate, often comprising proteins, fats, salts, acids, bases, and a multitude of food additives with diverse chemical properties. Among the various coating materials studied, PDMS,

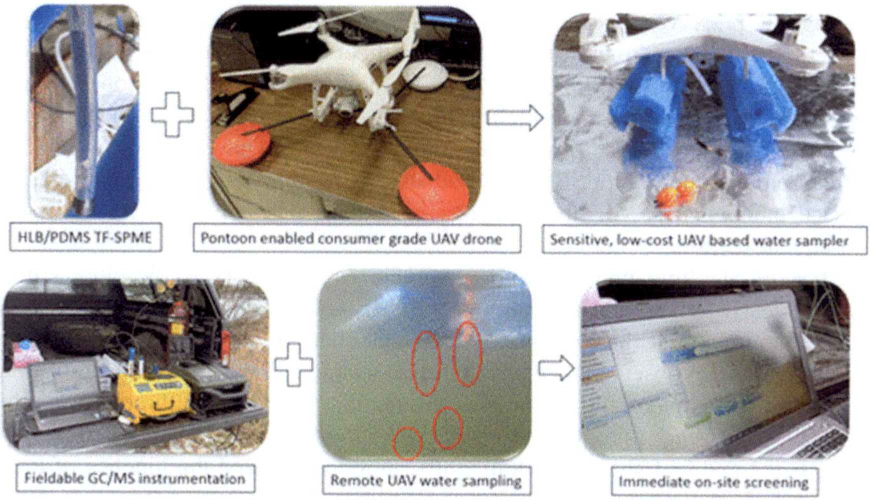

Fig. 4 Utilizing a drone-based TF-SPME system for water sampling. Reprinted from [121] with permission from American Chemical Society

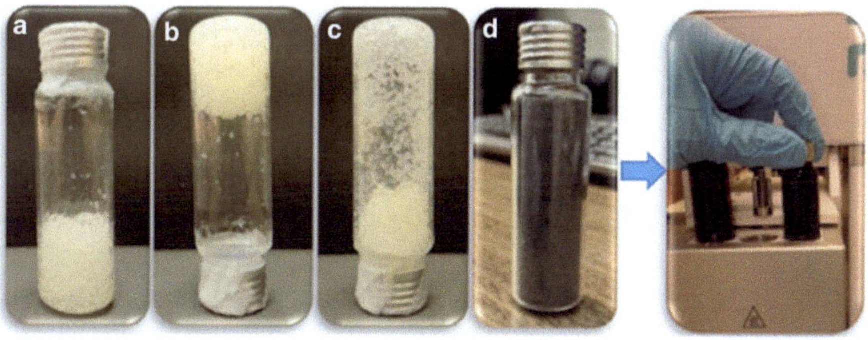

Fig. 5 Concept and rationale behind the development of dual-phase in-vial standard gas generation vials: **a** Utilization of a recently prepared silicone oil, PS/DVB vial. **b** Assessment of vial stability suitable for laboratory applications. **c** Evaluation of vial contents following agitation or transportation. **d** Deployment of DVB/PDMS-coated carbon fiber fabric as a sorbent, along with the integration of new vials into an autosampler unit. Reprinted from [44] with permission from Elsevier

characterized by its liquid nature and smooth, uniform surface, stands out for its remarkable resilience to irreversible fouling effects triggered by matrix components (as compared to solid coatings) [127], Consequently, it emerges as the most robust choice for the direct analysis of food samples. Nevertheless, PDMS's susceptibility to analytes of interest has posed a significant challenge. To address this limitation, researchers have explored the enhancement of conventional commercial SPME fiber coatings by incorporating a thin PDMS layer, thus creating a novel matrix-compatible coating that preserves the original coating's sensitivity to the target analytes [128]. As depicted in Fig. 6, these modified SPME fiber coatings, such as PDMS/DVB, DVB/Car/PDMS, and PDMS/DVB/PDMS, exhibit exceptional extraction efficiency and durability, rendering them highly effective for the direct analysis of complex matrices [128]. As a result, these PDMS-modified coatings have risen to prominence as the preferred choices for SPME in food analysis. Moreover, researchers have also devised innovative SPME fiber coatings, which we will delve into further in the subsequent section.

Chen et al. drew upon sampling rate correction theory to develop a non-invasive *in-vivo* sampling-rate-calibrated SPME-GC/MS method for the accurate quantification of target analytes [129]. The researchers employed their methodology directly on-site to observe and analyze the environmental dynamics, encompassing absorption, enrichment, migration, and elimination processes, of three insecticides (hexachlorobenzene, fipronil, and chlorfenapyr) within edible plants, specifically garlic bulbs and leaf sheaths. Additionally, they investigated the kinetics of these insecticides' elimination within living garlic plants. Figure 7 illustrates a schematic representation of Chen et al.'s in vivo SPME procedure. In this in vivo SPME

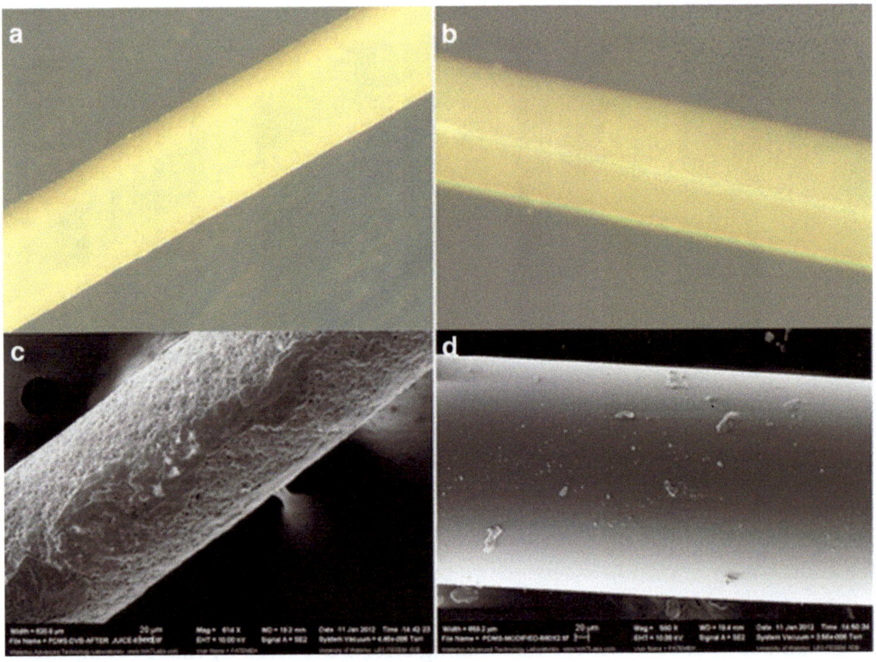

Fig. 6 **a** Microscopic image depicting a commercial PDMS/DVB coating in its pre-extraction state. **b** Microscopic image showing a PDMS/DVB/PDMS coating before any extraction. **c** Scanning Electron Microscope (SEM) image revealing the surface of a PDMS/DVB coating after undergoing 20 extraction cycles within grape juice. **d** SEM image depicting the surface morphology of a PDMS/DVB/PDMS coating after enduring more than 130 extraction cycles in grape juice, observed at 580× magnification. Reprinted from [128] with permission from the American Chemical Society

sampling method, an SPME fiber was inserted directly into the garlic bulbs or leaf sheaths, reaching a depth of approximately 1.5 cm while being shielded by an external cannula. Following a static extraction period of 25 min at a temperature of 25 °C, the probe was carefully withdrawn, washed with deionized water, wiped clean with Kimwipes, and promptly prepared for analysis using GC/MS.

The ability to detect spoilage and nutrient content in salmon is critical for ensuring it is safe to consume and determining its market value. To this end, Yu et al. [39] developed and fabricated an innovative SPME arrow coated with HLB/PDMS, which was subsequently integrated with GC-MS for the untargeted assessment of volatile metabolites and unsaturated fatty acids within fresh salmon samples. This newly developed device was effectively employed in two distinct operational modes, specifically headspace (HS) and direct immersion (DI) (as depicted in Fig. 8). As a result, it emerged as an excellent solution for real-time monitoring of salmon spoilage mechanisms and the comprehensive analysis of essential nutrients present in salmon fillets.

The application of ambient mass spectrometry techniques for pesticide analysis in produce, along with the validation of these techniques through chromatographic separation, has not received extensive exploration. In one of the few existing studies, Kasperkiewicz and Pawliszyn developed a coated blade spray (CBS) protocol to quantitate multiresidue pesticide levels in various fruit matrices [130]. In CBS, sampling, sample preparation, and introduction to analytical instrumentation is consolidated into a single device consisting of a polymeric sorbent coated onto a conductive support [131, 132]. The use of CBS allowed the authors to couple

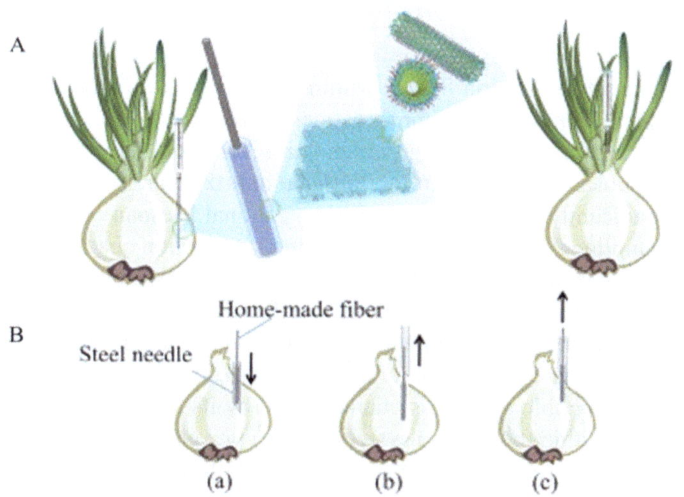

Fig. 7 **A** Real-time sampling within garlic bulbs and leaf sheaths. **B** In vivo SPME procedure: (*a*) Gently introduce the custom-made fiber into the garlic with the safeguard of a steel needle, (*b*) Extract analytes in vivo by carefully withdrawing the steel needle, (*c*) Retrieve the extracted fiber afterward. Reprinted from [129] with permission from Elsevier

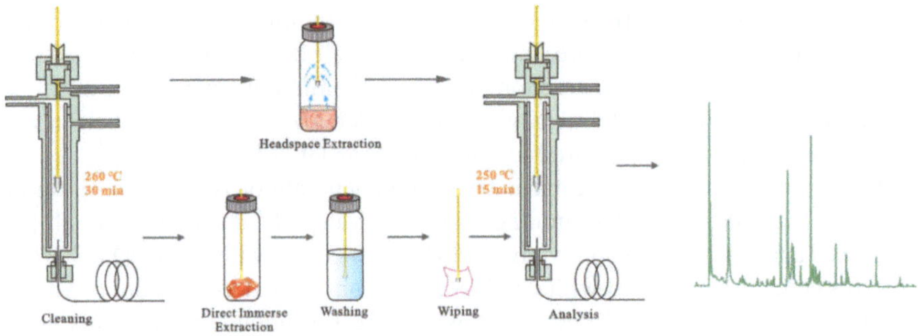

Fig. 8 Headspace and direct immersion SPME protocols. Reprinted from [39] with permission from Elsevier

the devices directly to mass spectrometry (MS) and liquid chromatography (LC) to perform multiresidue (e.g., organophosphates, organonitrogen, carbamates, neonicotinoids, strobilurins, triazines, spinosyns) analysis for a panel of pesticides in apple, blueberry, grape, and strawberry samples. Figure 9 shows the CBS-MS/MS and SPME-LC–MS/MS workflows to quantitatively assess 126 pesticides in apples, 139 pesticides in blueberries, 136 pesticides in grapes, and 135 pesticides in strawberries, as well as their analytical figures of merit, analytical properties (e.g., solvent usage, analysis time), and real-world sample quantification.

4.3 Biological Applications

Conducting direct-immersion SPME within complex matrices can be challenging. Generally, some form of sample pre-treatment is required to safeguard the coating and avert extraction phase fouling, which can result from the irreversible adsorption of large molecules present in the intricate matrix. This irreversible adsorption not only significantly shortens the fiber's operational lifespan (often limiting it to just a few samplings) but also alters the coating's extraction characteristics. Consequently, researchers are persistently exploring novel SPME coatings with enhanced performance capabilities for direct extractions from complex matrices.

The development of biocompatible coatings was a major breakthrough with respect to biological applications coupling SPME and LC/MS analysis. Within SPME, a biocompatible coating is characterized by its ability to (i) avoid eliciting toxic responses within the studied system and (ii) prevent the attachment of macromolecules, such as proteins, onto its surface [133]. To address the aforementioned biocompatibility issues, researchers have investigated coatings based on polyethylene glycol (PEG) [134], polypyrrole (PPY) [134, 135], restricted access materials (RAM) [136], and mixtures of SPE sorbents (coated silica particles) and biocompatible polymers [137]. As an illustration, Musteata et al. [135] utilized SPME fibers

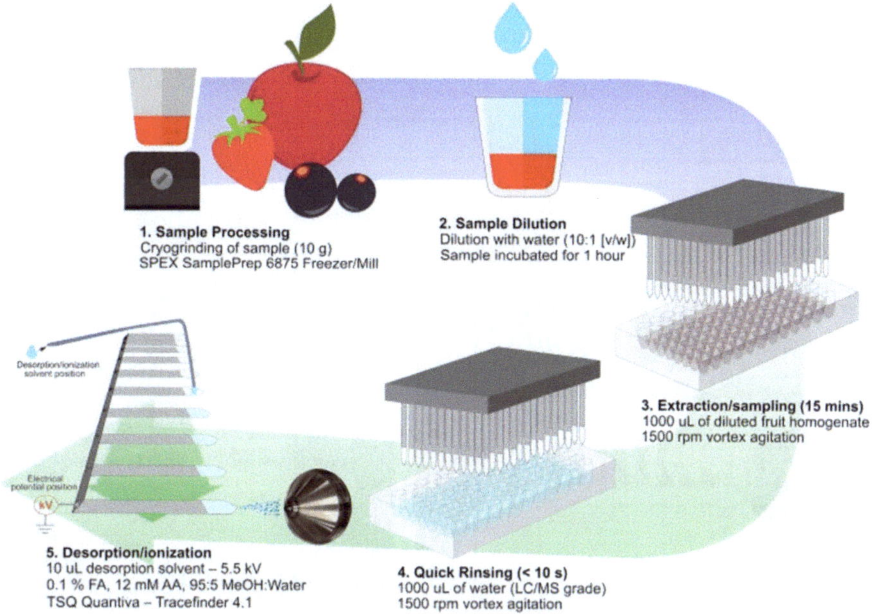

Fig. 9 The process for analyzing pesticides in fruit matrices (steps 1–4) was also utilized for CBS-MS/MS analysis (step 5). The LC-MS/MS protocol followed the identical sample-preparation workflow (steps 1–4). Reprinted from [130] with permission from Elsevier

featuring PPY/PEG and PEG/C_{18}-bonded coatings to directly extract diazepam and its metabolite from the circulating blood of beagle dogs.

In other work, scientists created novel biocompatible SPME fiber coatings through the amalgamation of polyacrylonitrile (PAN) with diverse extraction particles (including C_{18}, RP-amide-silica, HS-F5-silica, 5 μm). They subsequently assessed the efficiency of these coatings in extracting five distinct drugs from human plasma. [137]. In addition to biocompatibility and improved durability, the developed fibers offered significantly better extraction efficiency for the targeted drugs compared to PPY, RAM, and commercial CW/TPR coatings. Similarly, Mirnaghi et al. developed a method for preparing biocompatible C_{18}-PAN (polyacrylonitrile) thin-film coatings ("blades") for the direct extraction of small molecules from biological fluids [138]. Elsewhere, Sinha Roy et al. [42] developed a protocol enabling the high-throughput analysis of free concentrations of a panel of drugs in plasma, as well as the protein binding of a selection of substances with wide-ranging properties in order to elucidate the underlying principles of SPME technology. The chosen microsampling preparation device supported by plastic comprised 96 pins (as depicted in Fig. 10), each coated with a minimal quantity of matrix-compatible C_{18} extraction phase. This coating was employed to facilitate the extraction of small analytes of interest, even in the presence of macromolecules.

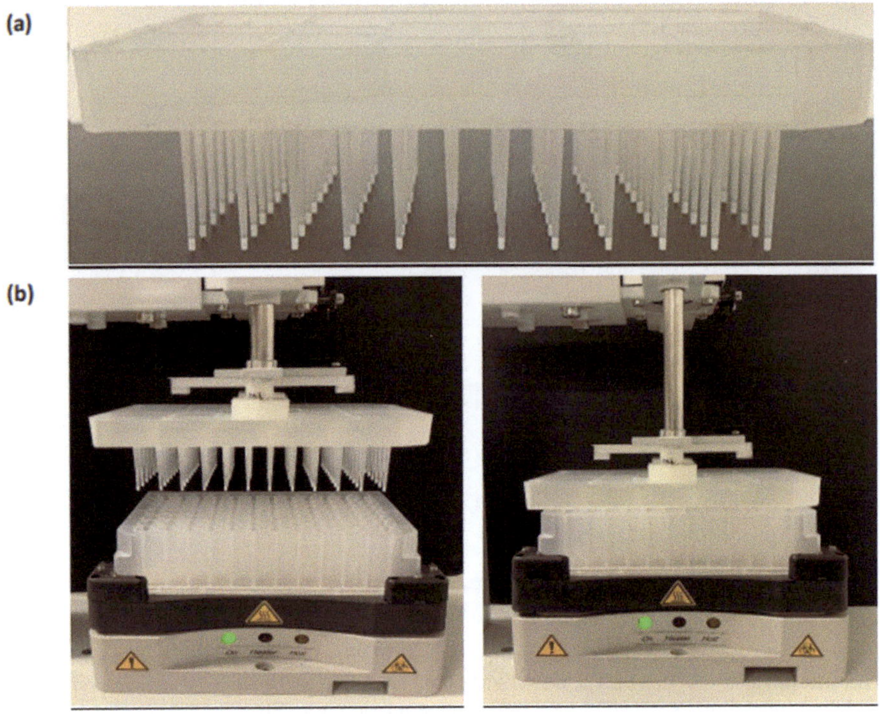

Fig. 10 **a** Supel™ BioSPME 96-Pin device. **b** BioSPME device coupled with a Concept96 automated system (PAS Technologies GmbH, Germany). Reprinted from [42] with permission from the American Chemical Society

Rocío-Bautista et al. directly coupled BioSPME to liquid electron ionization-tandem mass spectrometry (LEI-MS/MS) via a microfluidic open interface (MOI) to create a sensitive technique that eliminates matrix effects (ME) and enables the direct analysis of biological samples without necessitating sample purification or chromatographic separations [139]. In this protocol, the authors used C_{18} Bio-SPME fibers for direct immersion analysis of fentanyl compounds in urine and plasma. A schematic of the modified MOI-LEI-MS/MS system is shown in Fig. 11.

4.4 *In vivo Applications*

The application of in vivo SPME has found extensive use in numerous research investigations aiming to analyze organic analytes within intricate matrices [140–142]. Substances such as persistent organic pollutants (POPs), endocrine-disrupting compounds (EDCs), pesticides, disinfection byproducts (DBPs), and heavy metals have the potential to leach into environmental matrices like soil, air, water, and

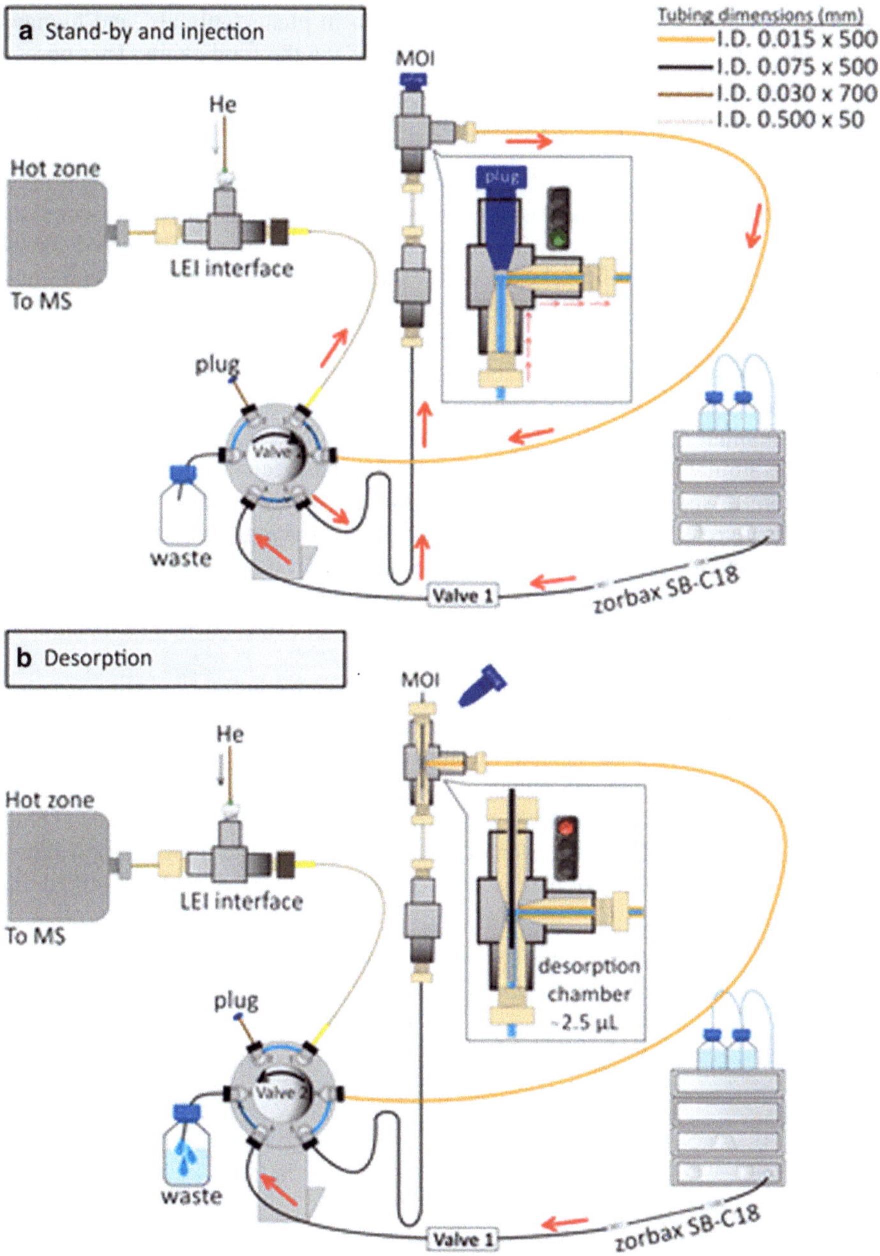

Fig. 11 Diagram illustrating the fluid dynamics of the MOI-LEI-MS/MS system. **a** Standby and injection configuration; **b** Desorption configuration. Reprinted from [139] with permission the from American Chemical Society

sediments, which can lead to their accumulation in plants, animals, and humans, either through direct contact with the matrices or via the food chain. This poses a risk to living organisms, as the accumulation of such compounds can lead to toxicity impairing different cellular processes at the genomic, proteomic, and metabolomic levels.

Napylov et al. [143] employed in vivo SPME sampling to assess oxylipin profiles within the brains of live, conscious rats. This groundbreaking and inventive technique circumvented alterations in oxylipin concentrations post-mortem, allowed for the real-time tracking of oxylipin levels with exceptional spatial precision, and could be executed using the identical experimental apparatus as in vivo microdialysis, a well-regarded standard in neuroscience research. Elsewhere, Musteata et al. [144] developed a fast in vivo microextraction technique with the potential to replace (at least in part) current sampling techniques based on blood drawing, especially in the case of small animals (Fig. 12). In this method, the sampling process does not require the animal to be handled once the interface has been installed, thus reducing its exposure to stress. This is a significant improvement, as lower levels of stress result in more relevant pharmacokinetic data, thus reducing the number of animals required to obtain reproducible data. In this study, Musteata et al. effectively utilized sampling devices founded on hypodermic tubes coupled with SPME fibers for the in vivo analysis of both free and total concentrations of diazepam and its metabolites within whole blood in rats.

Yuan et al. [145] developed a breath collection device utilizing a daily wearable face mask. In this approach, one or more SPME fibers are integrated into the face mask, leading to substantial selectivity and analyte enrichment through both specific and nonspecific adsorption mechanisms. Consequently, the proposed SPME-in-mask device proves well-suited for the ongoing collection of analytes from exhaled breath aerosols over extended periods, even in real-world settings, spanning multiple hours. After the exhaled breath samples were acquired, they were directly desorbed and

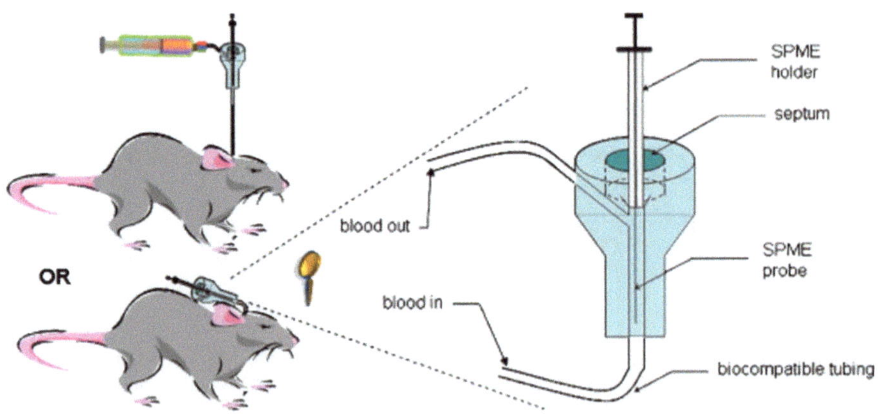

Fig. 12 In vivo SPME investigation involving rats: positioning of SPME devices and their connection interface to the carotid artery. Reprinted from [144] with permission from Elsevier

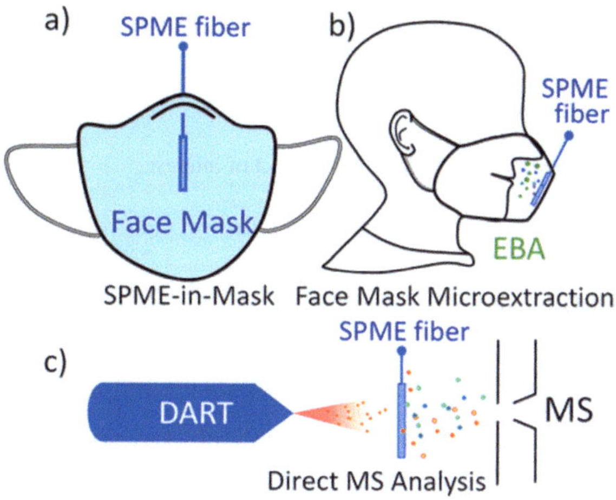

Fig. 13 Illustration depicting SPME-in-mask sampling and the subsequent direct mass spectrometry analysis of exhaled breath aerosol. **a** Insertion of the SPME fiber into a KN95 face mask; **b** Wearing of the SPME-in-mask for the collection of exhaled breath aerosol; **c** Direct linkage of SPME with DART-MS. Reprinted from [145] with permission from the American Chemical Society

ionized from the fibers via direct analysis in real time mass spectrometry (DART-MS) without further sample pretreatment. The SPME-in-mask concept is illustrated in Fig. 13.

5 Conclusions and Future Trends

Since its introduction in 1990, solid-phase microextraction has established itself as a highly popular microextraction approach for the analysis of a wide range of compounds in biological, food, and environmental samples. Indeed, SPME's numerous benefits endow it with great potential for several analytical applications. As an example, SPME proves highly advantageous for the analysis of volatile compounds found in exceedingly low concentrations within diverse food and environmental samples. In addition, SPME's status as a reliable and high-quality sample-preparation tool has further contributed to its widespread application for the identification and quantitation of myriad chemical compounds and biological substances. SPME's competitive edge over other extraction methods is primarily rooted in the large selection of available coating sorbents including polar, medium polar, non-polar, ion-exchange, and mixed-mode sorbents. Despite this variety, selecting an appropriate SPME fiber still requires detailed knowledge of the properties of the sample matrix under study. Finally, in addition to the large variety of fiber coatings, the availability of many different extraction and desorption techniques and derivatization procedures has also enabled the development of selective, sensitive, and repeatable SPME methods, especially for the analysis of food and environmental matrices.

Acknowledgements The authors would like to thank the Natural Sciences and Engineering Research Council of Canada (NSERC) Discovery Grant program and Canada Research Chairs program for their support.

Conflict of Interest
The authors have declared no conflict of interest.

References

1. Pawliszyn J (2003) Sample preparation: quo vadis? Anal Chem 75:2543–2558. https://doi.org/10.1021/ac034094h
2. Ramos L (2012) Critical overview of selected contemporary sample preparation techniques. J Chromatogr A 1221:84–98. https://doi.org/10.1016/j.chroma.2011.11.011
3. López-Lorente ÁI, Pena-Pereira F, Pedersen-Bjergaard S, et al (2022) The ten principles of green sample preparation. TrAC—Trends Anal Chem 148. https://doi.org/10.1016/j.trac.2022.116530
4. Varga R, Somogyvári I, Eke Z, Torkos K (2011) Determination of antihypertensive and anti-ulcer agents from surface water with solid-phase extraction-liquid chromatography-electrospray ionization tandem mass spectrometry. Talanta 83:1447–1454. https://doi.org/10.1016/j.talanta.2010.11.030
5. Ribeiro C, Tiritan ME, Rocha E, Rocha MJ (2007) Development and validation of a HPLC-DAD method for determination of several endocrine disrupting compounds in estuarine water. J Liq Chromatogr Relat Technol 30:2729–2746. https://doi.org/10.1080/10826070701560652
6. Abdulra'uf LB, Chai MK, Tan GH (2012) Applications of solid-phase microextraction for the analysis of pesticide residues in fruits and vegetables: a review. J AOAC Int 95:1272–1290. https://doi.org/10.5740/jaoacint.SGE_Abdulrauf
7. Wang H, Yan H, Qiu M et al (2011) Determination of dicofol in aquatic products using molecularly imprinted solid-phase extraction coupled with GC-ECD detection. Talanta 85:2100–2105. https://doi.org/10.1016/j.talanta.2011.07.061
8. Murtada K (2020) Trends in nanomaterial-based solid-phase microextraction with a focus on environmental applications—a review. Trends Environ Anal Chem 25:e00077. https://doi.org/10.1016/j.teac.2019.e00077
9. Reyes-Garcés N, Gionfriddo E, Gómez-Ríos GA et al (2018) Advances in solid phase microextraction and perspective on future directions. Anal Chem 90:302–360. https://doi.org/10.1021/acs.analchem.7b04502
10. Murtada K, Bowman D, Edwards M, Pawliszyn J (2022) Thin-film microextraction combined with comprehensive two-dimensional gas chromatography time-of-flight mass spectrometry screening for presence of multiclass organic pollutants in drinking water samples. Talanta 242:123301. https://doi.org/10.1016/j.talanta.2022.123301
11. Murtada K, Galpin V, Grandy JJ et al (2021) Development of porous carbon/polydimethylsiloxane thin-film solid-phase microextraction membranes to facilitate on-site sampling of volatile organic compounds. Sustain Chem Pharm 21:100435. https://doi.org/10.1016/j.scp.2021.100435
12. Risticevic S, Niri VH, Vuckovic D, Pawliszyn J (2009) Recent developments in solid-phase microextraction. Anal Bioanal Chem 393:781–795. https://doi.org/10.1007/s00216-008-2375-3
13. Risticevic S, Lord H, Górecki T et al (2010) Protocol for solid-phase microextraction method development. Nat Protoc 5:122–139. https://doi.org/10.1038/nprot.2009.179
14. Piri-Moghadam H, Alam MN, Pawliszyn J (2017) Review of geometries and coating materials in solid phase microextraction: Opportunities, limitations, and future perspectives. Anal Chim Acta 984:42–65. https://doi.org/10.1016/j.aca.2017.05.035

15. Duan C, Shen Z, Wu D, Guan Y (2011) Recent developments in solid-phase microextraction for on-site sampling and sample preparation. TrAC—Trends Anal Chem 30:1568–1574. https://doi.org/10.1016/j.trac.2011.08.005
16. Theodoridis G, Koster EH, de Jong GJ (2000) Solid-phase microextraction for the analysis of biological samples. J Chromatogr B Biomed Sci Appl 745:49–82. https://doi.org/10.1016/S0378-4347(00)00203-6
17. Xu J, Chen G, Huang S et al (2016) Application of in vivo solid-phase microextraction in environmental analysis. TrAC—Trends Anal Chem 85:26–35. https://doi.org/10.1016/j.trac.2016.03.003
18. Kataoka H, Ishizaki A, Saito K (2016) Recent progress in solid-phase microextraction and its pharmaceutical and biomedical applications. Anal Methods 8:5773–5788. https://doi.org/10.1039/c6ay00380j
19. Llompart M, Celeiro M, García-Jares C, Dagnac T (2019) Environmental applications of solid-phase microextraction. TrAC—Trends Anal Chem 112:1–12. https://doi.org/10.1016/j.trac.2018.12.020
20. Prosen H, Zupančič-Kralj L (1999) Solid-phase microextraction. TrAC—Trends Anal Chem 18:272–282. https://doi.org/10.1016/S0165-9936(98)00109-5
21. De Fátima AM (2000) Solid-phase microextraction: A promising technique for sample preparation in environmental analysis. J Chromatogr A 889:3–14. https://doi.org/10.1016/S0021-9673(00)00453-2
22. Lord H, Pawliszyn J (2000) Evolution of solid-phase microextraction technology
23. Olcer YA, Tascon M, Eroglu AE, Boyacı E (2019) Thin film microextraction: towards faster and more sensitive microextraction. TrAC—Trends Anal Chem 113:93–101. https://doi.org/10.1016/j.trac.2019.01.022
24. Wardencki W, Michulec M, Curyło J (2004) A review of theoretical and practical aspects of solid-phase microextraction in food analysis. Int J Food Sci Technol 39:703–717. https://doi.org/10.1111/j.1365-2621.2004.00839.x
25. Marrubini G, Dugheri S, Cappelli G et al (2020) Experimental designs for solid-phase microextraction method development in bioanalysis: A review. Anal Chim Acta 1119:77–100. https://doi.org/10.1016/j.aca.2020.04.012
26. Louch D, Motlagh S, Pawliszyn J (1992) Dynamics of organic compound extraction from water using liquid-coated fused silica fibers. Anal Chem 64:1187–1199. https://doi.org/10.1021/ac00034a020
27. Semenov SN, Koziel JA, Pawliszyn J (2000) Kinetics of solid-phase extraction and solid-phase microextraction in thin adsorbent layer with saturation sorption isotherm. J Chromatogr A 873:39–51. https://doi.org/10.1016/S0021-9673(99)01338-2
28. Pawliszyn J (1997) Solid phase microextraction: theory and practice. Wiley-VCH Publishers, Inc., New York
29. Paschke A, Popp P, Schüürmann G (1998) Water solubility and octanol/water-partitioning of hydrophobic chlorinated organic substances determined by using SPME/GC. Fresenius J Anal Chem 360:52–57. https://doi.org/10.1007/s002160050641
30. Langenfeld JJ, Hawthorne SB, Miller DJ (1996) Quantitative analysis of fuel-related hydrocarbons in surface water and wastewater samples by solid-phase microextraction. Anal Chem 68:144–155. https://doi.org/10.1021/ac950862w
31. Dean JR, Tomlinson WR, Makovskaya V et al (1996) Solid-phase microextraction as a method for estimating the octanol-water partition coefficient. Anal Chem 68:130–133. https://doi.org/10.1021/ac950778g
32. Pörschmann J, Kopinke FD, Pawliszyn J (1998) Solid-phase microextraction for determining the binding state of organic pollutants in contaminated water rich in humic organic matter. J Chromatogr A 816:159–167. https://doi.org/10.1016/S0021-9673(98)00525-1
33. Górecki T, Pawliszyn J (1997) Effect of sample volume on quantitative analysis by solid-phase microextraction: Part 1. Theoretical considerations. Analyst 122:1079–1086. https://doi.org/10.1039/a701303e

34. Górecki T, Khaled A, Pawliszyn J (1998) The effect of sample volume on quantitative analysis by solid phase microextraction. Part 2. Experimental verification. Analyst 123:2819–2824. https://doi.org/10.1039/a806788k
35. Nazdrajić E, Murtada K, Pawliszyn J (2021) The effect of sorbent particles in a binder on the mass transfer kinetics in separation media. In silico study and experimental verification. Anal Chem 93:14764–14772. https://doi.org/10.1021/acs.analchem.1c03373
36. Knox JH, Scott HP (1983) B and C terms in the Van Deemter equation for liquid chromatography. J Chromatogr A 282:297–313. https://doi.org/10.1016/S0021-9673(00)91609-1
37. Zhang YP, Luan CC, Lu ZY, et al (2022) Brass wires with different surface wettability used for in-tube solid-phase microextraction. J Chromatogr A 1670:462948. https://doi.org/10.1016/j.chroma.2022.462948
38. Xie W, Mullett WM, Miller-Stein CM, Pawliszyn J (2009) Automation of in-tip solid-phase microextraction in 96-well format for the determination of a model drug compound in human plasma by liquid chromatography with tandem mass spectrometric detection. J Chromatogr B Anal Technol Biomed Life Sci 877:415–420. https://doi.org/10.1016/j.jchromb.2008.12.036
39. Yu J, Xu XB, Murtada K, Pawliszyn J (2022) Untargeted analysis of microbial metabolites and unsaturated fatty acids in salmon via hydrophilic-lipophilic balanced solid-phase microextraction arrow. Food Chem 380:132219. https://doi.org/10.1016/j.foodchem.2022.132219
40. Zhang M, Cheng D, He X et al (2010) Magnetic silica-coated sub-microspheres with immobilized metal ions for the selective removal of bovine hemoglobin from bovine blood. Chem Asian J 5:1332–1340. https://doi.org/10.1002/asia.200900463
41. Yiantzi E, Murtada K, Terzidis K, et al (2022) Vacuum-assisted headspace thin-film microextraction: theoretical formulation and method optimization for the extraction of polycyclic aromatic hydrocarbons from water samples. Anal Chim Acta 1189:339217. https://doi.org/10.1016/j.aca.2021.339217
42. Roy KS, Nazdrajić E, Shimelis OI et al (2021) Optimizing a high-throughput solid-phase microextraction system to determine the plasma protein binding of drugs in human plasma. Anal Chem 93:11061–11065. https://doi.org/10.1021/acs.analchem.1c01986
43. Xu XB, Murtada K, Pawliszyn J (2021) Determination of selected volatile terpenes in fish samples via solid phase microextraction arrow coupled with GC-MS. Talanta 221:121446. https://doi.org/10.1016/j.talanta.2020.121446
44. Grandy JJ, Murtada K, Belinato JR et al (2020) Development and validation of an improved, thin film solid phase microextraction based, standard gas generating vial for the repeatable generation of gaseous standards. J Chromatogr A 1632. https://doi.org/10.1016/j.chroma.2020.461541
45. Murtada K, Pawliszyn J (2022) Protocol for the development of TFME-GC methods for analyzing multiclass organic constituents in water samples. Green Anal Chem 2:100016. https://doi.org/10.1016/j.greeac.2022.100016
46. Khaled A, Gionfriddo E, Acquaro V et al (2019) Development and validation of a fully automated solid phase microextraction high throughput method for quantitative analysis of multiresidue veterinary drugs in chicken tissue. Anal Chim Acta 1056:34–46. https://doi.org/10.1016/j.aca.2018.12.044
47. Khaled A, Gómez-Ríos GA, Pawliszyn J (2020) Optimization of coated blade spray for rapid screening and quantitation of 105 VETERINARY DRUGS IN BIOLOGICAL TISSUE SAmples. Anal Chem 92:5937–5943. https://doi.org/10.1021/acs.analchem.0c00093
48. Xu J, Zheng J, Tian J et al (2013) New materials in solid-phase microextraction. TrAC—Trends Anal Chem 47:68–83. https://doi.org/10.1016/j.trac.2013.02.012
49. Zheng J, Huang J, Yang Q et al (2018) Fabrications of novel solid phase microextraction fiber coatings based on new materials for high enrichment capability. TrAC—Trends Anal Chem 108:135–153. https://doi.org/10.1016/j.trac.2018.08.021
50. Feng J, Sun M, Li J et al (2011) A novel silver-coated solid-phase microextraction metal fiber based on electroless plating technique. Anal Chim Acta 701:174–180. https://doi.org/10.1016/j.aca.2011.05.040

51. Sun M, Feng J, Bu Y et al (2015) Palladium-coated stainless-steel wire as a solid-phase microextraction fiber. J Sep Sci 38:1584–1590. https://doi.org/10.1002/jssc.201401283
52. Xu HL, Li Y, Jiang DQ, Yan XP (2009) Hydrofluoric acid etched stainless steel wire for solid-phase microextraction. Anal Chem 81:4971–4977. https://doi.org/10.1021/ac900743s
53. Yazdi MN, Yamini Y, Asiabi H (2018) Fabrication of polypyrrole-silver nanocomposite for hollow fiber solid phase microextraction followed by HPLC/UV analysis for determination of parabens in water and beverages samples. J Food Compos Anal 74:18–26. https://doi.org/10.1016/j.jfca.2018.08.006
54. Yang Y, Li Y, Liu H et al (2014) Electrodeposition of gold nanoparticles onto an etched stainless steel wire followed by a self-assembled monolayer of octanedithiol as a fiber coating for selective solid-phase microextraction. J Chromatogr A 1372:25–33. https://doi.org/10.1016/j.chroma.2014.10.095
55. Lu Q, Lin R, Du C et al (2020) Metal probe microextraction coupled to dielectric barrier discharge ionization-mass spectrometry for detecting drug residues in organisms. Anal Chem 92:5921–5928. https://doi.org/10.1021/acs.analchem.0c00004
56. Jiang N, Wang J, Li W et al (2019) Silver nanoparticles-coated monolithic column for in-tube solid-phase microextraction of monounsaturated fatty acid methyl esters. J Chromatogr A 1585:19–26. https://doi.org/10.1016/j.chroma.2018.11.059
57. Gutiérrez-Serpa A, Napolitano-Tabares PI, Pino V et al (2018) Silver nanoparticles supported onto a stainless steel wire for direct-immersion solid-phase microextraction of polycyclic aromatic hydrocarbons prior to their determination by GC-FID. Microchim Acta 185:. https://doi.org/10.1007/s00604-018-2880-9
58. Karimi M, Aboufazeli F, Zhad HRLZ et al (2013) Determination of polycyclic aromatic hydrocarbons in Persian gulf and Caspian sea: Gold nanoparticles fiber for a head space solid phase micro extraction. Bull Environ Contam Toxicol 90:291–295. https://doi.org/10.1007/s00128-012-0906-2
59. Zhang Y, Yang Y, Li Y et al (2015) Growth of cedar-like Au nanoparticles coating on an etched stainless steel wire and its application for selective solid-phase microextraction. Anal Chim Acta 876:55–62. https://doi.org/10.1016/j.aca.2015.03.044
60. Djozan D, Abdollahi L (2003) Anodized zinc wire as a solid-phase microextraction fiber. Chromatographia 57:799–804. https://doi.org/10.1007/BF02491768
61. Gholivand MB, Piryaei M, Abolghasemi MM (2011) Anodized aluminum wire as a solid-phase microextraction fiber for rapid determination of volatile constituents in medicinal plant. Anal Chim Acta 701:1–5. https://doi.org/10.1016/j.aca.2011.05.046
62. Zhang G, Zou L, Xu H (2015) Anodic alumina coating for extraction of volatile organic compounds in human exhaled breath vapor. Talanta 132:528–534. https://doi.org/10.1016/j.talanta.2014.09.035
63. Zhang H, Chingin K, Li J et al (2018) Selective enrichment of phosphopeptides and phospholipids from biological matrixes on TiO_2 nanowire arrays for direct molecular characterization by internal extractive electrospray ionization mass spectrometry. Anal Chem 90:12101–12107. https://doi.org/10.1021/acs.analchem.8b03022
64. Song W, Guo M, Zhang Y et al (2015) Fabrication and application of zinc-zinc oxide nanosheets coating on an etched stainless steel wire as a selective solid-phase microextraction fiber. J Chromatogr A 1384:28–36. https://doi.org/10.1016/j.chroma.2015.01.059
65. Wang F, Zheng J, Qiu J et al (2017) In situ hydrothermally grown TiO_2@C core−shell nanowire coating for highly sensitive solid phase microextraction of polycyclic aromatic hydrocarbons. ACS Appl Mater Interfaces 9:1840–1846. https://doi.org/10.1021/acsami.6b14748
66. Budziak D, Martendal E, Carasek E (2007) Application of NiTi alloy coated with ZrO_2 as a new fiber for solid-phase microextraction for determination of halophenols in water samples. Anal Chim Acta 598:254–260. https://doi.org/10.1016/j.aca.2007.07.061
67. Ghasemi E, Farahani H (2012) Head space solid phase microextraction based on nano-structured lead dioxide: application to the speciation of volatile organoselenium in environmental and biological samples. J Chromatogr A 1258:16–20. https://doi.org/10.1016/j.chroma.2012.08.027

68. Gholivand MB, Shamsipur M, Shamizadeh M et al (2014) Cobalt oxide nanoparticles as a novel high-efficiency fiber coating for solid phase microextraction of benzene, toluene, ethylbenzene and xylene from aqueous solutions. Anal Chim Acta 822:30–36. https://doi.org/10.1016/j.aca.2014.02.032
69. Song A, Wang J, Lu G et al (2018) Oxidized multiwalled carbon nanotubes coated fibers for headspace solid-phase microextraction of amphetamine-type stimulants in human urine. Forensic Sci Int 290:49–55. https://doi.org/10.1016/j.forsciint.2018.06.031
70. Chen T, Xu H (2019) In vivo investigation of pesticide residues in garlic using solid phase microextraction-gas chromatography-mass spectrometry. Anal Chim Acta 1090:72–81. https://doi.org/10.1016/j.aca.2019.09.011
71. Naeemullah TM (2019) Development of tetraethylene pentamine functionalized multi-wall carbon nanotubes as a new adsorbent in a syringe system for removal of bisphenol A by using multivariate optimization techniques. Microchem J 147:1147–1154. https://doi.org/10.1016/j.microc.2019.04.040
72. Riahi-Zanjani B, Balali-Mood M, Asoodeh A et al (2018) Developing a new sensitive solid-phase microextraction fiber based on carbon nanotubes for preconcentration of morphine. Appl Nanosci 8:2047–2056. https://doi.org/10.1007/s13204-018-0882-x
73. Ma X, Huang P, Dang X et al (2019) MWCNTs/MnO_2 nanocomposite-based polythiophene coating for solid-phase microextraction and determination of polycyclic aromatic hydrocarbons in soil. Microchem J 146:1026–1032. https://doi.org/10.1016/j.microc.2019.02.031
74. Zhang S, Yang Q, Li Z et al (2018) Solid phase microextraction of phthalic acid esters from vegetable oils using iron (III)-based metal-organic framework/graphene oxide coating. Food Chem 263:258–264. https://doi.org/10.1016/j.foodchem.2018.04.132
75. Li JH, Xu H (2017) A novel polyaniline/polypyrrole/graphene oxide fiber for the determination of volatile organic compounds in headspace gas of lung cell lines. Talanta 167:623–629. https://doi.org/10.1016/j.talanta.2017.03.005
76. Lv F, Gan N, Cao Y et al (2017) A molybdenum disulfide/reduced graphene oxide fiber coating coupled with gas chromatography–mass spectrometry for the saponification-headspace solid-phase microextraction of polychlorinated biphenyls in food. J Chromatogr A 1525:42–50. https://doi.org/10.1016/j.chroma.2017.10.026
77. Xu L, Huang S, Liu Y et al (2020) Hollow carbon nanobubbles-coated solid-phase microextraction fibers for the sensitive detection of organic pollutants. Anal Chim Acta 1097:85–93. https://doi.org/10.1016/j.aca.2019.10.056
78. Ji R, Wu Y, Bian Y, et al (2021) Nitrogen-doped porous biochar derived from marine algae for efficient solid-phase microextraction of chlorobenzenes from aqueous solution. J Hazard Mater 407:124785. https://doi.org/10.1016/j.jhazmat.2020.124785
79. Dong ZM, Cheng L, Sun T et al (2021) Carboxylation modified meso-porous carbon aerogel templated by ionic liquid for solid-phase microextraction of trace tetracyclines residues using HPLC with UV detection. Microchim Acta 188. https://doi.org/10.1007/s00604-021-04707-2
80. Jiang H, Hu X, Li Y et al (2019) Large-pore ordered mesoporous carbon as solid-phase microextraction coating for analysis of polycyclic aromatic hydrocarbons from aqueous media. Talanta 195:647–654. https://doi.org/10.1016/j.talanta.2018.11.090
81. Zheng J, Chen L, Xie X et al (2020) Polydopamine modified ordered mesoporous carbon for synergistic enhancement of enrichment efficiency and mass transfer towards phenols. Anal Chim Acta 1095:109–117. https://doi.org/10.1016/j.aca.2019.10.036
82. Pang Y, Zang X, Li H et al (2020) Solid-phase microextraction of organophosphorous pesticides from food samples with a nitrogen-doped porous carbon derived from g-C3N4 templated MOF as the fiber coating. J Hazard Mater 384:121430. https://doi.org/10.1016/j.jhazmat.2019.121430
83. Wei F, He Y, Qu X et al (2019) In situ fabricated porous carbon coating derived from metal-organic frameworks for highly selective solid-phase microextraction. Anal Chim Acta 1078:70–77. https://doi.org/10.1016/j.aca.2019.05.061

84. Zhang N, Huang C, Tong P et al (2018) Moisture stable Ni-Zn MOF/g-C3N4 nanoflowers: A highly efficient adsorbent for solid-phase microextraction of PAHs. J Chromatogr A 1556:37–46. https://doi.org/10.1016/j.chroma.2018.04.066
85. Li YA, Yang F, Liu ZC et al (2014) A porous Cd(ii)-MOF-coated quartz fiber for solid-phase microextraction of BTEX. J Mater Chem A 2:13868–13872. https://doi.org/10.1039/c4ta01940g
86. Wei S, Kou X, Liu Y, et al (2020) Facile construction of superhydrophobic hybrids of metal-organic framework grown on nanosheet for high-performance extraction of benzene homologues. Talanta 211:120706. https://doi.org/10.1016/j.talanta.2019.120706
87. Liu G, Liu H, Tong Y, et al (2020) Headspace solid-phase microextraction of semi-volatile ultraviolet filters based on a superhydrophobic metal-organic framework stable in high-temperature steam. Talanta 219:121175. https://doi.org/10.1016/j.talanta.2020.121175
88. Suwannakot P, Lisi F, Ahmed E et al (2020) Metal-organic framework-enhanced solid-phase microextraction mass spectrometry for the direct and rapid detection of perfluorooctanoic acid in environmental water samples. Anal Chem 92:6900–6908. https://doi.org/10.1021/acs.analchem.9b05524
89. Niu J, Li Z, Yang H et al (2016) A water resistant solid-phase microextraction fiber with high selectivity prepared by a metal organic framework with perfluorinated pores. J Chromatogr A 1441:16–23. https://doi.org/10.1016/j.chroma.2016.02.076
90. Jia Y, Su H, Wang Z et al (2016) Metal-organic framework@microporous organic network as adsorbent for solid-phase microextraction. Anal Chem 88:9364–9367. https://doi.org/10.1021/acs.analchem.6b03156
91. Bagheri H, Javanmardi H, Abbasi A, Banihashemi S (2016) A metal organic framework-polyaniline nanocomposite as a fiber coating for solid phase microextraction. J Chromatogr A 1431:27–35. https://doi.org/10.1016/j.chroma.2015.12.077
92. Kong J, Zhu F, Huang W et al (2019) Sol–gel based metal-organic framework zeolite imidazolate framework-8 fibers for solid-phase microextraction of nitro polycyclic aromatic hydrocarbons and polycyclic aromatic hydrocarbons in water samples. J Chromatogr A 1603:92–101. https://doi.org/10.1016/j.chroma.2019.06.063
93. Lin C, Qi G, Wang L et al (2020) Facile preparation of stainless steel microextraction fiber via in situ growth of metal–organic framework UiO-66 and its application to sensitive analysis of polycyclic musks. J Sep Sci 43:2240–2246. https://doi.org/10.1002/jssc.201901118
94. Hu Y, Lian H, Zhou L, Li G (2015) In situ solvothermal growth of metal-organic framework-5 supported on porous copper foam for noninvasive sampling of plant volatile sulfides. Anal Chem 87:406–412. https://doi.org/10.1021/ac502146c
95. Sun S, Huang L, Xiao H et al (2019) In situ self-transformation metal into metal-organic framework membrane for solid-phase microextraction of polycyclic aromatic hydrocarbons. Talanta 202:145–151. https://doi.org/10.1016/j.talanta.2019.04.063
96. Qiu J, Zhang T, Wang F, et al (2020) Sheathed in situ heteroepitaxial growth metal-organic framework probe for detection of polycyclic aromatic hydrocarbons in river water and living fish. Sci Total Environ 729:138971. https://doi.org/10.1016/j.scitotenv.2020.138971
97. Guo JX, Qian HL, Zhao X et al (2019) In situ room-temperature fabrication of a covalent organic framework and its bonded fiber for solid-phase microextraction of polychlorinated biphenyls in aquatic products. J Mater Chem A 7:13249–13255. https://doi.org/10.1039/c9ta02974e
98. Gao W, Li G, Liu H, et al (2021) Covalent organic frameworks with tunable pore sizes enhanced solid-phase microextraction direct ionization mass spectrometry for ultrasensitive and rapid analysis of tetrabromobisphenol A derivatives. Sci Total Environ 764:144388. https://doi.org/10.1016/j.scitotenv.2020.144388
99. Sun X, Ji W, Hou S, Wang X (2020) Facile synthesis of trifluoromethyl covalent organic framework for the efficient microextraction of per-and polyfluorinated alkyl substances from milk products. J Chromatogr A 1623:461197. https://doi.org/10.1016/j.chroma.2020.461197
100. Ji W, Guo YS, Xie HM, et al (2020) Rapid microwave synthesis of dioxin-linked covalent organic framework for efficient micro-extraction of perfluorinated alkyl substances from water. J Hazard Mater 397:122793. https://doi.org/10.1016/j.jhazmat.2020.122793

101. Wang M, Zhou X, Zang X et al (2018) Determination of pesticides residues in vegetable and fruit samples by solid-phase microextraction with a covalent organic framework as the fiber coating coupled with gas chromatography and electron capture detection. J Sep Sci 41:4038–4046. https://doi.org/10.1002/jssc.201800644
102. Liu L, Meng WK, Li L et al (2019) Facile room-temperature synthesis of a spherical mesoporous covalent organic framework for ultrasensitive solid-phase microextraction of phenols prior to gas chromatography-tandem mass spectrometry. Chem Eng J 369:920–927. https://doi.org/10.1016/j.cej.2019.03.148
103. Wen L, Wu P, Wang LL et al (2020) Solid-phase microextraction using a β-ketoenamine-linked covalent organic framework coating for efficient enrichment of synthetic musks in water samples. Anal Methods 12:2434–2442. https://doi.org/10.1039/c9ay02755f
104. Liu L, Meng WK, Zhou YS et al (2019) B-Ketoenamine-linked covalent organic framework coating for ultra-high-performance solid-phase microextraction of polybrominated diphenyl ethers from environmental samples. Chem Eng J 356:926–933. https://doi.org/10.1016/j.cej.2018.09.081
105. Ma TT, Shen XF, Yang C et al (2019) Covalent immobilization of covalent organic framework on stainless steel wire for solid-phase microextraction GC-MS/MS determination of sixteen polycyclic aromatic hydrocarbons in grilled meat samples. Talanta 201:413–418. https://doi.org/10.1016/j.talanta.2019.04.031
106. Wang W, Wang J, Zhang S et al (2016) A novel Schiff base network-1 nanocomposite coated fiber for solid-phase microextraction of phenols from honey samples. Talanta 161:22–30. https://doi.org/10.1016/j.talanta.2016.08.009
107. Zhao X, Cui H, Wang Y et al (2018) Development Strategies and Prospects of Nano-based Smart Pesticide Formulation. J Agric Food Chem 66:6504–6512. https://doi.org/10.1021/acs.jafc.7b02004
108. Wu M, Chen G, Liu P et al (2016) Polydopamine-based immobilization of a hydrazone covalent organic framework for headspace solid-phase microextraction of pyrethroids in vegetables and fruits. J Chromatogr A 1456:34–41. https://doi.org/10.1016/j.chroma.2016.05.100
109. Khataei MM, Yamini Y, Ghaemmaghami M (2020) Reduced graphene-decorated covalent organic framework as a novel coating for solid-phase microextraction of phthalate esters coupled to gas chromatography-mass spectrometry. Microchim Acta 187. https://doi.org/10.1007/s00604-020-4224-9
110. Guo H, Chen G, Ma J, Jia Q (2019) A triazine based organic framework with micropores and mesopores for use in headspace solid phase microextraction of phthalate esters. Microchim Acta 186:4–10. https://doi.org/10.1007/s00604-018-3060-7
111. Zhang S, Yang Q, Li Z et al (2017) Covalent organic frameworks as a novel fiber coating for solid-phase microextraction of volatile benzene homologues. Anal Bioanal Chem 409:3429–3439. https://doi.org/10.1007/s00216-017-0286-x
112. Merkle S, Kleeberg K, Fritsche J (2015) Recent developments and applications of solid phase microextraction (SPME) in food and environmental analysis—a review. Chromatography 2:293–381. https://doi.org/10.3390/chromatography2030293
113. Xu CH, Chen GS, Xiong ZH et al (2016) Applications of solid-phase microextraction in food analysis. TrAC—Trends Anal Chem 80:12–29. https://doi.org/10.1016/j.trac.2016.02.022
114. Volante M, Cattaneo M, Bianchi M, Zoccola G (1998) Some applications of solid phase micro extraction (SPME) in the analysis of pesticide residues in food. J Environ Sci Heal—Part B Pestic Food Contam Agric Wastes 33:279–292. https://doi.org/10.1080/03601239809373144
115. Kumar A, Gaurav MAK, Matysik F-M (2009) Analysis of biological samples using solid-phase microextraction. Bioanal Rev 1:35–55. https://doi.org/10.1007/s12566-009-0004-z
116. Moein MM, Said R, Bassyouni F, Abdel-Rehim M (2014) Solid phase microextraction and related techniques for drugs in biological samples. J Anal Methods Chem. https://doi.org/10.1155/2014/921350
117. Lei XL, Wang JD (2002) Solid-phase microextraction for the analysis of drugs and biological samples. Se Pu 20:210–215

118. Kori AH, Jagirani MS, Soylak M (2023) Graphene-Based Nanomaterials: A Sustainable Material for Solid-Phase Microextraction (SPME) for Environmental Applications. Anal Lett 1–16. https://doi.org/10.1080/00032719.2023.2173221
119. Zheng J, Kuang Y, Zhou S et al (2023) Latest improvements and expanding applications of solid-phase microextraction. Anal Chem 95:218–237. https://doi.org/10.1021/acs.analchem.2c03246
120. Wu J, Luo S, Huang X (2023) Fabrication of monolith-based electrodes for simultaneous solid-phase microextraction of phenylurea and sulfonylurea herbicides assisted by electric field. Chem Eng J 455:140786. https://doi.org/10.1016/j.cej.2022.140786
121. Grandy JJ, Galpin V, Singh V, Pawliszyn J (2020) Development of a drone-based thin-film solid-phase microextraction water sampler to facilitate on-site screening of environmental pollutants. Anal Chem 92:12917–12924. https://doi.org/10.1021/acs.analchem.0c01490
122. Ouyang G, Pawliszyn J (2008) A critical review in calibration methods for solid-phase microextraction. Anal Chim Acta 627:184–197. https://doi.org/10.1016/j.aca.2008.08.015
123. Barroso MF, Noronha JP, Delerue-Matos C, Oliveira MBPP (2011) Flavored waters: Influence of ingredients on antioxidant capacity and terpenoid profile by HS-SPME/GC-MS. J Agric Food Chem 59:5062–5072. https://doi.org/10.1021/jf1048244
124. Liu XJ, Jin QZ, Liu YF et al (2011) Changes in volatile compounds of peanut oil during the roasting process for production of aromatic roasted peanut oil. J Food Sci 76:404–412. https://doi.org/10.1111/j.1750-3841.2011.02073.x
125. Melgarejo P, Calín-Sánchez Á, Vázquez-Araújo L et al (2011) Volatile composition of pomegranates from 9 Spanish cultivars using headspace solid phase microextraction. J Food Sci 76:114–120. https://doi.org/10.1111/j.1750-3841.2010.01945.x
126. Sabik H, Fortin J, Martin N (2012) Identification of pyrazine derivatives in a typical maple syrup using headspace solid-phase microextraction with gas chromatography-mass spectrometry. Food Chem 133:1006–1010. https://doi.org/10.1016/j.foodchem.2011.07.132
127. Beltran J, López FJ, Hernández F (2000) Solid-phase microextraction in pesticide residue analysis. J Chromatogr A 885:389–404. https://doi.org/10.1016/S0021-9673(00)00142-4
128. Souza Silva ÉA, Pawliszyn J (2012) Optimization of fiber coating structure enables direct immersion solid phase microextraction and high-throughput determination of complex samples. Anal Chem 84:6933–6938. https://doi.org/10.1021/ac301305u
129. Chen T, Yu X, Tian X, et al (2022) Study on the environmental fate of three insecticides in garlic by in vivo sampling rate calibrated-solid phase microextraction-gas chromatography-mass spectrometry. Food Chem 367:130740. https://doi.org/10.1016/j.foodchem.2021.130740
130. Kasperkiewicz A, Pawliszyn J (2021) Multiresidue pesticide quantitation in multiple fruit matrices via automated coated blade spray and liquid chromatography coupled to triple quadrupole mass spectrometry. Food Chem 339:127815. https://doi.org/10.1016/j.foodchem.2020.127815
131. Gómez-Ríos GA, Pawliszyn J (2014) Development of coated blade spray ionization mass spectrometry for the quantitation of target analytes present in complex matrices. Angew Chemie—Int Ed 53:14503–14507. https://doi.org/10.1002/anie.201407057
132. Kasperkiewicz A, Gómez-Ríos GA, Hein D, Pawliszyn J (2019) Breaching the 10 Second Barrier of Total Analysis Time for Complex Matrices via Automated Coated Blade Spray. Anal Chem 91:13039–13046. https://doi.org/10.1021/acs.analchem.9b03225
133. Ouyang G, Vuckovic D, Pawliszyn J (2011) Nondestructive sampling of living systems using in vivo solid-phase microextraction. Chem Rev 111:2784–2814. https://doi.org/10.1021/cr100203t
134. Es-Haghi A, Zhang X, Musteata FM et al (2007) Evaluation of bio-compatible poly(ethylene glycol)-based solid-phase microextraction fiber for in vivo pharmacokinetic studies of diazepam in dogs. Analyst 132:672–678. https://doi.org/10.1039/b701423f
135. Musteata FM, Musteata ML, Pawliszyn J (2006) Fast in vivo microextraction: A new tool for clinical analysis. Clin Chem 52:708–715. https://doi.org/10.1373/clinchem.2005.064758
136. Mullett WM, Pawliszyn J (2002) Direct determination of benzodiazepines in biological fluids by restricted-access solid-phase microextraction. Anal Chem 74:1081–1087. https://doi.org/10.1021/ac010747n

137. Musteata ML, Musteata FM, Pawliszyn J (2007) Biocompatible solid-phase microextraction coatings based on polyacrylonitrile and solid-phase extraction phases. Anal Chem 79:6903–6911. https://doi.org/10.1021/ac070296s
138. Mirnaghi FS, Chen Y, Sidisky LM, Pawliszyn J (2011) Optimization of the coating procedure for a high-throughput 96-blade solid phase microextraction system coupled with LC-MS/MS for analysis of complex samples. Anal Chem 83:6018–6025. https://doi.org/10.1021/ac2010185
139. Rocío-Bautista P, Famiglini G, Termopoli V et al (2021) Direct coupling of Bio-SPME to liquid electron ionization-MS/MS via a modified microfluidic open interface. J Am Soc Mass Spectrom 32:262–269. https://doi.org/10.1021/jasms.0c00303
140. Yu M, Roszkowska A, Pawliszyn J (2022) In vivo solid-phase microextraction and applications in environmental sciences. ACS Environ Au 2:30–41. https://doi.org/10.1021/acsenvironau.1c00024
141. Queiroz MEC, Souza ID de, Oliveira IG de, Grecco CF (2022) In vivo solid phase microextraction for bioanalysis. TrAC—Trends Anal Chem 153:. https://doi.org/10.1016/j.trac.2022.116656
142. Riboni N, Fornari F, Bianchi F, Careri M (2020) Recent advances in in vivo SPME sampling. Separations 7. https://doi.org/10.3390/separations7010006
143. Napylov A, Reyes-Garces N, Gomez-Rios G et al (2020) In vivo solid-phase microextraction for sampling of oxylipins in brain of awake, moving rats. Angew Chemie—Int Ed 59:2392–2398. https://doi.org/10.1002/anie.201909430
144. Musteata FM, de Lannoy I, Gien B, Pawliszyn J (2008) Blood sampling without blood draws for in vivo pharmacokinetic studies in rats. J Pharm Biomed Anal 47:907–912. https://doi.org/10.1016/j.jpba.2008.03.028
145. Yuan ZC, Li W, Wu L et al (2020) Solid-phase microextraction fiber in face mask for in vivo sampling and direct mass spectrometry analysis of exhaled breath aerosol. Anal Chem 92:11543–11547. https://doi.org/10.1021/acs.analchem.0c02118

Stir Bar Sorptive Extraction

Juan L. Benedé, Rafael Lucena, Soledad Cárdenas, and Alberto Chisvert

Abstract From its introduction until now, more than 20 years ago, stir bar sorptive extraction (SBSE) has been consolidated as sample preparation technique. This chapter revisits the fundamentals of this solventless technique and discusses the different aspects affecting its performance, with special emphasis on working under non-equilibrium conditions. Special attention is focused on its limitations, mainly those derived for the extraction of non-polar compounds, and how researchers try to solve them by resorting to derivatization strategies, by developing new workflows and approaches, and/or by proposing new sorbents and synthetic procedures. Those SBSE-derived extraction techniques and the advantages they present are also described and deeply discussed. An exhaustive revision of those published papers just applying these techniques are not described considering they have been extensively compiled in recent published review articles, but those contributing with some of the above-mentioned developments are commented on.

Keywords Microextraction · Sample preparation · Solventless extraction · Stir bar

Abbreviations

AμE	Adsorptive microextraction
BAμE	Bar adsorptive μ-extraction
CD	Cyclodextrin

J. L. Benedé · A. Chisvert (✉)
GICAPC Research Group, Department of Analytical Chemistry, University of Valencia, 46100 Burjassot, Valencia, Spain
e-mail: alberto.chisvert@uv.es

R. Lucena · S. Cárdenas (✉)
Affordable and Sustainable Sample Preparation (AS2P) Research Group, Departamento de Química Analítica, Instituto Universitario de Investigación en Química Fina y Nanoquímica IUNAN, Universidad de Córdoba, Campus de Rabanales, Edificio Marie Curie (anexo), 14071 Córdoba, Spain
e-mail: scardenas@uco.es

M. Á. Rodríguez-Delgado et al. (eds.), *Microextraction Techniques*, Integrated Analytical Systems, https://doi.org/10.1007/978-3-031-50527-0_5

CE	Capillary electrophoresis
COF	Covalent organic framework
DE	Desorption efficiency
DμSPE	Dispersive microsolid-phase extraction
DS	Desorption solvent
DVB	Divinylbenzene
EDMA	Ethylene dimethacrylate
EE	Extraction efficiency
EF	Enrichment factor
EG	Ethylene glycol
FPSE	Fabric phase sorptive extraction
GC	Gas chromatography
GO	Graphene oxide
HFμE	Hollow fiber microextraction
HLB	Hydrophilic-lipophilic balance
HS	Headspace
HSSE	Headspace sorptive extraction
ICECLES	Ice concentration linked with extractive stirrer
ICP	Inductive coupled plasma
LC	Liquid chromatography
LD	Liquid desorption
LDH	Layered double hydroxide
MAOE	Octyl methacrylate
MAX	Mixed mode anion exchange
MECs	Microextraction capsules
MI-FPSE	Magnet-integrated-FPSE
MIP	Molecularly-imprinted polymer
MOF	Metal-organic framework
MPS	3-(Trimethoxysilyl)propyl methacrylate
MSAμE	Multi-spheres adsorptive μ-extraction
MTES	Methyltrimethoxysilane
MWCNTs	Multiwalled carbon nanotubes
o-SWNHs	Oxidized single-walled carbon nanohorns
PA	Polyacrylate
PANI	Polyaniline
PCB	Polychlorinated biphenyl
PDMS	Polydimethylpolysiloxane
PEEK	Poly(ether ether ketone)
PEG	Polyethylene glycol
PPESK	Poly(phthalazine ether sulfone ketone)
PTFE	Polytetrafluoroethylene
PTV	Programmable temperature vaporizer
RAM	Restricted access material
RDSE	Rotating disk sorptive extraction
rGO	Reduced graphene oxide

SA-SBSE	Solvent-assisted stir bar sorptive extraction
SBME	Solvent bar microextraction
SBSDME	Stir bar sorptive dispersive microextraction
SBSE	Stir bar sorptive extraction
SCSE	Stir cake sorptive extraction
SME	Stir membrane extraction
SPME	Solid-phase microextraction
TD	Thermal desorption
ZIF	Zeolitic imidazole framework

1 Introduction

Stir bar sorptive extraction, abbreviated as SBSE, is a solventless extraction technique invented in 1999, and patented, by Prof. Sandra and co-workers [1]. In its original format, it consists of the partition of analytes between an aqueous sample (or solution) and polydimethylsiloxane (PDMS) (used as sorbent) immobilized on a magnetic-core bar (typically 10–40 mm length) immersed into the aqueous phase, in such a way the bar is stirred by using a laboratory magnetic stirrer. After a defined period of time, the agitation is stopped and the stir bar containing the analytes is taken out and rinsed with deionized water. Afterwards, it is carefully dried with a paper tissue or under a nitrogen stream, and subsequently either it is stirred into an appropriate solvent to back-extract the analytes by liquid desorption (LD), or it is subjected to high temperatures, in case of (semi)volatile and thermally-stable compounds, to desorb them by thermal desorption (TD). LD allows multiple analysis of the extract and it is the preferred option for the subsequent measurement by liquid chromatography (LC), capillary electrophoresis (CE) or inductively coupled plasma (ICP), among others. On the contrary, TD is the preferred option when gas chromatography (GC) is used, and it allows achieving higher sensitivity than with LD, since all the extracted amount is transferred to the measuring instrument. Figure 1 shows a schematic representation of the experimental procedure of SBSE.

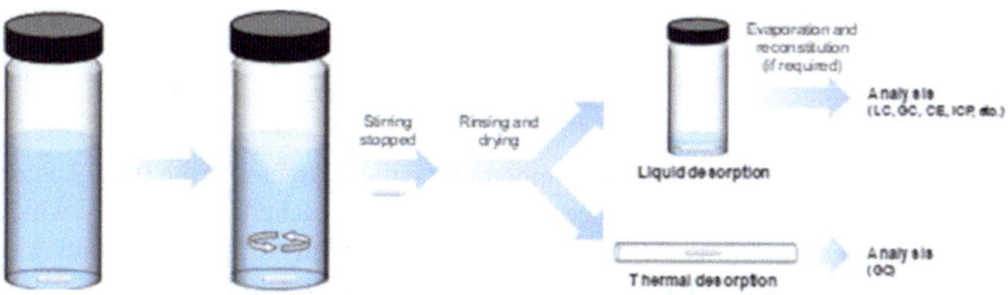

Fig. 1 Schematic representation of the experimental procedure of SBSE

From an operational point of view, SBSE is very simple and easy to carry it out, without requiring supervision, so it can be working overnight to compensate the long extraction times often required. Before their use, the stir bar needs to be conditioned by cleaning with suitable solvents (e.g., acetonitrile) or through thermal treatment (e.g., 320 °C) to minimize interferences and memory effects [2].

Although nowadays the stir bar is made up of different materials and forms, as it will be discussed later, the original stir bar consisted of three parts, i.e., a magnetic rod that enables the rotating movement, a glass jacket coating the magnetic rod, and a thin layer of PDMS coating the glass jacket where the analytes are really extracted by means of hydrophobic interactions through Van der Waals forces, although hydrogen bonds can also be stablished [3]. Although the intermediate glass jacket could seem unnecessary, it was essential to prevent decomposition of PDMS catalyzed by the metallic rod [4]. These devices have been, and still are, for many years marketed by Gerstel GmbH & Co. KG under the trade name of Twister® [5, 6]. This information will be completed later.

Much has been written about SBSE, as evidenced by the countless review articles that can be found in the bibliography describing the principles and applications of this technique [3, 7–11]. The objective of this chapter is not to repeat once again what has already been published but to revisit the fundamentals and describe them from a more didactic perspective, while describing the evolution of this technique through novel sorbents, instrumental developments, and derived techniques.

2 Fundamentals

SBSE emerged as a way to enhance the extraction efficiency (EE), and thus the sensitivity, achieved by solid-phase microextraction (SPME) (described in this chapter), the unique sorbent-based microextraction technique that existed at that moment.

Both microextraction techniques are based on the partition of the target compound (e.g., A) between the aqueous sample and a small amount of PDMS immobilized in an inert support, either a stir bar in SBSE or a fiber in SPME. The equilibrium constant governing this equilibrium, i.e., the partitioning coefficient ($K_{PDMS/water}$) can be defined as:

$$K_{PDMS/water_A} = \frac{[A]_{extracted}}{[A]_{remaining}} = \frac{\frac{m_{extracted_A}}{V_{PDMS}}}{\frac{m_{remaining_A}}{V_{water}}} \quad (1)$$

where [A] is the concentration of the compound A either extracted in the PDMS phase or remained in the aqueous phase once the equilibrium is reached, respectively, which in turn can be expressed as the ratio between the mass (m) of the compound A in each phase with respect to the volume (V) of each phase, respectively.

The EE for this compound A is defined as the ratio between the amount extracted of this compound in the PDMS layer ($m_{extracted}$) with respect to the amount of the same compound that was initially present in the aqueous sample ($m_{initial}$). After equilibrium is reached, the initial amount is distributed between the PDMS layer and the aqueous solution ($m_{remaining}$), thus:

$$EE\ (\%) = \frac{m_{extracted_A}}{m_{initial_A}} \times 100 = \frac{m_{extracted_A}}{m_{remaining_A} + m_{extracted_A}} \times 100 \quad (2)$$

Combining both equations, it is easy to see that EE depends on $K_{PDMS/water}$ and volumes ratio. Since the $K_{o/water}$ values are usually accessible unlike $K_{PDMS/water}$ ones, and PDMS behaves similarly to octanol, it could be assumed that $K_{PDMS/water} \sim K_{o/water}$, and thus:

$$EE\ (\%) = \frac{1}{\frac{\frac{V_{water}}{V_{PDMS}}}{K_{PDMS/water_A}} + 1} \times 100 \approx \frac{1}{\frac{\frac{V_{water}}{V_{PDMS}}}{K_{o/water_A}} + 1} \times 100 \quad (3)$$

In a partition equilibrium, unlike an adsorption equilibrium where the compounds are adsorbed in the active sites on the surface, the total amount of the extraction phase has a high influence on the EE. According to Eq. 3, for the same compound under the same extraction conditions, the lower the amount of PDMS is, the lower the EE is. This is the reason why EE values are low in SPME, since the extremely thin PDMS-coated fused silica or stainless-steal fibers limit the amount of available PDMS (typically around 0.5 μL for a 100-μm film thickness [1]). On the contrary, the PDMS amount is much higher in SBSE due to the higher surface area of the stir bar, which depending on its length (10–20 mm) and thickness (0.5–1.0 mm) can reach more than 120 μL [5], i.e., more than 240 times compared to SPME. As EE does not depend linearly on V_{PDMS}, the sensitivity does not increase in the same way as V_{PDMS}, as wrongly stated elsewhere, but it does increase notably. The difference between both approaches can be visualized by representing Eq. 3 as in Fig. 2, where, as a practical example, 25 mL of aqueous sample is extracted with SPME ($V_{PDMS} = 0.25$ and 0.5 μL) or with SBSE ($V_{PDMS} = 60$ and 120 μL). As it can be seen, the theoretical EE increases with $K_{o/water}$, or what is the same, the extraction is more favourable the more non-polar the compound is (i.e., higher $K_{o/water}$). As also predicted by Eq. 3, the EE is higher as the amount of PDMS increases, and thus the EE is always superior for SBSE than for SPME. It should be noticed that for moderately non-polar compounds ($K_{o/water} \sim 10^3$–10^4), which represent most of applications, EE > 70% is achieved by SBSE, whereas it does not reach to 20% for SPME. To a lesser extent, SBSE allows the extraction of polar and medium-polar compounds ($K_{o/water} < 10^3$), which are hardly achievable by SPME. For SPME, quantitative EE values are just achieved for extremely non-polar compounds ($K_{o/water} > 10^6$) and thus quantitative extractions are hardly encountered by employing this microextraction technique.

If we move our attention from the V_{PDMS} to the V_{water}, lower theoretical EE but higher theoretical extracted amount ($m_{extracted}$) are achieved for higher sample volumes maintaining the same PDMS-coated stir bar. This can be seen by plotting

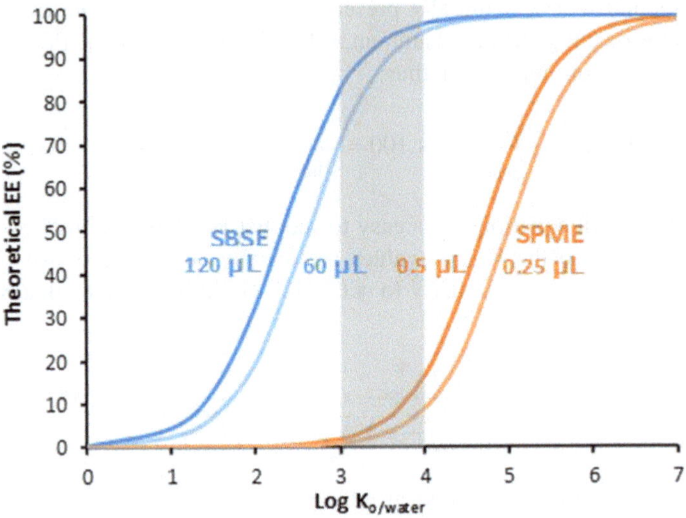

Fig. 2 Comparison of theoretical EE values obtained by SPME (V_{PDMS} = 0.25 and 0.5 µL) and SBSE (V_{PDMS} = 60 and 120 µL) in the extraction of 25 mL of aqueous sample

EE throughout Eq. 3 and $m_{extracted}$ throughout Eq. 2 for different sample volumes. Figure 3 shows, as a practical example, the extraction of a compound with $K_{o/water}$ = 10^4 from a water sample containing 10 ng mL^{-1} by means of a PDMS-coated stir bar with 60 µL PDMS.

From these results, and taking into account that the desorption is accomplished under the same conditions, large sample volumes are the best option to achieve higher sensitivity, obviously if there are not problems related to sample availability.

Apart from volume ratio, all those experimental variables affecting the $K_{o/water}$ also affect the EE. Temperature increases the solubility of solutes in a solvent, thus temperature can affect differently the solubility of the target compound in both phases, and then is compound-dependant. However, the increase in solubility is often more accentuated in the aqueous phase and thus it causes a decrease in $K_{o/water}$ and as a consequence it can be concluded that EE often decreases with temperature. Nevertheless, the effect of the temperature is often ignored and people work at room temperature [9]. In the case of ionisable compounds, such as weak acids and bases, pH plays a key role, since the neutral form is more extractable than the ionic one, thus those values that favour the formation of the neutral form increase the EE. The addition of ion-pair reagents also facilitates the extraction of ionized acids and bases by formation of a neutral adduct. The ionic strength also affects the $K_{o/w}$ and thus the EE, in such a way, an increase in the salt content forces the solutes to move to the organic phase ('salting-out effect'). Sometimes, a small amount of polar organic solvent (modifier), such as methanol, is added to the bulk aqueous sample solution to avoid the adsorption of the target compounds onto glass vessels and thus it prevents

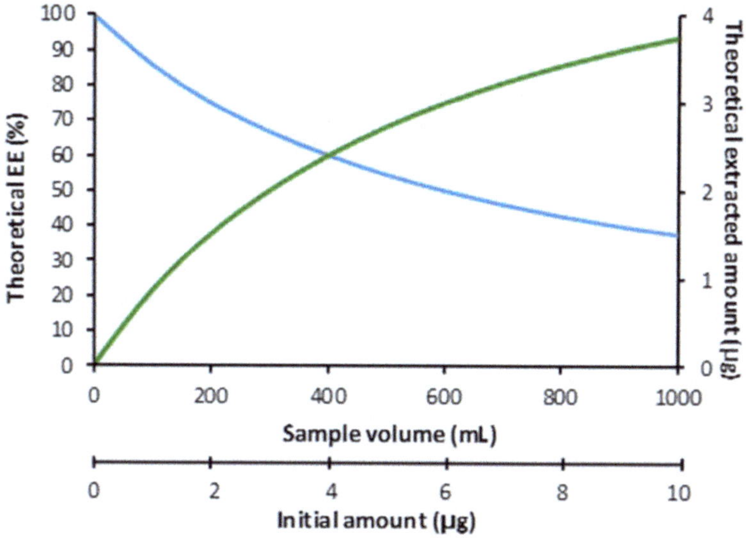

Fig. 3 Variation of theoretical EE (blue line) and theoretical extracted amount (green line) with sample volume for a compound with $K_{o/water} = 10^4$ extracted from a water sample containing 10 ng mL^{-1} by means of a PDMS-coated stir bar with 60 μL PDMS

analyte losses ('wall-effect'), but it also increases the solubility in the aqueous phase and thus it reduces collaterally the $K_{o/water}$ and the EE.

Through Eq. 3, researchers can predict the theoretical EE for a specific compound with known $K_{o/water}$ and for a given volume ratio. However, these theoretical values are rarely achieved experimentally owing mainly to three reasons: on the one hand, it is assumed that $K_{PDMS/water} \sim K_{o/water}$, which could be slightly different; on the other hand, for very non-polar compounds, the above mentioned 'wall-effect' could be significant; and last but not least, the equilibrium might not have been reached after a defined period of extraction time [1]. Figure 4 shows the results experimentally obtained for the SBSE of a group of semivolatile compounds compared with the predicted theoretical values.

Not reaching the equilibrium state is a consequence of the existence, in addition to those variables that affect thermodynamics mentioned above, of other variables involved in the extraction that exert a kinetic control over it. These variables are the extraction time, the stirring rate, the temperature (again) and the surface area of the stir bar.

With the aim to contextualize the discussion on the kinetic control in SPME and SBSE, it should be said that, unlike those microextraction techniques where the sorbent is dispersed (see chapter "Dispersive-Micro-Solid Phase Extraction (d-μSPE)") [12], the kinetics in both SPME and SBSE are slow. So, longer extraction times are needed to reach the equilibrium state and thus to obtain the maximum thermodynamic EE for a target compound under these conditions. Figure 5 shows a

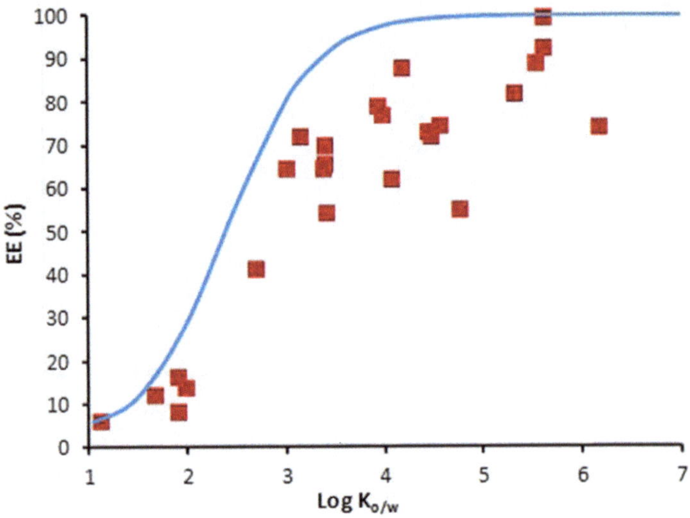

Fig. 4 Experimental EE (red squares) compared with theoretical EE (blue line) values obtained for different semivolatile compounds from a 10 mL aqueous sample by using SBSE ($V_{PDMS} = 55$ μL) (adapted with permission from [1])

real case obtained in the extraction of polychlorinated biphenyls (PCB) from a water sample [13], where the experimental EE gradually increases with extraction time. This behavior has been reported by other authors.

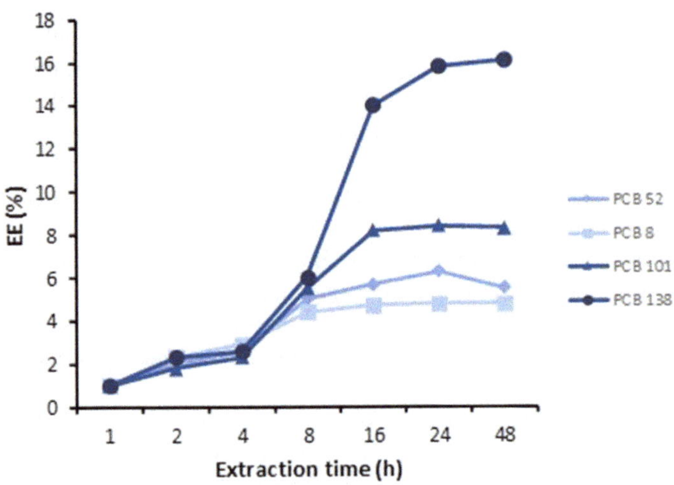

Fig. 5 Experimental evolution of EE with extraction time observed in the extraction of different PCB from water samples (adapted with permission from [13])

In practice, to avoid extending the total analysis time too much, it is usual to shorten the extraction time and to work under non-equilibrium conditions, or if time is not a critical parameter to get the results, the extraction can be left overnight. If we opt by working under non-equilibrium conditions, we should take into account that it does not affect the accuracy of the determination if calibration is conducted with standards extracted under the same working conditions than samples. However, this way of proceeding can jeopardize the precision and sensitivity. Regarding precision, it is affected as a consequence of not working in a plateau, thus the extraction time needs to be strictly controlled. In any case, the precision can be improved by working with surrogates [11]. Regarding sensitivity, it is lowered as a consequence of not achieving the maximum thermodynamic EE. Thus, a compromise situation is usually looked for, so that the selected extraction time is increased if it compensates the increase in the EE, what depends on the region EE (%) *vs* time in which the system is. To this regard, SBSE, like SPME, is considered as non-exhaustive microextraction technique.

Bearing in mind that extraction occurs under diffusion-controlled conditions, diffusion through the boundary layer between the bulk aqueous solution and PDMS is rate controlling, and this can be enhanced by an efficient stirring. However, a vigorous uncontrolled agitation could physically damage the stir bar and also cause bubble formation, which in both cases negatively affects the EE [8]. Unlike SPME, SBSE integrates the extraction and stirring elements in the same device, which, in addition to simplifying the extraction, reduces the thickness of the boundary layer and thus speeds the diffusion of the extractable compounds from the bulk sample to the stir bar [14, 15].

Apart from the thermodynamic effect that exerts the temperature over the $K_{o/water}$ discussed before, temperature per se usually accelerates the kinetics, but also diminishes the viscosity of the liquid bulk sample enhancing the mass transfer and thus decrease the time required to reach the equilibrium state. To this regard, as described before, the addition of salt improves the thermodynamics by the ‘salting out’ effect but it could damage the kinetics by increasing the viscosity.

Finally, the higher the surface area of the stir bar exposed to the bulk sample solution, the faster the extraction process is, since the sorbent is more accessible to entrap the target compounds.

With regard to the desorption conditions, we should distinguish between LD or TD, as commented before. In case of LD, a new partitioning equilibrium is now stablished between the PDMS and the desorption solvent (DS), so it is governed by a new partitioning coefficient ($K_{DS/PDMS}$). In this sense, the nature and volume of this solvent exert a great influence in the thermodynamics of the desorption efficiency (DE), which similarly as discussed early, it could be defined as:

$$DE\ (\%) = \frac{1}{\frac{\frac{V_{PDMS}}{V_{DS}}}{K_{DS/PDMS_A}} + 1} \times 100 \quad (4)$$

The solvent is chosen from the wide range of options that can be found in the laboratory compatible with the extraction phase. Aqueous solutions buffered at an appropriate pH to cause ionization of the analytes and their back-extraction could also be used, if it is the case. Its compatibility with the subsequent analytical instrument should be considered. To this regard, evaporation and reconstitution in an appropriate solvent could also be implemented despite increasing the total analysis time. Regarding the desorption volume, at least it should totally cover the stir bar to efficiently wet it, whereas, as it can be seen from Eq. 4, the higher the volume, the higher the DE. Nevertheless, as desorption volume increase, the enrichment factor (EF) for a compound A, defined as the ratio of the concentration in the desorption solvent ($[A]_{DS}$) with respect to the initial concentration in the sample ($[A]_{initial}$) (Eq. 5), would worsen as a consequence of collateral dilution, thus decreasing the sensitivity of the overall procedure. Nevertheless, at this stage, evaporation and reconstitution in a less amount of this same or another solvent could be carried out.

$$EF_A = \frac{[A]_{DS}}{[A]_{initial}} \quad (5)$$

Here again, the stirring rate and temperature enhance the kinetics [8], but it is usual to work at room temperature conditions.

Regarding TD, it is used in the case of (semi)volatile and thermally-stable compounds, so their boiling points and vapor pressures are the thermodynamic parameters inherently associated to the compounds themselves. For TD, a dedicated unit is required, consisting of two programmable temperature vaporizers (PTV). The first one heats a glass tube containing the stir bar, in such a way the retained compounds are vaporized and transferred to a cold trap (i.e., a quartz tube packed with a sorbent or a series of sorbents of increasing strength) with the aid of the carrier gas (Fig. 6a). There, the compounds are cryofocused to avoid the excessive peak broadening caused if they were transferred directly to the GC instrument. The desorption temperature and the carrier gas flow play a crucial role in this step. After the required desorption time and once the compounds are in the cold trap, the carrier gas flow reverses and the second PTV is ballistically heated to transfer rapidly and efficiently the compounds to the GC instrument (Fig. 6b).

At this point it should be said that SBSE can also operate in headspace (HS) mode rather than in the immersion mode discussed up to now, which is particularly interesting to extract (semi)volatile compounds. In this mode, also known as headspace sorptive extraction (HSSE), the stir bar is held in the HS of the vial by using special devices (Fig. 7), and it remains static in such a way the target compounds are partitioned between the aqueous sample and HS, and then between HS and PDMS. An additional magnetic stir bar immersed into the bulk sample (or solution) is used to facilitate the mass transfer from it to the HS. Compared to the immersion mode, HS-SBSE requires longer extraction times since the kinetics are more limited, whereas it is more selective since non-volatile potentially interfering compounds are avoided.

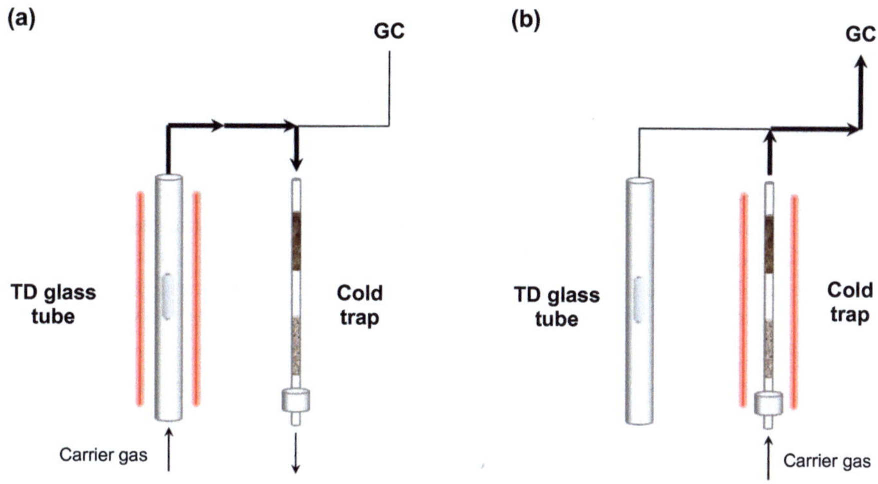

Fig. 6 Schematic diagram of thermal desorption process. **a** glass tube heating; **b** cold trap heating

The boiling points and vapor pressures of the target compounds control the thermodynamics, besides the pH, the salinity, and the organic modifier content, as discussed before for the immersion mode.

To conclude with this section, it is obvious to emphasize that the influence of both thermodynamic and kinetic variables needs to be considered in both extraction/desorption steps with the aim of reaching an efficient extraction/desorption in a reasonable time.

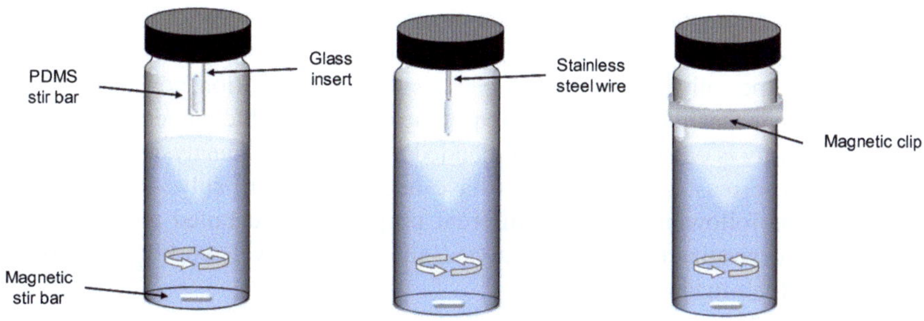

Fig. 7 Different devices for holding the extraction stir bar in the HS

3 Main Limitations and Solutions

As said before, PDMS was the unique and commercially available coating material for SBSE for a long time, and therefore it has been the most used and discussed in the literature for a wide range of applications. Despite this, due to the hydrophobic nature of this polymer, SBSE was initially limited to the unspecific extraction of non-polar compounds (generally for those with log $K_{o/w} > 3$), so the researchers had to resort to different strategies if the target compounds were of polar nature.

In this sense, a new stir bar with more polar features was later marketed, also by Gerstel GmbH & Co. KG. This alternative stir bar is made up of PDMS and ethylene glycol (EG)-modified silicone (EG/Silicone Twister®), which allows unspecific sorption of non-polar compounds and additional specific binding of polar hydrogen bond donors, such as phenols, demonstrating greater affinity compared to PDMS Twister® [16, 17]. Against, it has been verified that the stability of this coating material is inferior due to the softer nature of the polymer, and for this reason it is covered with an inert supporting grid for mechanical stabilization. Likewise, numerous scratches were observed on the surface of the grid when reused several times. Because of this, its use is recommended either in the HS mode, or immobilized in the aqueous donor phase and being stirred with an additional inert stir bar.

It is worth mentioning that a third stir bar coated with an alternative sorbent was also commercialized for a time by the same company, namely, polyacrylate (PA) with a proportion of polyethylene glycol (PEG) (Acrylate Twister®), but it is no longer available since robustness and applicability were limited [6].

Despite the commercial availability of these less non-polar coatings, their extraction mechanisms are still mainly based on hydrophobic interactions due to the presence, to a greater or lesser extent, of PDMS in its composition, and non-polar and medium polar compounds are by far the most extracted analytes. In fact, even EG/Silicone Twister® was shown to still be a challenge for the extraction of some polar compounds (especially for those with log $K_{o/w} < 2$) [18]. In addition, nowadays, multiresidue methods that allow the extraction of the widest possible range of analytes of different polarity is one of the most demanded needs by many analysis laboratories.

For all these reasons, great efforts have been directed to solve these limiting factors. In the following sections, different proposals are detailed and discussed, namely, the derivatization of the analytes, the use of novel workflows, and the fabrication of lab-made stir bars.

3.1 Derivatization

Different derivatization strategies carried out as an alternative to extract polar compounds more efficiently can be found in the literature [19]. In these cases, the polar functional groups of the analytes are converted to less polar derivatives, whose transfer to the PDMS stir bar is feasible. Some examples are alkylation, acylation, or silylation, among others.

In-situ derivatization of the analytes is the most common strategy, which occurs simultaneously with the extraction step in the aqueous donor phase. In this sense, the derivative is first formed within the solution in the presence of the derivatizing agent, which is subsequently extracted by the sorbent phase (Fig. 8a). This procedure reduces the number of steps compared to derivatization prior to extraction, and therefore the whole analysis time. Alternatively, the derivatization can also be on-stir bar, i.e., by previously loading the derivatization agent on the PDMS phase and then the analytes are incorporated into it, thus derivatization and extraction being also simultaneous (Fig. 8b). However, derivatization reactions that can be performed in aqueous solutions are limited, even some of them, such as silylation, does not occur since silylating agents are very sensitive to water and other protic solvents. In addition, derivatizing agents may be a source of interferences and errors.

Apart from these simultaneous derivatization strategies to extract polar analytes on PDMS stir bars, post-extraction derivatization strategies have also been reported to enhance the volatilization of the analytes and thus the chromatographic performance for GC analysis. In-tube derivatization (or in-port derivatization) occurs in the glass TD tube, where a few microliters of the derivatization reagent are added in a capillary tube or glass wool alongside the post-extraction stir bar containing the analytes (desorption and derivatization are simultaneous) (Fig. 8c). However, this on-line derivatization is limited to the fact that the non-derivatized polar compounds have been efficiently extracted on the stir bar, and more so if it is from PDMS phase. If LD is carried out, the derivatization reagent may be added to the desorption solvent after or during the desorption (in-extract derivatization) (Fig. 8d). In this case, the silylation reaction is possible if the solvent is not protic.

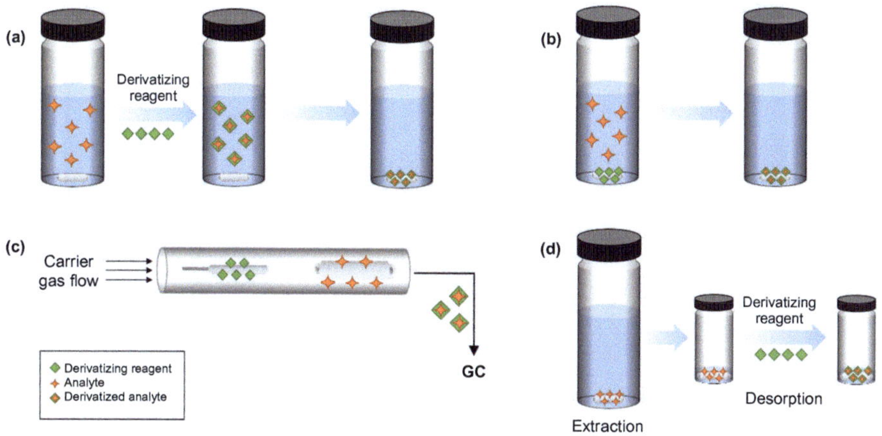

Fig. 8 Schematic diagram of: **a** in-situ derivatization; **b** on-stir bar derivatization; **c** in-tube derivatization; **d** in-extract derivatization

3.2 Novel Workflows in SBSE

When multi-residue analyzes are needed for the simultaneous determination of a large number of compounds covering a wide range of polarities, the problem arises when the extraction conditions for the analytes are quite different. For this reason, several alternatives to the conventional SBSE workflow were presented.

Ochiai et al. proposed back in 2006 dual SBSE [20], where two sample aliquots are subjected parallelly to different extractions conditions with separate stir bars, using the optimal conditions for the analytes in each case. After extraction, the stir bars are desorbed together, mainly by TD in the same glass tube (in one or two steps), and consequently only one chromatogram is obtained, reducing overall analysis time (Fig. 9a). It should be noted that what changes in each aliquot are the extraction conditions (pH, derivatization agent, ionic strength, extraction time, etc.), but they should not necessarily be two stir bars with different characteristics.

Although for other purposes, at this point, it should be mentioned that some authors employed the so-called multi-shot SBSE in order to obtain a higher sensitivity. In this methodology, several sample aliquots were extracted under the same extraction conditions using a stir bar per aliquot, and then desorbed together. When the results of this procedure were compared with those obtained using a single stir bar in a sample volume equivalent to the sum of the aliquots, an enhancement in sensitivity was verified [21].

The combination of stir bars with different polarities may expand the range of compounds to be extracted. In sequential SBSE [22], the same sample aliquot is

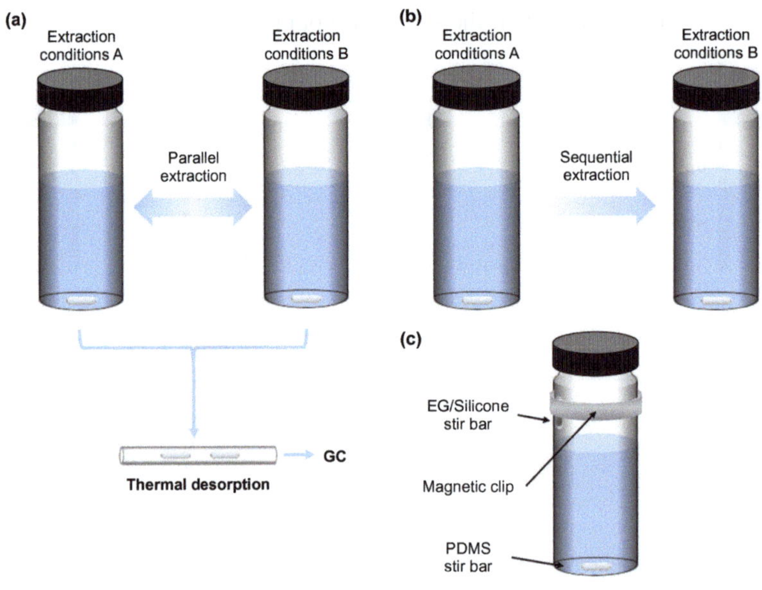

Fig. 9 Schematic diagram of: **a** dual SBSE; **b** sequential SBSE; **c** multi SBSE

subjected to different extraction conditions, even using different types of stir bars, in a sequential manner (Fig. 9b). It is also possible to modify the extraction modality (immersion and HS) between extractions, which is useful if volatile compounds are determined, as it has been proven that a high temperature may decrease the EE due to the volatilization up to the HS. After extraction, both stir bars are simultaneously desorbed for a single analysis. This workflow usually requires less sample volume than dual SBSE, but as the extractions are not carried out simultaneously, the analysis time increases.

In 2013, Ochiai et al. presented multi SBSE (^{m}SBSE) [23], which enables the extraction of a single sample aliquot using simultaneously both PDMS- and EG-Silicone-coated stir bars, and the simultaneous desorption of both (Fig. 9c). The superior extraction capacity of this workflow to cover a wide polarity range was demonstrated [24]. Although the extraction device can be made in the laboratory [25], Gerstel GmbH & Co. KG commercializes a device under the tradename of Twicester® specifically designed for ^{m}SBSE. Up to three stir bars can be used, two of them magnetically positioned with a clip on the inner wall of the vial for HS and the third being stirred in the bottom. This arrangement prevents damage to the EG-Silicone-coated stir bars due to mechanical stirring.

A novel extraction technique that relies on stir bars that have been swollen with solvent was presented in 2016 under the term solvent-assisted SBSE (SA-SBSE) [26] to extend the applicability of conventional SBSE to more polar compounds. In this approach, a small volume of solvent (e.g., ethyl acetate, acetone, acetonitrile, methanol) is added to the conventional PDMS-coated stir bar before the extraction step leading to a swelling of the sorbent phase (Fig. 10a). Thereby, compared to conventional SBSE, the SA-SBSE phase volume is significantly increased (thus reducing the phase ratio), and, at the same time, it modifies its polarity depending on the solvent used, leading to improved extraction not only for polar compounds within the range of log $K_{o/water}$ values between 1 and 2, but also for non-polar compounds. Gerstel GmbH & Co. KG developed and commercializes a stir bar specifically designed for this approach named as Flex Twister®).

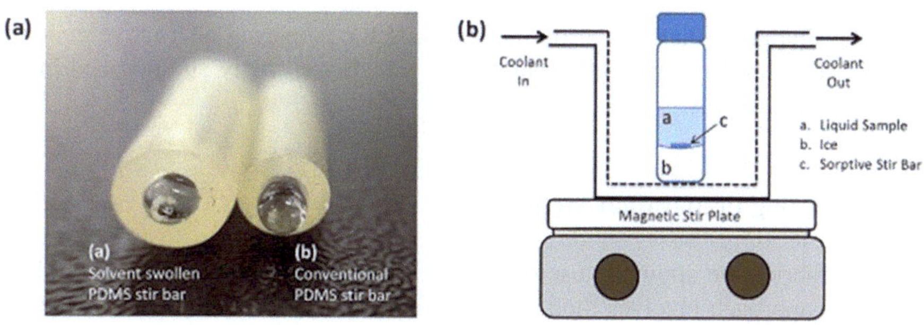

Fig. 10 **a** Comparison between a solvent swollen PDMS stir bar and a conventional PDMS stir bar in SA-SBSE (reproduced with permission from [26]. **b** Schematic diagram of ICECLES apparatus (reproduced with permission from [27])

Another alternative to overcome the drawback of poor extraction of polar compounds by conventional SBSE was proposed in the same year by Maslamani et al. [27]. Ice concentration linked with extractive stirrer (ICECLES) is based on the gradual freezing of the aqueous solution during SBSE from the bottom of the vial to the top using a double-walled beaker and a circulating chiller. As the donor phase freezes the analytes are gradually concentrated into the PDMS stir bar that remains in the upper liquid phase (Fig. 10b). It was demonstrated its higher performance for the extraction of polar compounds compared to conventional SBSE [28]. The main drawback was the limited sample volume (up to 10 mL), since it is moved away from the magnetic field as the ice front moves towards the top of the vial.

Recently, the concept of sequential SBSE was extended for a two-step fractionation of compounds with different polarities, by using a combination of ^{m}SBSE and SA-^{m}SBSE. This new workflow, termed fractionated SBSE (Fr-SBSE) [29], consists of introducing first a set of three PDMS stir bars in a sample volume to extract the non-polar compounds. Then, after removing these stir bars, three PDMS stir bars swollen with solvent are introduced in the same sample for the extraction of polar compounds (Fig. 11). This extraction procedure provides two fractions with different polarities, which are either thermally desorbed or back-extracted in an organic solvent. A similar but simpler methodology was recently proposed by the same authors, combining SBSE and SA-SBSE with in-situ derivatization [30].

3.3 *Alternative Coatings for SBSE*

The development and manufacture of new lab-made coatings has been one of the major aims of researchers working in the field to expand the potential and versatility of SBSE [31]. In this way, there is no dependency on commercial availability, which can limit the application, as previously stated. Beyond the ability to efficiently extract the analytes, the mechanical and chemical stability of the coating are two of the most sought properties when preparing alternative coatings. To obtain an increase of the extraction efficiency, a thick coating layer is preferable.

In the literature, there are different methods for preparing alternative coatings on stir bars with different sorbent materials, which are briefly summarized in the following subsections.

3.3.1 Coating Preparation Methods

The first fabrication approach used for SBSE-stir bars was the sol–gel technology [32]. This approach involves the transformation of a liquid colloidal solution (sol) to a solid matrix (gel). The most typical procedure consists of the hydrolysis of the coating precursors (e.g., methyltrimethoxysilane (MTES)) followed by the polycondensation of the hydroxylated species (i.e., inorganic network growth), incorporation of active organic ligands (e.g., hydroxy-terminated PDMS) into the network, and

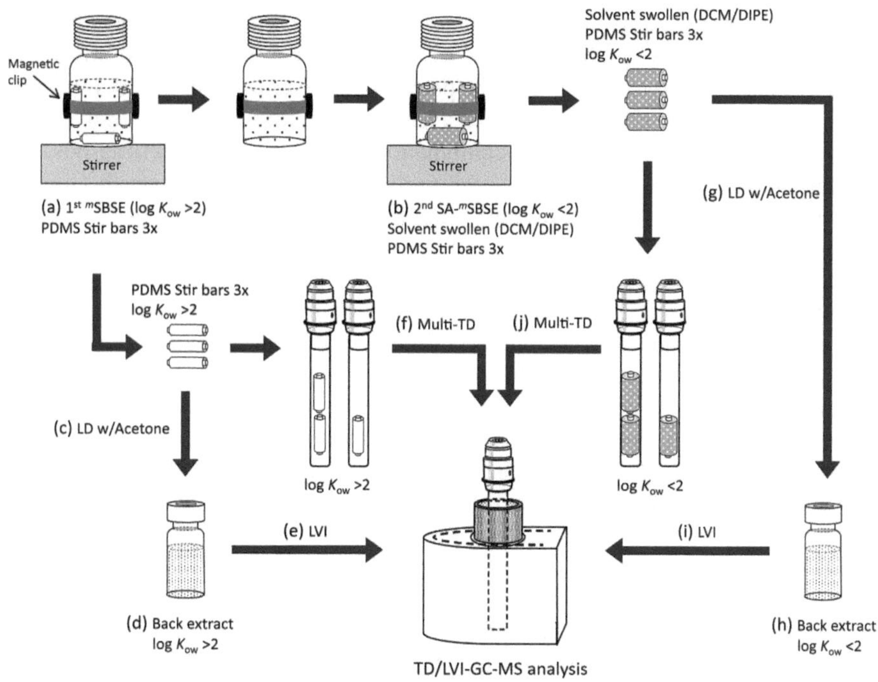

Fig. 11 Schematic diagram of Fractionated SBSE (Fr-SBSE) (reproduced with permission from [29])

finally chemical bonding of the coating on the previously treated glass stir bar to generate silanol groups on its surface.

Several other precursors are available and different functional groups (i.e., modifiers) can be easily introduced into the three-dimensional network structure during its growth to provide them with the desired polarity, such as β-cyclodextrin (β-CD) [33], polyaniline (PANI) [34], and other materials detailed in Sect. 3.3.2. Thereby, the inclusion of all these modifiers enhances the extraction of polar compounds compared to PDMS-only stir bars.

These stir bars present a good chemical, thermal and mechanical stability, and thus a long lifetime, since there is a strong chemical bonding between the coating and the glass surface of the stir bar. Additionally, the coating obtained is usually thick and uniform due to the good reproducibility in the preparation. Against, as the typical sol–gel coatings are based on non-polar PDMS, they may still lack selectivity for the most polar compounds, and also the pretreatment of the glass surface may be laborious.

The monolithic fabrication consists of the in-situ polymerization of a monomer and cross-linker mixture in the presence of a porogen solvent and a radical initiator. The polymerization is then thermally- or photo-initiated and lasts for a period of several hours.

Monoliths are porous materials containing a network of interconnected micro-sized pores, and as a result they possess very good permeability and adsorption capacity. Moreover, it is easy to tune the polarity of the resulting monolith by simply selecting the appropriate monomer from a large availability, depending on the chemical properties of the analytes. A combination of various monomers is possible, and the ratio between them and the composition of the porogen affect the rigidity, porosity, and polarity of the resulting monolith.

The fixation of the monolith on the glass surface of the stir bar can be physical or chemical. For the former, the immobilization of the coating is achieved just by simply immersing the glass stir bar in the polymerization mixture inside a mold with the desired dimensions and then starting the polymerization. For the chemical attachment, the coating fabrication involves the pretreatment of the surface of the stir bar by silylation to create double bonds, for example with 3-(trimethoxysilyl)propyl methacrylate (MPS), and subsequent polymer growth on it. Although the physical attachment is significantly easier than the chemical one since the previous step is avoided, the latter presents a higher chemical and mechanical stability due to the chemical bonding.

In a similar way, molecular imprinting technology involves the fabrication of polymers with molecule-specific cavities to recognize a target molecule (i.e., molecularly-imprinted polymer [MIP]), thus enhancing the selectivity of the material. The synthesis of the coating is carried out in the same way as explained above, but in presence of a template molecule (i.e., the target analyte or an analogous compound) in the polymerization mixture, and its subsequent removal at the end by washing steps. The formed cavity complements in size, shape, and chemical environment to the template. The ratio of crosslinker and porogen plays an important role in increasing the recognition capacity of the MIP (i.e., imprinting factor).

On the other hand, the main limitations of MIP fabrication are the need to ensure the complete removal of the template (otherwise it could provide false positives), which lengthens the synthesis time, and its better extraction efficiency in an organic medium rather than in an aqueous one. The latter can be improved with the incorporation of hydrophilic monomers.

Adhesion techniques are efficient alternatives to immobilize the sorbent materials directly on the stir bar surface, either by physical or chemical coating. For the physical coating, different methodologies have been proposed. The first proposal, and one of the most widely used in this context, is to cover the stir bar with an adhesive film (i.e., epoxy glue or a PDMS sol), followed by the attachment of the solid material (e.g., rolling the stir bar in the material), and subsequent incubation and drying [35]. Alternatively, a pretreated stir bar can be placed in an organic solution containing the extraction phase for a period of time. Once removed, the solvent is evaporated, and the coating remains on the surface. Other more sophisticated alternatives are magnetic adhesion [36], if magnetic sorbents are used, or flame deposition [37], among others. On the other hand, chemical adhesion involves the previous modification of the glass stir bar and the subsequent covalent immobilization of the material. Unlike the sol–gel technology, these stir bars are broadly not based on PDMS or the in-situ growth of a three-dimensional network.

Solvent exchange (or immersion precipitation) consists of dissolving or dispersing the material (e.g., a polymer) in a suitable solvent (e.g., formic acid), immerse first the stir bar in the solution to adhere the material onto the surface of the stir bar, and finally immerse the stir bar in water for a period of time to allow diffusion of the solvent out and leaving the film of the material on the surface. The first application of this procedure was reported by Guan et al. [38] who deposited poly(phthalazine ether sulfone ketone) (PPESK) on the stir bar, presenting a good mechanical stability.

3.3.2 Sorbent Materials

As previously indicated, the use of a wide variety of materials as the extraction phase in SBSE has been one of the main focuses of attention of researchers, since it is essential to broaden the applicability of the technique. There are different interesting review articles in the literature that have already addressed this issue [39–42], so these materials and their applications will only be briefly summarized here. It is important to note at this point that the same material can be immobilized on the stir bar by the different methods described in the previous section. Thus, the selected preparation method will affect the morphology, thickness, and stability of the coating, among others.

Different carbon-based materials, such as graphene oxide (GO) [43], reduced graphene oxide (rGO) [44] or multiwalled carbon nanotubes (MWCNTs) [45, 46], have been shown to be effective in extracting the compounds of interest in SBSE. They present high surface area, thermal and chemical stability, and their ability to have hydrophobic, $\pi - \pi$ and/or electrostatic interactions.

Huang et al. [47] were the first to introduce the monolithic materials in SBSE. They prepared an octyl methacrylate (MAOE)-ethylene dimethacrylate (EDMA) monolith for the extraction of non-polar polycyclic aromatic hydrocarbons in water samples and polar steroids in urine samples. Since then, a wide variety of monomers have been used to fabricate monoliths [48, 49]. The selection of the monomers is made based on the properties of the analytes, so that they interact through hydrophobic, hydrophilic, hydrogen bonding and/or electrostatic interactions.

Regarding the use of the selectivity provided by the MIPs, Zhu et al. [50] were the first to report the use of a MIP-based film for SBSE, which was prepared by precipitation of the polymer (nylon-6) containing the template molecule onto the surface of a commercial PDMS-coated stir bar. On the other hand, the first application of a MIP chemically attached to the surface of the stir bar was reported by Xu et al. [51]. Over the years, alternatives have been proposed to avoid the use of expensive or toxic molecules as a template and the problem of possible residual template leakage. In this sense, dummy templates (i.e., molecules similar in shape and interactions to the analytes) have been used [52].

Metal–organic frameworks, widely known as MOFs, are hybrid inorganic–organic microporous crystalline materials with a three-dimensional network by the assembly of metal ions and organic ligands by coordinative bonds. For the last decade, they have been widely used as extraction phases due to their high chemical and thermal stability, large porosity, and huge surface area. Their use in SBSE has not been an exception [53, 54]. Covalent organic frameworks (COFs) are similar to MOFs in chemical structure and properties, but their assembly is between different units by covalent bonds, and they have gained interest as extraction phase for the last years [55–57].

Some other sorbent materials have also been used as coatings in SBSE but to a much lesser extent. In this sense, polyurethane foams are polymers produced by the reaction of polyisocyanate with polyols and water in the presence of specific catalyst. These materials offer high chemical stability, flexibility, and the ability to cut them to the desired size. Although this material presented a promising future in SBSE [58], its use has not been exploited in recent years. Layered double hydroxides (LDH) are two-dimensional nanosorbents composed of two layers of divalent and trivalent cations with an anionic interphase [59]. Restricted access materials (RAMs) are biocompatible particles that enable the direct extraction of analytes from biological fluids since are able to fractionate the protein component. However, this material has been scarcely used in SBSE [60]. Immunoaffinity materials such as aptamers have been used since they present a high selectivity degree [61].

In addition, the combination of various materials in the same coating (i.e., hybrid materials) has also been proposed for SBSE purposes. For instance, a novel glycidyl methacrylate (GMA)-based polymer with an amino-modified MOF was recently developed for the first time [62], allowing to incorporate the best features of both materials in the same sorbent.

3.3.3 Stir Bar Geometry and Coating Support

One of the main drawbacks of the stir bars with conventional geometry (i.e., uniform elongated bars) completely covered by the coating material is its direct contact with the bottom of the extraction vessel in immersion mode, which may cause its damage and/or loss due to the high stirring rate. To solve these problems, alternative stir bar geometries with better mechanical resistance than the conventional one have been proposed along the last two decades.

In a first attempt, in 2007, Yu et al. [33] prepared a stir bar from the combination of two glass-coated bars with different diameters placed in parallel, with a long steel wire sealed inside one of the glass bars. One side of the combined stir bar was coated with PDMS/β-CD by sol–gel method, and no coating on the other side. The authors demonstrated that this stirring device was durable to withstand frictional forces at high stirring speed and could be reused at least 100 times with minimum loss in EE.

Two years later, the same authors presented a dumbbell-shaped stir bar to prevent the friction loss [63]. Specifically, a capillary glass bar with an iron wire inside was sealed at both ends in the shape of a spherical bubble (with an internal diameter larger than the glass bar) by alcohol flame (Fig. 12a). Then, the bars were immersed in a sol solution of PDMS/β-CD/divinylbenzene (DVB) to physically adhere the coating. They concluded that the dumbbell-shaped stir bar presented a longer lifetime since it

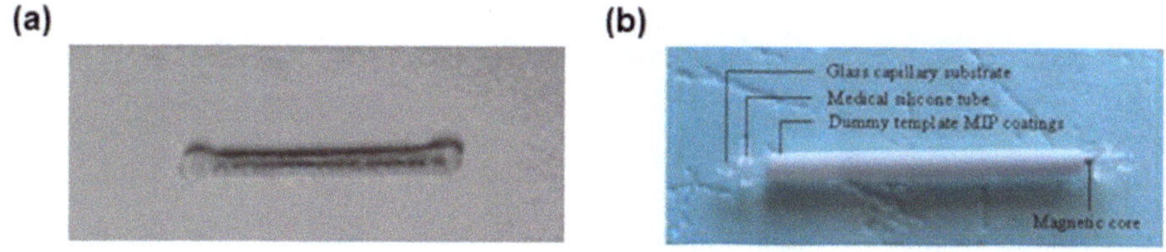

Fig. 12 **a** Photograph of dumbbell-shaped stir bar (adapted with permission from [63]). **b** Photograph of barbell-shaped stir bar (reproduced with permission from [52])

was able to be reused 40 times, while a normal-shaped stir bar was able for 30 times under the same operating conditions.

With a similar setup, Liu et al. [52] proposed in 2016 a MIP-coated glass stir bar that was sleeved by silicone wheels at both edges. In this case, they termed it "barbell-shaped" stir bar (Fig. 12b). In the same way as the dumbbell-shaped, the friction between the coating and the extraction vessel was avoided.

Moreover, although glass-coated stir bars are the most used coating supports in SBSE, other less fragile materials than glass have also been used, with the additional advantage that they can be used directly, avoiding the pretreatment of the surface of the glass stir bars before the immobilization of the extraction material.

Zhang et al. [43] were the first to report a chemically-bonded coating on a stainless-steel wire as jacket-free device for SBSE in 2014. Therein, they modified the wire first with polydopamine and then with GO, resulting in a good stability of the stir bar. Compared to conventional glass-coated stir bar, it avoided the pretreatment, thus saving operational time. Against, the metal rod suffered from corrosion when exposed long time under acidic conditions.

Fan et al. [34] proposed a stainless-steel spring as coating support (Fig. 13a). Its spiral structure presented two advantages. On the one hand, compared to a stainless-steel wire, more extraction phase can be physically fixed on it, which favored the extraction efficiency of the analytes, as demonstrated by the authors. On the other hand, it prevents the friction of the coating with the bottom of the extraction vessel, thus prolonging its lifetime.

An easier-to-prepare dumbbell-shaped stir bar was proposed by Sukree et al. [64], where a stainless-steel net is rolled into a tube, and filled with the sorbent and a metal rod to allow the stirring. Then, the two ends of the tube were closed with Teflon caps with larger diameter than the resulting stainless-steel tube (Fig. 13b). The greater advantage of this stirring device is the possibility of easily changing the sorbent material inside the tube depending on the analytes to be extracted.

In 2018, Zhou et al. [65] applied a etched poly(ether ether ketone) (PEEK) jacket stir bar. As PEEK presents high chemical resistance and smooth surface, it was treated with sulfuric acid before functionalization. Two lollipop-shaped stainless-steel needles prepared by burning polypropylene at one end were inserted into the PEEK tube for the construction of a facile detachable dumbbell-shaped stir bar (Fig. 13c). During the elution, one of the needles was detached in such a way can be

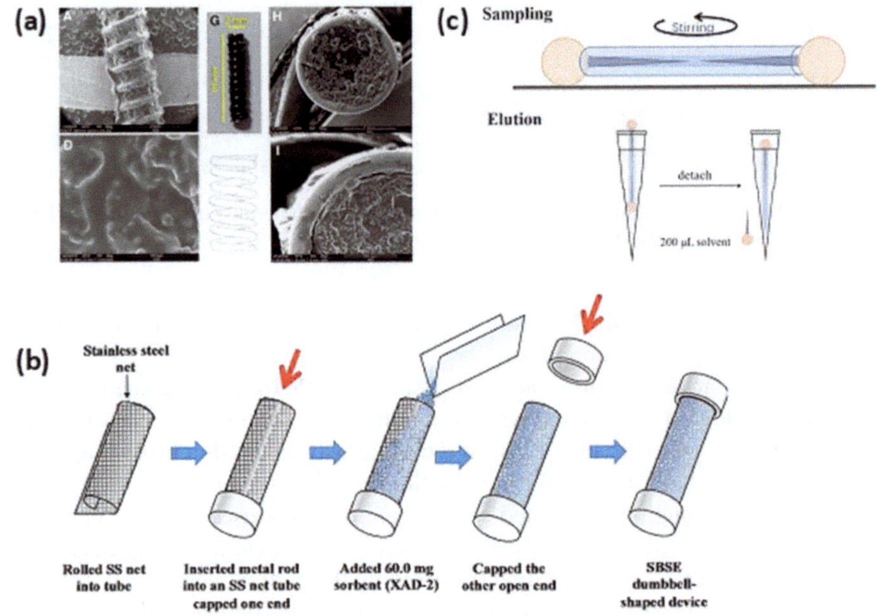

Fig. 13 **a** Images of stainless-steel spring (adapted with permission from [34]). **b** Schematic diagram of the preparation of dumbbell-shaped stir bar (reproduced with permission from [64]). **c** Schematic diagram of the detachable dumbbell-shaped PEEK jacket stir bar (reproduced with permission from [65])

easily inserted into a pipette tip. This setup was also recently applied to a polypropylene hollow fiber as the jacket for stir bar [56]. In this case, the porous structure of the bare hollow fiber avoided the tedious etching process with sulfuric acid.

Commercial polytetrafluoroethylene (PTFE) jacketed stir bar has also been employed as coating support for SBSE as it presents affordability and low cost. However, the modification of PTFE can be complicated due to its chemically resistant surface. Zhang et al. [66] immobilized graphene onto the surface previously modified with polydopamine.

Mirzaee et al. [59] proposed in 2020 the in-tube SBSE for the first time. Specifically, they immobilized the extraction material on the inner surface of a small piece of an aluminum tube, which itself participates in the fabrication of the sorbent. As the coating is not in direct contact with the extraction vessel, it prevents its deterioration.

An anodized aluminum wire was electrochemically prepared and used as nanoporous substrate for in-situ growth of a zeolitic imidazole framework (ZIF) by Ghani et al. [67]. The anodized aluminum presents a porous layer on the metal surface, and the stir bar was mechanically stable. Later, this substrate was also used for the in-situ growth of a zeolite imidazolate framework on the surface of a LDH [68].

On the other hand, to avoid interferences from the sample matrix that could affect the lifetime of the stir bar, such as the macromolecules, Mao et al. [69] used a PTFE membrane-protected stir bar, by encapsulating the coated stir bar in a porous membrane. In this way, according to author's words, the high molecular weight interference compounds would be blocked by the protective porous membrane and the lifetime of the stir bar was prolonged.

In any case, further exploring the suitable support material with porous structure and making a suitable structure design remains one of the goals at SBSE.

4 Novel Developments

The advantages of integrating extraction and stirring in the same unit have propitiated the development of similar formats that share this principle. The most outstanding characteristics of these microextraction devices raised in an increased versatility of both formats, sorbent phases and stirring mode (magnetic or mechanical). The most relevant approaches are detailed in this section, focused on the description of the principle behind the development and its main favourable features rather than in the specific application or analytical figures that can be easily found in the specific reference.

4.1 Magnetically Stirred Units

4.1.1 Stir Membrane Extraction

The use of membranes as active elements for analyte isolation presents several advantages over other configurations. They especially refer to their planar nature, which results in a high surface-to-volume ratio. Moreover, their porous structure permits the flow of the liquid or gaseous samples through them, which improves the kinetics. The incorporation of a membrane in a stirring device synergically combines the benefits of both. Stir membrane extraction (SME) was proposed by Alcudia et al. in 2009 [70]. The device consists of the use of a commercial polypropylene unit as membrane holder (Fig. 14a). It was pierced by a protected iron wire which provided the stirring of the unit under magnetic agitation. Two windows were opened on the plastic holder to allow the flow of the sample through the membrane. This first design was evaluated for the extraction of polycyclic aromatic hydrocarbons from waters. The extraction only required 15 min while elution was accomplished by face-down immersion of the membrane in the desorption solvent therefore, there was no need to remove the membrane from the holder while higher enrichment factors were obtained. The better performance of this configuration over conventional SBSE using PDMS stir bars of different surfaces was also demonstrated by the authors, which was justified by the

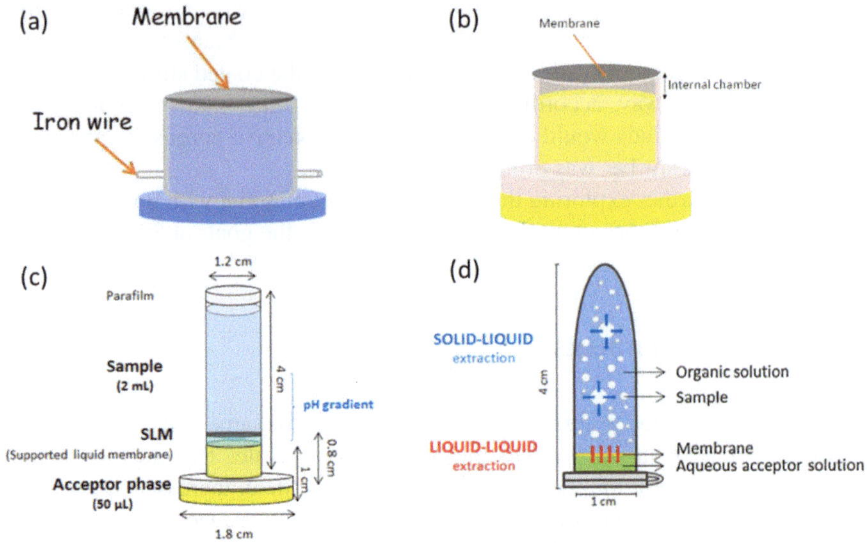

Fig. 14 Different configurations of stir membrane extraction. **a** Stir membrane extraction. **b** SME for LPME. **c** SME for processing limited-volume samples. **d** SME for the analysis of solid samples. Panels a–c reproduced under Creative Commons Attribution (CC BY) license from reference [15]. Panel d reproduced with permission from [74]

enhanced transference of the analytes from the sample to the membrane thanks to the stirring and the permeability of the extractant phase.

This basic configuration can be adapted to other microextraction modalities. The first modification consists of closing the membrane holder with a plastic cap (Fig. 14b). In this way a small chamber is created over the membrane, which can be filled with an organic (two-phase) [71] or aqueous (three-phase) solution [72]. The versatility of the SME is dramatically increased as, in the first situation, organic compounds are extracted based on the partitioning equilibrium between the sample (aqueous) and the extractant (organic solvent). The second alternative increases the selectivity of the extraction because the transference of the analyte between the two aqueous phases in favored by their intermediate solubilization into the organic phase filling the pores of the membrane (supported liquid membrane) and driven via a pH gradient established between the donor and acceptor aqueous phase. This approach is suitable for the extraction of ionizable polar analytes.

One of the main advantages of miniaturized extraction techniques is the possibility of facing new analytical problems, for example, those involving limited-volume samples such as saliva and related biofluids. In this case, the SME configuration can be modified by increasing the volume of the upper chamber that can be filled with the sample instead of the liquid extractant (Fig. 14c). This system works under the three-phase format and the stirring is accomplished by means of a vortex [73].

Another example of the versatility of the SME is that it also works processing solid samples as indicated in Fig. 14d. An Eppendorf is used as extraction device and the membrane is used to confine the extractant aqueous phase in the cap [74]. The extraction is carried out in the body of the Eppendorf where the sample is in deep contact with the organic extractant. Once isolated from the sample, the target analytes are re-extracted in the aqueous phase by passing through the supported liquid membrane.

Similarly, the polymeric membrane can be substituted by a borosilicate disk [75] or a series of small magnets [76] to broaden the application field through the possibility of using different coatings for analytes extraction.

4.1.2 Stir Cake and Rotating Disk Sorptive Extraction

The substitution of the membrane by a monolithic sorbent resulted in the so-called stir cake sorptive extraction (SCSE) [77]. In this configuration, the sorbent is synthesized inside a home-made plastic holder, which is also fitted with a protected iron wire, responsible for the stirring of the unit under a magnetic field. The advantages of SCSE over SBSE are the higher extraction capacity of the sorbent phase and its longer lifetime (reusability of 300 times *versus* 60 reuses). Since its proposal in 2011, several sorbents have been prepared and used in this format, including polymeric ionic liquids [78] and organic-phase monoliths [79]. Also, the plastic holder can be modified in terms of size and geometry to contribute to the sustainability of the synthesis, for example, reusing plastic bottle caps [80].

Rotating disk sorptive extraction (RDSE) can also be described in this section as it consists of a PTFE disk (diameter ca. 1.5 cm) with an embedded miniaturized magnet (for rotation by a laboratory magnetic stirrer) and one of its sides coated with the sorbent phase in the form of a thin film [81]. Since the sorbent phase is not in contact with the bottom of the vessel, high stirring speed can be applied without damage. Notwithstanding this, the disk can also be rotated by using a rotary rod connected to an electric stirrer [82]. The authors demonstrated that this last configuration reduces the time needed to reach the extraction equilibrium. This was ascribed to the fact that the movement of the disk reduces the boundary layer, and the transference of the analytes is, therefore, faster. As it is the case with SCSE, the variety of sorbent phases that can be deposited over the disk surface clearly increases the number of different families of compounds that can be extracted. A detailed description of the analytes and samples that can be processed has been recently reviewed [83].

4.1.3 Adsorptive Microextraction

Despite developing novel sorbent phases to increase the versatility of SBSE during the last decades, as described previously, Nogueira et al. proposed in 2010 the decoupling of the sorbent phase from the magnetic stirring unit. The so-called adsorptive microextraction (AμE) was proposed to afford the challenge of extracting polar

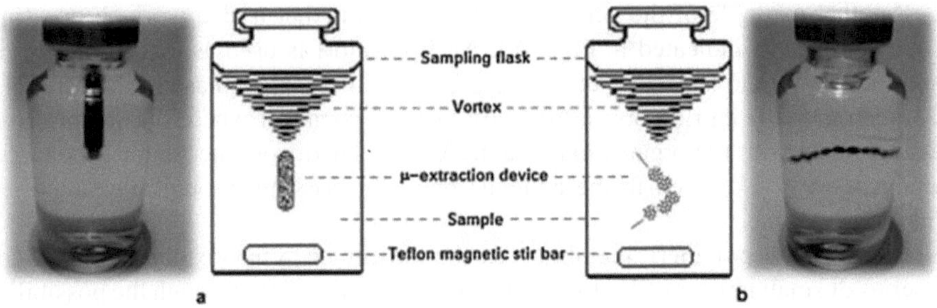

Fig. 15 Schematic representation and images of BAμE (**a**) and MSAμE (**b**) during the μ-extraction process (reproduced with permission from [84])

compounds from waters [84]. For the aim, two different configurations, namely bar adsorptive μ-extraction (BAμE) and multi-spheres adsorptive μ-extraction (MSAμE), were evaluated. Two representative examples are given in Fig. 15. The preparation of BAμE units involved the retention of a powdered sorbent over polypropylene hollow cylindrical substrates by adhesive forces while in MSAμE the spherical particles of sorbent were attached to a threat and covered by the powdered sorbent, which was fixed by thermal curing. Their application in microextraction techniques requires using a conventional Teflon magnetic stirring bar to promote the agitation of the sample and thus the migration of the target analyte to the sorbent phase. As both, BAμE and MSAμE substrates, were lighter than water, they remained below the vortex, under the so-called floating sampling technology.

An interesting issue with this configuration is the stability of the coating during the extraction and thus, the potential reusability of the extraction units. In this first study, the authors concluded that the thermal curing confers the sorbent with higher stability in the organic media (direct immersion in pure solvent for 60 min under sonication), temperature (20–50 °C, 3 h, sonication) and pH (1–14, 3 h, sonication). The instability of the adhesive supporting film would be the reason behind this behavior. Nevertheless, both configurations can be used for analyte isolation under standard operational conditions and solvents, although activated carbon and polystyrene-DVB performed better in terms of stability, robustness, and μ-extraction efficiency.

As it was described for SCSE and RDSE, this configuration minimizes the friction with the vessel walls. Moreover, both BAμE and MSAμE require less extraction time and lower agitation speed to reach similar performance than SBSE.

BAμE has evolved following sustainability criteria and in 2018 an eco-friendly alternative was developed [85]. The authors proposed the use of a flexible nylon support of reduced dimensions (7.5 × 1.0 mm) that is coated with the appropriate sorbent. Analyte elution is carried out in a glass vial insert which, on the one hand makes it compatible with automatic instrumental analysis and, on the other hand increases the preconcentration factor that can be achieved and hence, the sensitivity of the analytical method.

A hollow fiber filled with the most convenient organic solvent can also be used as floating extraction unit [86]. The so-called hollow fiber microextraction (HFμE) uses a polypropylene membrane of 10 mm in length which is immersed for a few seconds in the organic solvent of choice, being 20 μL embedded in the pores. Next, the unit is immersed into the sample and the agitation of the vial allows its free floating below the vortex created by the stirring magnetic bar. Once finished, the unit is withdrawn by means of clean tweezes and transferred to a glass insert for liquid desorption with the help of ultrasounds. The whole vial is transferred to an LC autosampler for instrumental analysis. The preparation of this type of extraction unit is quite simple and rapid, using negligible amounts of organic solvent for both, extraction, and elution steps.

The versatility of this miniaturized extraction technique can be improved if two hollow fibers are added to the sample, giving rise to what is known dual-HFμE. In this case, the different nature of the organic solvents broadens the chemical nature of the analytes to be extracted. If it is combined with large volume injection, the sensitivity is dramatically enhanced [87].

4.1.4 Stir Bar Sorptive Dispersive Microextraction

Dispersive microextraction techniques exhibit better performance than non-dispersed miniaturized approaches thanks to the higher contact between the sorbent/solvent phase and the analyte distributed within the sample matrix [12]. However, its main disadvantage is the collection of the extractant phase enriched with the analyte after the extraction step. It usually requires filtration or centrifugation. The inclusion of magnetic materials in these dispersive techniques facilitates the procedure as the extractant is isolated from the sample matrix or eluent by means of an external magnet (see chapter "Dispersive-Micro-Solid Phase Extraction (d-μSPE)"). However, in some cases, the recovery of the solid or liquid extractant phase is not complete or requires and excessive time to occur. In 2014, a new microextraction technique named stir bar sorptive dispersive microextraction (SBSDME) was developed as an elegant combination of SBSE and dispersive microsolid-phase extraction (DμSPE) [88, 89]. The fundamentals of SBSDME are found in the use of a strong permanent magnet over which a thin layer of a magnetic sorbent is deposited. Playing with the stirring rates, the solid phase is retained (lower speeds) or detached (higher speeds) from the magnetic support. Therefore, as indicated in Fig. 16a, in the first step, the magnet coated with the sorbent phase is introduced in the vial containing the sample. Then, the system is stirred at a high-speed provoking the dispersion of the sorbent phase into the liquid sample for a given time. Next, the speed is reduced to zero and the magnetism of the bar retrieves the magnetic sorbent enriched with the analytes without needing an external magnetic field. Finally, the bar is withdrawn from the sample and the target compounds are liquid or thermally desorbed for further instrumental analysis.

Automation of SBSDME has been proposed using a lab-in-syringe manifold on-line connected to a spectrophotometer [90]. The configuration developed all the steps of the process in an on-line fashion: sorbent dispersion, magnetic collection, elution, and detection. The main shortcoming of this configuration is that only 5 mL of

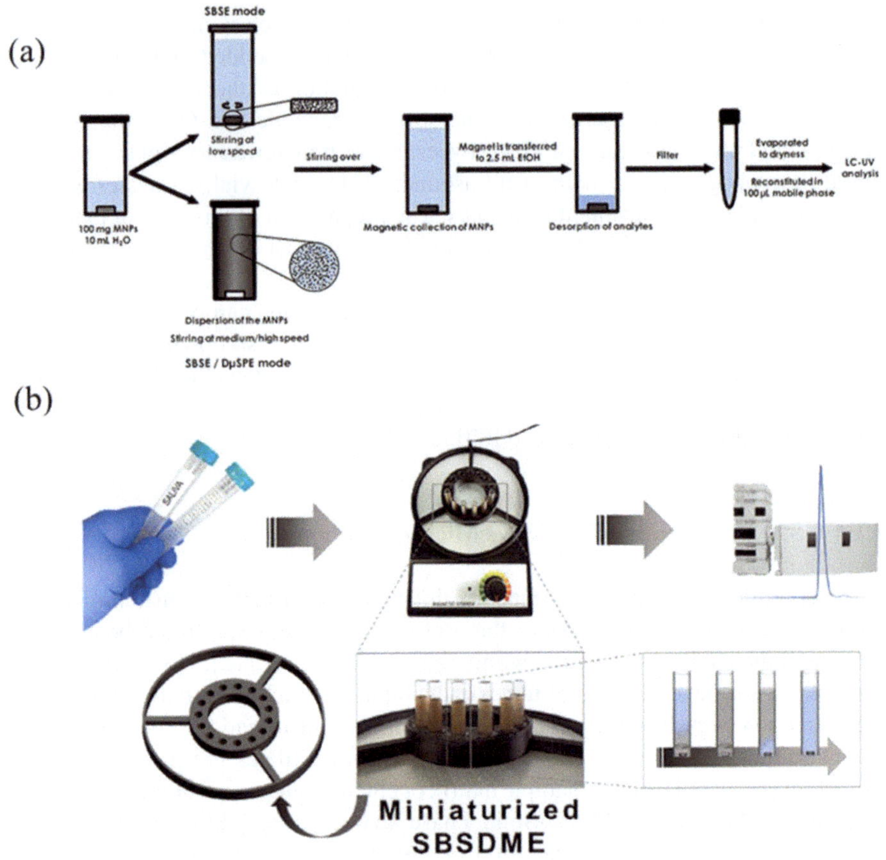

Fig. 16 Configurations of: **a** conventional (reproduced with permission from [88]); **b** miniaturized SBSDME (reproduced with permission from [91])

sample can be processed. Therefore, to increase the sensitivity of the measurements, the processes was repeated up to eight times using fresh aliquots of sample prior to analytes elution and determination.

Very recently, SBSDME was miniaturized to face the processing of low availability samples [91]. A dedicated device was constructed to hold 400 μL glass vial as sample containers and 3 mm × 2 mm bar shaped magnetic as stirring elements. As can be seen in the Fig. 16b, up to 15 samples can be simultaneously processed. In addition to the low sample volume, the amounts of sorbent and eluent are also reduced thus contributing to the sustainability of the sample preparation.

4.1.5 Solvent Bar Microextraction

Solvent bar microextraction (SBME) was proposed in 2004 by Jiang and Kee-Lee [92]. It consists of the confinement of few microliters of octanol into the lumen of a hollow fiber membrane followed by sealing of both ends. The resulting solvent bar was then added to a liquid sample and stirred by means of an additional magnet. As a main advantage, its use in "dirty" samples (e.g., soil slurries) can be highlighted as the hollow fiber acts as a filter of the particulate matter that can eventually be found dispersed in the matrix.

Since its proposal, several configurations have been developed, including the two-phase and three-phase modes. In the first case, the organic solvent fills the lumen and the pores of the hollow fiber while in the tree-phase the organic medium impregnates the pores and separate two aqueous phases, thus acting as a liquid membrane.

The basis arrangement of SBME has been adapted to increase its performance [93]. For example, one of both ends can remain unreached for compatibility with volatile organic solvents, which are easily evaporated during the seal of the second fiber end [94]. Efficiency of the extraction can be improved by keeping the fiber at the bottom of the sample vessel either using a pipette-tip [95] or a stainless-steel wire [96] Also, a dual solvent-stir bar microextraction has been designed where a stainless-steel bar with four fixing positions was used to hold two SBME [97]. The magnetic bar can also be externally stuck to the hollow fiber although this alternative considerably reduces the surface area available for analytes diffusion [98]. All these approaches are represented in Fig. 17.

4.1.6 Capsuled Microextraction

Extraction phases can also be stirred in the sample solution without the need of a plastic holder. Unlike SME and related techniques, the magnetic bar is attached to the sorbent element in different ways, thus allowing the device to spin itself for analytes diffusion. Georgiadis et al. coined in 2019 the term microextraction capsules (MECs) to describe a device consisting of a built-in magnet, a cellulose fiber substrate coated with a sol–gel organic–inorganic sorbent and porous membrane [99] As it can be seen in Fig. 18, the magnet is introduced in a polypropylene membrane which is joined to the extractant phase protected by a polypropylene membrane. The built-in magnet avoids the need for an additional, independent magnetic bar, while the porous polypropylene membrane allows sample permeation, protecting the sorbent from impurities coming from the matrix, thus extending its reusability. The high porosity of the capsule facilitates analytes diffusion for both isolation and elution steps. Also, it provides many interaction chemistries (polar, non-polar, anion-exchange of mixed mode) which also extend its applicability to a wide variety of analytes.

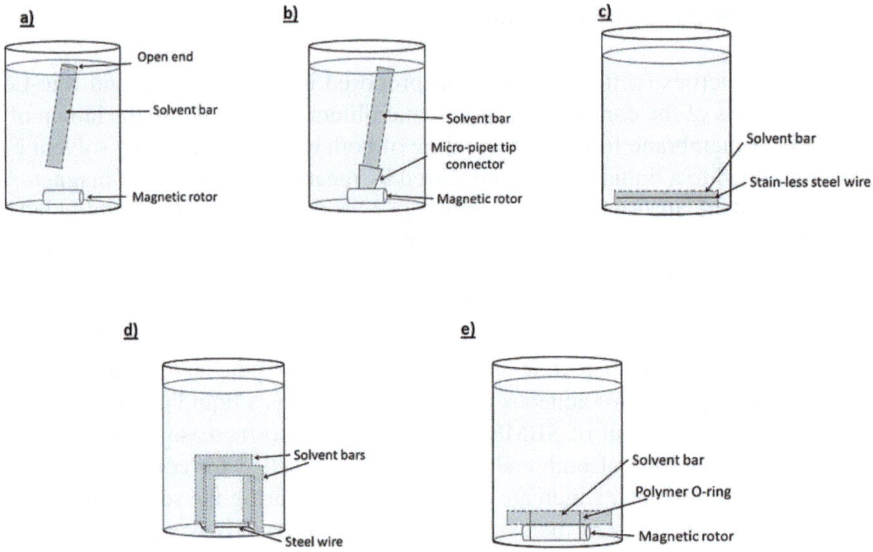

Fig. 17 Modifications of the conventional SBME setup: **a** "Cone" SBME; **b** solvent stir bar SBME; **c** magnetic bar SBME; **d** dual solvent stir bar micro-extraction; **e** magnetic support SBME (reproduced with permission from [93])

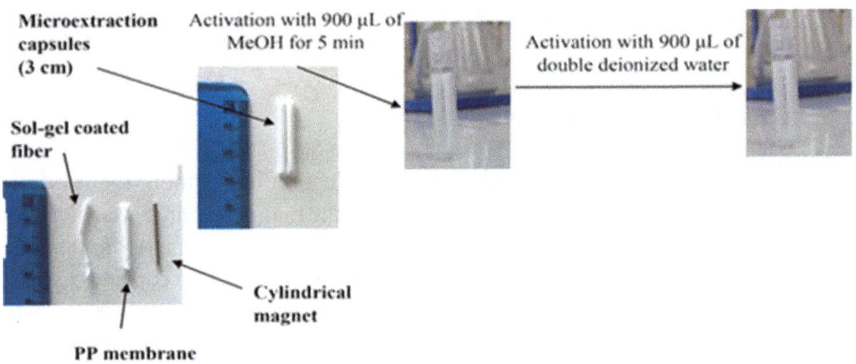

Fig. 18 Preparation of a capsule phase microextraction unit (adapted with permission from [99])

4.1.7 Flat-Shaped Self-Rotating Devices

The use of planar substrates presents several advantages in microextraction. Among them, the larger superficial area, thanks to the fact that the two-sorbent sides are available for extraction, can be highlighted. If they are used under self-rotating configuration, the need for a holder is obviated as well as the need for an additional magnetic bar. These facts make this approach more environmental and user-friendly and, at the same time, reduces costs and time.

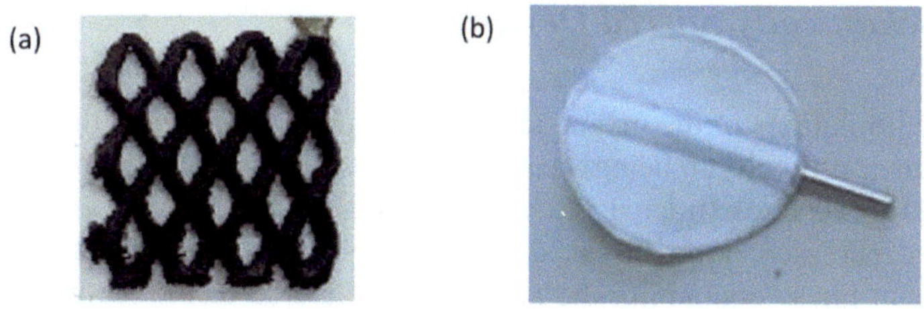

Fig. 19 Flat-shaped self -rotating devices. **a** Iron-mess screen (reproduced with permission from [100]). **b** Magnet-integrated fabric phase sorptive extraction (reproduced with permission from [102])

Two flat elements have been proposed. Kerman et al. synthesized an iron mesh screen that was electrochemically coated with polypyrrole. The device can rotate in the presence of an external magnetic field and the presence of apertures in the design increases the sorptive phase available for analyte interaction, which promotes the diffusion of the sample through the extractant phase (in comparison to bar or plates) [100]. To demonstrate the advantages of this configuration over other geometries, the authors compared the efficiency of the mesh screen, cylindrical and solid-plate layouts (see Fig. 19a) under standard extraction conditions (temperature 30 °C, stirring rate 1000 rpm, extraction time 30 min and no salt addition). Quantitative extractions were obtained for the mesh screen, followed by the solid plate coated with the same polymeric phase (ca. 55%) and cylindrical geometry performed the worst (ca. 25%). The authors attributed this enhanced performance to the extra stream pathways provided by the open structure of the mesh that facilitate the sample diffusion through the sorbent phase. The cylindrical (rod wound on the support) and the solid plate allow only part of the sorbent to the exposed to interaction with the analyte.

Fabric phase sorptive extraction (FPSE) consists of using a natural or synthetic fabric substrate which is subsequently modified with a hybrid organic–inorganic polymeric phase [101]. The stability of the thin film coating and its porous structure results in extremely high extraction efficiency. The integration of a stirring element allows the FP to freely rotate in the sample [102]. According to the authors' description, the magnet-integrated-FPSE (MI-FPSE) was constructed using two circular membranes ($r = 0.75$ cm) integrated with a metallic magnetic stir bar (see Fig. 19b). As it was the case with the previously described approach, this MI-FPSE is easy to handle with better reproducibility, faster extraction equilibrium and shorter extraction times. As an upgrade regarding other alternatives, FPs with different interaction chemistries can be used to fabricate the MI-FPSE thus broadening the range of polarity of the analytes that can be simultaneously extracted.

4.2 Mechanically Stirred Devices

One of the limitations of the magnetically stirred devices and even the self-rotating layouts is their difficulty in being adapted to perform on-site extraction. If the target compounds are on-site extracted, the sampling logistics are reduced, and the analytes integrity is increased during transportation and storage. In these strategies, only the extractant phase containing the analytes is delivered to the laboratory and the absence of the aqueous matrix eliminates undesired secondary reactions. The only requirement is the stability of the analytes on the extraction unit during the storage. All the devices that are described below share the simplicity of installation, removal, and replacement of the extraction units. They also avoid coating deterioration due to the friction with the bottom of the sample vessels.

In a first approach a home-mase portable electric stirrer was coupled to an SBSE, working under the off/on-site modes [103]. The bar was fixed to the stirrer through a magnetic stir rod welded onto the bottom of the mini-electro motor (stirring speed 2000–6000 rpm). The stir bar used was lab-made by coating a glass-coated iron stir bar with a thin film of PDMS. The portable SBSE can work under the HS, direct immersion, or continuous flow modes.

Qin et al. demonstrated the advantages of using a PDMS thin film coupled to an electric drill for on-site extraction in waters [104]. Interestingly, the authors compared the performance of the planar substrate with a SPME fiber. As expected, the PDMS thin film resulted in a better efficiency thanks to the most favorable surface-to-volume ratio and larger extractant phase (ca. 100 times higher).

Borosilicate disks can also be used as planar support of the sorbent phase. They exhibit a higher mechanical stability while being easily functionalized. Roldán-Pijuán et al. modified these disks with oxidized single-walled carbon nanohorns (o-SWNHs) [105]. The o-SWNHs disks were fixed to a screw of a portable drill to develop on-site extractions. The rotation of the disk homogenized a defined volume of sample around it and therefore the extraction can be considered almost independent of sample volume. The methodology was robust and highly reproducible among different synthesized o-SWNHs disks.

Despite the high reproducibility reported for the laboratory-made extraction units, the use of standardized and commercial elements helps to increase this analytical properly and is less time-consuming. Casado-Carmona et al. presented a portable stir membrane device that can be used with commercial nylon membranes to carry out the on-site extraction of target compounds from environmental water samples [106]. A countersunk pot magnet permits the attachment of the nylon membrane using a metallic washer. The system is coupled to a wireless electric drill using a screw of variable length (depending on the sampling site requirements) and a nut. The membrane can be easily removed after each extraction for analyte elution and quantification.

This configuration can be simplified by substituting the membrane by a magnetic paper thanks to the minimization of the diffusion boundary layer. The flat support was prepared by immersion of a piece of paper in a dispersion containing nylon-6 (dissolved) and magnetic nanoparticles (dispersed). The magnetic paper is directly attached to the magnet, avoiding the need of metal washers. In addition, an improved blade is fixed over the magnet to promote mass transference [107].

Commercial particulate particles with different interaction chemistries are widely used in environmental analysis thanks to their high efficiency compared to polymeric sorbents. This material can also be used in the previously described device combining the advantages of a flat extraction unit with the integrated mechanical stirring [108]. The preparation of the sorbent phase consists of the deposition of hydrophilic-lipophilic balance (HLB) particles over a magnetic tape. It was then fixed to a screw by means of a countersunk pot magnet, as previously described. A small blade was added to facilitate the analyte diffusion. The device was integrated into a glass bottle's cap fitted with a small electric motor. A portable power supply was used to facilitate the portability of the system. The extraction unit can be adapted to the sample volume that would be eventually needed to reach a given sensitivity level by changing the volume of the bottle used for the extraction. The authors used an internal standard to compensate the influence of the ionic strength of the sample on the analytical signal.

Open-sources technologies such Arduino can be used to automate these devices. Also, several sensors (temperature, conductivity) can also be added to enrich the sample information. Moreover, the planar sorbent phase can be attached to the stirring element by means of an alligator clip [109]. This configuration, shown in Fig. 20, maximized the surface available for analyte interaction. In this proposal, mixed mode anion exchange (MAX) particles to avoid pH adjustment were used for analytes isolation and they were achieved to the support by means of a double sided adhesive tape.

Using several extraction units simultaneously can increase the amount of analyte extracted or, if they are of different nature, expand the variety of compounds (hydrophobic, hydrophilic, charged) that can be isolated in a single extraction step. Makkiniang et al. developed a portable and miniaturized apparatus that can hold up to 6 miniaturized multi-stirred microextractors [110]. A monolithic polymeric phase containing carboxylated MWCNTs was prepared, and the rods were connected to the motor by using pipette tips of different volumes (see Fig. 21a). The low cost of the extraction units allows the simultaneous use of several units, therefore, a higher sample throughput is achieved.

Solvent bars can also be used in mechanical stirrer devices, using an electronic motor [111]. In this case, four hollow fibers were arranged in a cubic-like configuration between two polymeric disks which are connected to the motor (see Fig. 21b). The higher the number of hollow fibers, the better the extraction efficiency. In the case that only one hollow fiber was needed to reach the desired sensitivity, the device can be used to obtain replicate values of the analysis in a single step.

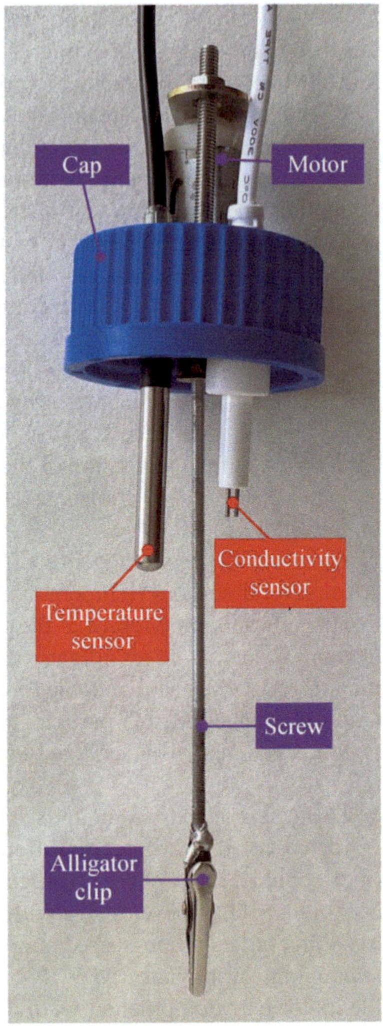

Fig. 20 Photograph resembling the elements used to construct the on-site extraction device integrating temperature and conductivity probes (adapted with permission from reference [109])

5 Future Remarks

This chapter provides a general overview of SBSE technique from a broad perspective since related techniques and new materials have been outlined. Initially, SBSE was proposed as the simple integration of the sorptive phase into a stirring bar, a common element in any microextraction technique to enhance mass transference. In the last few years, SBSE has experienced a remarkable evolution driven by resolving its initial shortcomings.

Compared to in-fiber SPME, the SBSE coatings are thicker, thus increasing the potential extraction capacity of the technique. However, thicker coatings restrict the

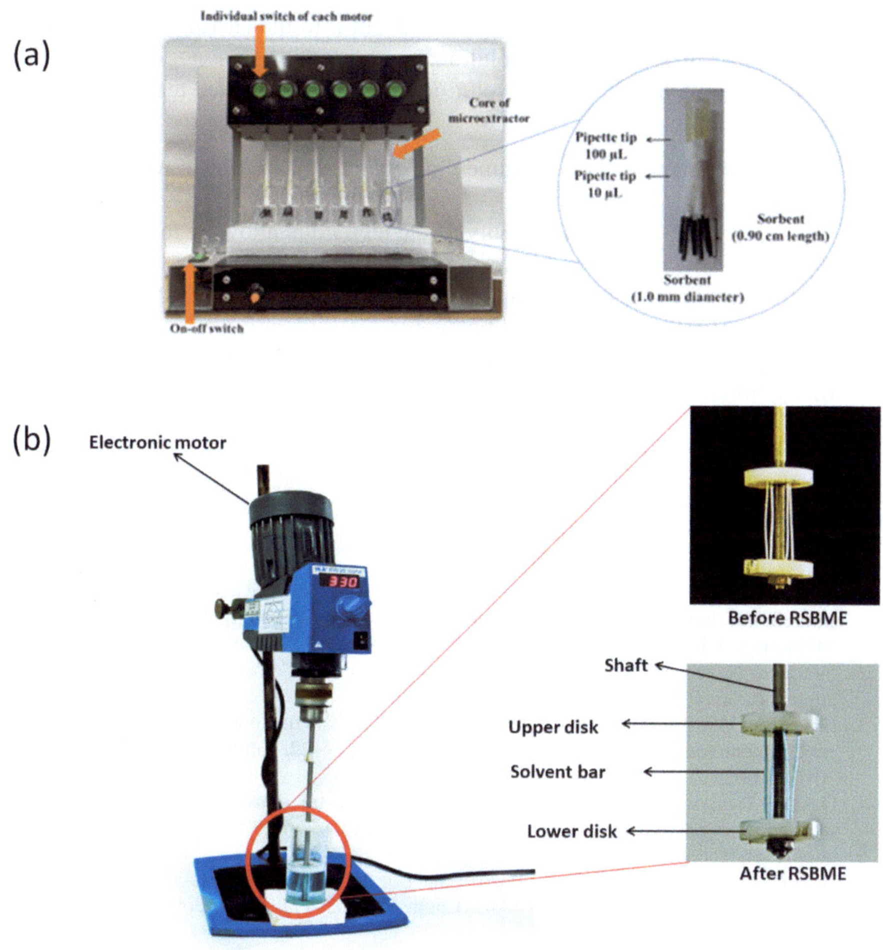

Fig. 21 Magnetically stirred devices using several extraction units simultaneously. **a** Multi stir rod microextractors (reproduced with permission from [110]). **b** Solvent bars (reproduced with permission from [111])

extraction kinetics, which are, sometimes only partially, compensated by the efficient stirring of the solution. New porous materials, such as monoliths and membranes, have been proposed to boost the contact area between the sorptive phase and the analytes. In most cases, the extraction units needed to be completely redesigned to deploy the new materials giving rise to new microextraction modalities.

PDMS was extensively used as the coating in the first SBSE approaches. This material has demonstrated an efficient extraction capacity. However, its non-hydrophobic nature somewhat limits the applicability of SBSE to the extraction of non-polar compounds. Developing new coatings covering a wider range of polarities has been vital to widening the technique's versatility. The development of new

LPME modalities based on integrating the solvent into the stirring element can be highlighted as a milestone in this evolution.

We foresee some trends for the evolution of SBSE and related techniques in the next years, including:

(a) The development of new commercial coatings covering a wider range of polarities.
(b) The improvement of the portability of the technique allowing the development of on-site extraction procedures.
(c) The evaluation of the direct coupling of the extraction devices with instrumental techniques for the sake of simplification.
(d) The implementation of open technologies, including 3D printing, improving the affordability and versatility of the technique.

References

1. Baltussen E, Sandra P, David F, Cramers C (1999) Stir bar sorptive extraction (SBSE), a novel extraction technique for aqueous samples: theory and principles. J Microcolumn Sep 11:737–747. https://doi.org/10.1002/(SICI)1520-667X(1999)11:10<737::AID-MCS7>3.0.CO;2-4
2. Florêncio Nogueira JM (2017) Stir bar sorptive extraction. Compr Anal Chem 76:463–481. https://doi.org/10.1016/bs.coac.2017.01.006
3. Nogueira JMF (2015) Stir-bar sorptive extraction: 15 years making sample preparation more environment-friendly. TrAC—Trends Anal Chem 71:214–223. https://doi.org/10.1016/j.trac.2015.05.002
4. David F, Tienpont B, Sandra P (2003) Stir-bar sorptive extraction of trace organic compounds from aqueous matrices. LC GC North Am 21:108–118
5. https://gerstel.com/en/twister-sbse
6. David F, Ochiai N, Sandra P (2019) Two decades of stir bar sorptive extraction: a retrospective and future outlook. TrAC—Trends Anal Chem 112:102–111. https://doi.org/10.1016/j.trac.2018.12.006
7. David F, Sandra P (2007) Stir bar sorptive extraction for trace analysis. J Chromatogr A 1152:54–69. https://doi.org/10.1016/j.chroma.2007.01.032
8. Prieto A, Basauri O, Rodil R et al (2010) Stir-bar sorptive extraction: a view on method optimisation, novel applications, limitations and potential solutions. J Chromatogr A 1217:2642–2666. https://doi.org/10.1016/j.chroma.2009.12.051
9. Zuloaga O, Etxebarria N, González-Gaya B, et al (2019) Stir-bar sorptive extraction. In: Poole C (ed) Handbooks in separation science, 1st edn. Elsevier, pp 493–530
10. Hasan CK, Ghiasvand A, Lewis TW et al (2020) Recent advances in stir-bar sorptive extraction: coatings, technical improvements, and applications. Anal Chim Acta 1139:222–240. https://doi.org/10.1016/j.aca.2020.08.021
11. Camino-Sánchez FJ, Rodríguez-Gómez R, Zafra-Gómez A et al (2014) Stir bar sorptive extraction: recent applications, limitations and future trends. Talanta 130:388–399. https://doi.org/10.1016/j.talanta.2014.07.022
12. Chisvert A, Cárdenas S, Lucena R (2019) Dispersive micro-solid phase extraction. TrAC—Trends Anal Chem 112:226–233. https://doi.org/10.1016/j.trac.2018.12.005
13. Camino-Sánchez FJ, Zafra-Gómez A, Cantarero-Malagón S, Vílchez JL (2012) Validation of a method for the analysis of 77 priority persistent organic pollutants in river water by stir bar sorptive extraction in compliance with the European Water Framework Directive. Talanta 89:322–334. https://doi.org/10.1016/j.talanta.2011.12.037

14. Lucena R (2012) Extraction and stirring integrated techniques: examples and recent advances. Anal Bioanal Chem 403:2213–2223. https://doi.org/10.1007/s00216-012-5826-9
15. Cárdenas S, Lucena R (2017) Recent advances in extraction and stirring integrated techniques. Separations 4:6. https://doi.org/10.3390/separations4010006
16. Sgorbini B, Cagliero C, Cordero C et al (2012) New medium-to-high polarity twister coatings for liquid and vapour phase sorptive extraction of matrices of vegetable origin. J Chromatogr A 1265:39–45. https://doi.org/10.1016/j.chroma.2012.09.097
17. Serrano de la Hoz K, Salinas MR, Ferrandino A (2016) Different coatings for the HS-SBSE grape volatile analysis in model solution: preliminary results. Food Chem 212:814–820. https://doi.org/10.1016/j.foodchem.2016.06.047
18. Gilart N, Miralles N, Marcé RM et al (2013) Novel coatings for stir bar sorptive extraction to determine pharmaceuticals and personal care products in environmental waters by liquid chromatography and tandem mass spectrometry. Anal Chim Acta 774:51–60. https://doi.org/10.1016/j.aca.2013.03.010
19. Bizkarguenaga E, Iparragirre A, Navarro P et al (2013) In-port derivatization after sorptive extractions. J Chromatogr A 1296:36–46. https://doi.org/10.1016/j.chroma.2013.03.058
20. Ochiai N, Sasamoto K, Kanda H, Nakamura S (2006) Fast screening of pesticide multiresidues in aqueous samples by dual stir bar sorptive extraction-thermal desorption-low thermal mass gas chromatography-mass spectrometry. J Chromatogr A 1130:83–90. https://doi.org/10.1016/j.chroma.2006.06.032
21. Kawaguchi M, Ishii Y, Sakui N et al (2004) Stir bar sorptive extraction with in situ derivatization and thermal desorption-gas chromatography-mass spectrometry in the multi-shot mode for determination of estrogens in river water samples. J Chromatogr A 1049:1–8. https://doi.org/10.1016/j.chroma.2004.08.013
22. Ubeda C, Callejón RM, Troncoso AM et al (2016) Volatile profile characterisation of Chilean sparkling wines produced by traditional and Charmat methods via sequential stir bar sorptive extraction. Food Chem 207:261–271. https://doi.org/10.1016/j.foodchem.2016.03.117
23. Ochiai N, Sasamoto K, Ieda T et al (2013) Multi-stir bar sorptive extraction for analysis of odor compounds in aqueous samples. J Chromatogr A 1315:70–79. https://doi.org/10.1016/j.chroma.2013.09.070
24. Berrou K, Dunyach-Remy C, Lavigne JP et al (2019) Multiple stir bar sorptive extraction combined with gas chromatography-mass spectrometry analysis for a tentative identification of bacterial volatile and/or semi-volatile metabolites. Talanta 195:245–250. https://doi.org/10.1016/j.talanta.2018.11.042
25. Vyviurska O, Thai HA, Garančovská D et al (2022) Enhanced multi-stir bar sorptive extraction for wine analysis: alteration in headspace mode. Food Res Int 158:111510. https://doi.org/10.1016/j.foodres.2022.111510
26. Ochiai N, Sasamoto K, David F, Sandra P (2016) Solvent-assisted stir bar sorptive extraction by using swollen polydimethylsiloxane for enhanced recovery of polar solutes in aqueous samples: application to aroma compounds in beer and pesticides in wine. J Chromatogr A 1455:45–56. https://doi.org/10.1016/j.chroma.2016.05.085
27. Maslamani N, Manandhar E, Geremia DK, Logue BA (2016) ICE concentration linked with extractive stirrer (ICECLES). Anal Chim Acta 941:41–48. https://doi.org/10.1016/j.aca.2016.09.003
28. Skaggs CS, Alluhayb AH, Logue BA (2020) Comparison of the extraction efficiency of ice concentration linked with extractive stirrer, stir bar sorptive extraction, and solid-phase microextraction for pesticides from drinking water. J Chromatogr A 1622:461102. https://doi.org/10.1016/j.chroma.2020.461102
29. Ochiai N, Sasamoto K, Sasaki T et al (2020) Fractionated stir bar sorptive extraction using conventional and solvent-assisted approaches for enhanced identification capabilities of aroma compounds in beverages. J Chromatogr A 1628:461475. https://doi.org/10.1016/j.chroma.2020.461475
30. Ochiai N, Sasamoto K, David F, Sandra P (2022) Dual stir bar sorptive extraction using conventional and solvent-assisted approaches including in-situ derivatization followed by

thermal desorption and gas chromatography–tandem mass spectrometry for analysis of polyfunctional thiols in white wine sample. Adv Sample Prep 3:100034. https://doi.org/10.1016/j.sampre.2022.100034
31. Gilart N, Marcé RM, Borrull F, Fontanals N (2014) New coatings for stir-bar sorptive extraction of polar emerging organic contaminants. TrAC - Trends Anal Chem 54:11–23. https://doi.org/10.1016/j.trac.2013.10.010
32. Liu W, Wang H, Guan Y (2004) Preparation of stir bars for sorptive extraction using sol-gel technology. J Chromatogr A 1045:15–22. https://doi.org/10.1016/j.chroma.2004.06.036
33. Yu C, Hu B (2007) Novel combined stir bar sorptive extraction coupled with ultrasonic assisted extraction for the determination of brominated flame retardants in environmental samples using high performance liquid chromatography. J Chromatogr A 1160:71–80. https://doi.org/10.1016/j.chroma.2007.05.042
34. Fan W, He M, You L et al (2020) Spiral stir bar sorptive extraction with polyaniline-polydimethylsiloxane sol-gel packings for the analysis of trace estrogens in environmental water and animal-derived food samples. J Sep Sci 43:1137–1144. https://doi.org/10.1002/jssc.201900819
35. Mao X, Fan W, He M et al (2015) C18-coated stir bar sorptive extraction combined with HPLC-ICP-MS for the speciation of butyltins in environmental samples. J Anal At Spectrom 30:162–171. https://doi.org/10.1039/c4ja00294f
36. Díaz-Álvarez M, Turiel E, Martín-Esteban A (2019) Molecularly imprinted polymer monolith containing magnetic nanoparticles for the stir-bar sorptive extraction of thiabendazole and carbendazim from orange samples. Anal Chim Acta 1045:117–122. https://doi.org/10.1016/j.aca.2018.09.001
37. Feng J, Sun M, Bu Y, Luo C (2016) Development of a carbon-nanoparticle-coated stirrer for stir bar sorptive extraction by a simple carbon deposition in flame. J Sep Sci 39:918–922. https://doi.org/10.1002/jssc.201501008
38. Guan W, Wang Y, Xu F, Guan Y (2008) Poly(phthalazine ether sulfone ketone) as novel stationary phase for stir bar sorptive extraction of organochlorine compounds and organophosphorus pesticides. J Chromatogr A 1177:28–35. https://doi.org/10.1016/j.chroma.2007.10.077
39. Hashemi SH, Kaykhaii M (2021) Nanoparticle coatings for stir bar sorptive extraction, synthesis, characterization and application. Talanta 221:121568. https://doi.org/10.1016/j.talanta.2020.121568
40. Maciel EVS, de Toffoli AL, Neto ES et al (2019) New materials in sample preparation: Recent advances and future trends. TrAC—Trends Anal Chem 119:115633. https://doi.org/10.1016/j.trac.2019.115633
41. He M, Chen B, Hu B (2014) Recent developments in stir bar sorptive extraction microextraction techniques. Anal Bioanal Chem 406:2001–2026. https://doi.org/10.1007/s00216-013-7395-y
42. He M, Wang Y, Zhang Q et al (2021) Stir bar sorptive extraction and its application. J Chromatogr A 1637:461810. https://doi.org/10.1016/j.chroma.2020.461810
43. Zhang W, Zhang Z, Zhang J et al (2014) Covalent immobilization of graphene onto stainless steel wire for jacket-free stir bar sorptive extraction. J Chromatogr A 1351:12–20. https://doi.org/10.1016/j.chroma.2014.05.038
44. Zhang Q, You L, Chen B et al (2021) Reduced graphene oxide coated nickel foam for stir bar sorptive extraction of benzotriazole ultraviolet absorbents from environmental water. Talanta 231:122332. https://doi.org/10.1016/j.talanta.2021.122332
45. Hu C, Chen B, He M, Hu B (2013) Amino modified multi-walled carbon nanotubes/polydimethylsiloxane coated stir bar sorptive extraction coupled to high performance liquid chromatography-ultraviolet detection for the determination of phenols in environmental samples. J Chromatogr A 1300:165–172. https://doi.org/10.1016/j.chroma.2013.05.004
46. Zou N, Yuan C, Liu S et al (2016) Coupling of multi-walled carbon nanotubes/polydimethylsiloxane coated stir bar sorptive extraction with pulse glow discharge-ion mobility spectrometry for analysis of triazine herbicides in water and soil samples. J Chromatogr A 1457:14–21. https://doi.org/10.1016/j.chroma.2016.06.043

47. Huang X, Yuan D (2007) Preparation of stir bars for sorptive extraction based on monolithic material. J Chromatogr A 1154:152–157. https://doi.org/10.1016/j.chroma.2007.03.132
48. Yao X, Zhou Z, He M et al (2018) One-pot polymerization of monolith coated stir bar for high efficient sorptive extraction of perfluoroalkyl acids from environmental water samples followed by high performance liquid chromatography-electrospray tandem mass spectrometry detection. J Chromatogr A 1553:7–15. https://doi.org/10.1016/j.chroma.2018.04.014
49. You L, He M, Chen B, Hu B (2017) One-pot synthesis of zeolitic imidazolate framework-8/poly (methyl methacrylate-ethyleneglycol dimethacrylate) monolith coating for stir bar sorptive extraction of phytohormones from fruit samples followed by high performance liquid chromatography-ultravi. J Chromatogr A 1524:57–65. https://doi.org/10.1016/j.chroma.2017.10.001
50. Zhu X, Cai J, Yang J et al (2006) Films coated with molecular imprinted polymers for the selective stir bar sorption extraction of monocrotophos. J Chromatogr A 1131:37–44. https://doi.org/10.1016/j.chroma.2006.07.041
51. Xu Z, Hu Y, Hu Y, Li G (2010) Investigation of ractopamine molecularly imprinted stir bar sorptive extraction and its application for trace analysis of β2-agonists in complex samples. J Chromatogr A 1217:3612–3618. https://doi.org/10.1016/j.chroma.2010.03.046
52. Liu R, Feng F, Chen G et al (2016) Barbell-shaped stir bar sorptive extraction using dummy template molecularly imprinted polymer coatings for analysis of bisphenol A in water. Anal Bioanal Chem 408:5329–5335. https://doi.org/10.1007/s00216-016-9628-3
53. Wang C, Zhou W, Liao X et al (2018) Covalent immobilization of metal organic frameworks onto chemical resistant poly(ether ether ketone) jacket for stir bar extraction. Anal Chim Acta 1025:124–133. https://doi.org/10.1016/j.aca.2018.04.056
54. Heidarbeigi M, Saraji M, Jafari MT (2021) In situ growth of copper-based metal-organic framework on a helical shape copper wire as a sorbent in stir-bar sorptive extraction of fenthion followed by corona discharge ion mobility spectrometry. J Chromatogr A 1651:462279. https://doi.org/10.1016/j.chroma.2021.462279
55. Zang L, He M, Wu Z et al (2021) Imine-linked covalent organic frameworks coated stir bar sorptive extraction of non-steroidal anti-inflammatory drugs from environmental water followed by high performance liquid chromatography-ultraviolet detection. J Chromatogr A 1659:462647. https://doi.org/10.1016/j.chroma.2021.462647
56. Liu Z, Zhou W, Hong Y et al (2022) Covalent organic framework-V modified porous polypropylene hollow fiber with detachable dumbbell-shaped structure for stir bar sorptive extraction of benzophenones. J Chromatogr A 1664:462798. https://doi.org/10.1016/j.chroma.2021.462798
57. Wang J, Feng J, Sun M et al (2023) Sulfonic acid-functionalized covalent organic frameworks as the coating for stir bar sorptive extraction of fluoroquinolones in milk samples. Microchim Acta 190:1–9. https://doi.org/10.1007/s00604-022-05534-9
58. Neng NR, Pinto ML, Pires J et al (2007) Development, optimisation and application of polyurethane foams as new polymeric phases for stir bar sorptive extraction. J Chromatogr A 1171:8–14. https://doi.org/10.1016/j.chroma.2007.09.033
59. Mirzaee MT, Seidi S, Razeghi Y et al (2020) In-tube stir bar sorptive extraction based on 3-aminopropyl triethoxysilane surface-modified Ce-doped ZnAl layered double hydroxide thin film for determination of nonsteroidal anti-inflammatory drugs in saliva samples. Microchim Acta 187:528. https://doi.org/10.1007/s00604-020-04489-z
60. Lambert JP, Mullett WM, Kwong E, Lubda D (2005) Stir bar sorptive extraction based on restricted access material for the direct extraction of caffeine and metabolites in biological fluids. J Chromatogr A 1075:43–49. https://doi.org/10.1016/j.chroma.2005.03.119
61. Wang M, Zeng J, Wang J et al (2021) Dual-mode aptasensor for simultaneous detection of multiple food-borne pathogenic bacteria based on colorimetry and microfluidic chip using stir bar sorptive extraction. Microchim Acta 188:244. https://doi.org/10.1007/s00604-021-04902-1
62. Zatrochová S, Martínez-Pérez-Cejuela H, Catalá-Icardo M et al (2022) Development of hybrid monoliths incorporating metal–organic frameworks for stir bar sorptive extraction coupled

with liquid chromatography for determination of estrogen endocrine disruptors in water and human urine samples. Microchim Acta 189:92. https://doi.org/10.1007/s00604-022-05208-6
63. Yu C, Yao Z, Hu B (2009) Preparation of polydimethylsiloxane/β-cyclodextrin/divinylbenzene coated "dumbbell-shaped" stir bar and its application to the analysis of polycyclic aromatic hydrocarbons and polycyclic aromatic sulfur heterocycles compounds in lake water and soil by hig. Anal Chim Acta 641:75–82. https://doi.org/10.1016/j.aca.2009.03.031
64. Sukree W, Sooksawat D, Kanatharana P et al (2020) A miniature stainless steel net dumbbell-shaped stir-bar for the extraction of phthalate esters in instant noodle and rice soup samples. J Environ Sci Heal—Part B Pestic Food Contam Agric Wastes 55:60–68. https://doi.org/10.1080/03601234.2019.1659053
65. Zhou W, Wang C, Wang X, Chen Z (2018) Etched poly(ether ether ketone) jacket stir bar with detachable dumbbell-shaped structure for stir bar sorptive extraction. J Chromatogr A 1553:43–50. https://doi.org/10.1016/j.chroma.2018.04.022
66. Zhang Z, Mwadini MA, Chen Z (2016) Polytetrafluoroethylene-jacketed stirrer modified with graphene oxide and polydopamine for the efficient extraction of polycyclic aromatic hydrocarbons. J Sep Sci 39:4011–4018. https://doi.org/10.1002/jssc.201600716
67. Ghani M, Ghoreishi SM, Azamati M (2018) In-situ growth of zeolitic imidazole framework-67 on nanoporous anodized aluminum bar as stir-bar sorptive extraction sorbent for determining caffeine. J Chromatogr A 1577:15–23. https://doi.org/10.1016/j.chroma.2018.09.049
68. Khoobi A, Salavati-Niasari M, Ghani M et al (2019) Multivariate optimization methods for in-situ growth of LDH/ZIF-8 nanocrystals on anodized aluminium substrate as a nanosorbent for stir bar sorptive extraction in biological and food samples. Food Chem 288:39–46. https://doi.org/10.1016/j.foodchem.2019.02.118
69. Mao X, He M, Chen B, Hu B (2016) Membrane protected C18 coated stir bar sorptive extraction combined with high performance liquid chromatography-ultraviolet detection for the determination of non-steroidal anti-inflammatory drugs in water samples. J Chromatogr A 1472:27–34. https://doi.org/10.1016/j.chroma.2016.10.051
70. Alcudia-León MC, Lucena R, Cárdenas S, Valcárcel M (2009) Stir membrane extraction: a useful approach for liquid sample pretreatment. Anal Chem 81:8957–8961. https://doi.org/10.1021/ac9016192
71. Alcudia-León MC, Lucena R, Cárdenas S, Valcárcel M (2011) Stir membrane liquid-liquid microextraction. J Chromatogr A 1218:869–874. https://doi.org/10.1016/j.chroma.2010.12.075
72. Alcudia-León MC, Lucena R, Cárdenas S, Valcárcel M (2011) Determination of phenols in waters by stir membrane liquid-liquid-liquid microextraction coupled to liquid chromatography with ultraviolet detection. J Chromatogr A 1218:2176–2181. https://doi.org/10.1016/j.chroma.2011.02.033
73. Roldán-Pijuán M, Alcudia-León MC, Lucena R et al (2013) Stir-membrane liquid microextraction for the determination of paracetamol in human saliva samples. Bioanalysis 5:307–315. https://doi.org/10.4155/bio.12.312
74. Rodríguez-Gómez R, Roldán-Pijuán M, Lucena R et al (2014) Stir-membrane solid-liquid-liquid microextraction for the determination of parabens in human breast milk samples by ultra high performance liquid chromatography-tandem mass spectrometry. J Chromatogr A 1354:26–33. https://doi.org/10.1016/j.chroma.2014.05.071
75. Roldán-Pijuán M, Lucena R, Alcudia-León MC et al (2013) Stir octadecyl-modified borosilicate disk for the liquid phase microextraction of triazine herbicides from environmental waters. J Chromatogr A 1307:58–65. https://doi.org/10.1016/j.chroma.2013.07.086
76. Alcudia-León MC, Lucena R, Cárdenas S, Valcárcel M (2013) Magnetically confined hydrophobic nanoparticles for the microextraction of endocrine-disrupting phenols from environmental waters. Anal Bioanal Chem 405:2729–2734. https://doi.org/10.1007/s00216-012-6683-2
77. Huang X, Chen L, Lin F, Yuan D (2011) Novel extraction approach for liquid samples: stir cake sorptive extraction using monolith. J Sep Sci 34:2145–2151. https://doi.org/10.1002/jssc.201100283

78. Huang X, Chen L, Yuan D, Bi S (2012) Preparation of a new polymeric ionic liquid-based monolith for stir cake sorptive extraction and its application in the extraction of inorganic anions. J Chromatogr A 1248:67–73. https://doi.org/10.1016/j.chroma.2012.06.004
79. Lin F, Nong S, Huang X, Yuan D (2013) Sensitive determination of organic acid preservatives in juices and soft drinks treated by monolith-based stir cake sorptive extraction and liquid chromatography analysis. Anal Bioanal Chem 405:2077–2081. https://doi.org/10.1007/s00216-012-6646-7
80. Oliva-Lamarca Y, Fresco-Cala B, Cárdenas S (2019) Synthesis, characterization, and application of chemically interconnected carbon nanotube monolithic sorbents by photopolymerization in polypropylene caps. Anal Bioanal Chem 411:3291–3299. https://doi.org/10.1007/s00216-019-01795-1
81. Richter P, Leiva C, Choque C et al (2009) Rotating-disk sorptive extraction of nonylphenol from water samples. J Chromatogr A 1216:8598–8602. https://doi.org/10.1016/j.chroma.2009.10.044
82. Jachero L, Ahumada I, Richter P (2014) Rotating-disk sorptive extraction: Effect of the rotation mode of the extraction device on mass transfer efficiency. Anal Bioanal Chem 406:2987–2992. https://doi.org/10.1007/s00216-014-7693-z
83. Richter P, Arismendi D, Becerra-Herrera M (2021) The fundamentals, chemistries and applications of rotating-disk sorptive extraction. TrAC—Trends Anal Chem 137:116209. https://doi.org/10.1016/j.trac.2021.116209
84. Neng NR, Silva ARM, Nogueira JMF (2010) Adsorptive micro-extraction techniques—novel analytical tools for trace levels of polar solutes in aqueous media. J Chromatogr A 1217:7303–7310. https://doi.org/10.1016/j.chroma.2010.09.048
85. Ide AH, Nogueira JMF (2018) New-generation bar adsorptive microextraction (BAμE) devices for a better eco-user-friendly analytical approach—application for the determination of antidepressant pharmaceuticals in biological fluids. J Pharm Biomed Anal 153:126–134. https://doi.org/10.1016/j.jpba.2018.02.001
86. Ide AH, Nogueira JMF (2018) Hollow fiber microextraction: a new hybrid microextraction technique for trace analysis. Anal Bioanal Chem 410:2911–2920. https://doi.org/10.1007/s00216-018-0971-4
87. Ide AH, Nogueira JMF (2020) Dual-hollow fiber microextraction (dual-HFμE)—application for monitoring trace levels of organochlorine pesticides in real matrices. Int J Environ Anal Chem 100:1402–1414. https://doi.org/10.1080/03067319.2019.1655006
88. Benedé JL, Chisvert A, Giokas DL, Salvador A (2014) Development of stir bar sorptive-dispersive microextraction mediated by magnetic nanoparticles and its analytical application to the determination of hydrophobic organic compounds in aqueous media. J Chromatogr A 1362:25–33. https://doi.org/10.1016/j.chroma.2014.08.024
89. Vállez-Gomis V, Grau J, Benedé JL et al (2021) Fundamentals and applications of stir bar sorptive dispersive microextraction: a tutorial review. Anal Chim Acta 1153:338271. https://doi.org/10.1016/j.aca.2021.338271
90. Maya F, Palomino Cabello C, Estela JM et al (2015) Automatic in-syringe dispersive microsolid phase extraction using magnetic metal-organic frameworks. Anal Chem 87:7545–7549. https://doi.org/10.1021/acs.analchem.5b01993
91. Azorín C, Benedé JL, Chisvert A (2023) New challenges in sample preparation: miniaturized stir bar sorptive dispersive microextraction as a high-throughput and feasible approach for low-availability sample preparation. Anal Chim Acta 1238:340627. https://doi.org/10.1016/j.aca.2022.340627
92. Jiang X, Lee HK (2004) Solvent bar microextraction. Anal Chem 76:5591–5596. https://doi.org/10.1021/ac040069f
93. López-López JA, Mendiguchía C, Pinto JJ, Moreno C (2019) Application of solvent-bar micro-extraction for the determination of organic and inorganic compounds. TrAC—Trends Anal Chem 110:57–65. https://doi.org/10.1016/j.trac.2018.10.034
94. López-López JA, Ogalla-Chozas E, Lara-Martín PA, Pintado-Herrera MG (2017) Solvent bar micro-extraction (SBME) based determination of PAHs in seawater samples. Sci Total Environ 598:58–63. https://doi.org/10.1016/j.scitotenv.2017.04.125

95. Liu W, Zhang L, Fan L et al (2012) An improved hollow fiber solvent-stir bar microextraction for the preconcentration of anabolic steroids in biological matrix with determination by gas chromatography-mass spectrometry. J Chromatogr A 1233:1–7. https://doi.org/10.1016/j.chroma.2012.01.064
96. Correa L, Fiscal JA, Ceballos S et al (2015) Hollow-fiber solvent bar microextraction with gas chromatography and electron capture detection determination of disinfection byproducts in water samples. J Sep Sci 38:3945–3953. https://doi.org/10.1002/jssc.201500324
97. Yu C, Liu Q, Lan L, Hu B (2008) Comparison of dual solvent-stir bars microextraction and U-shaped hollow fiber-liquid phase microextraction for the analysis of Sudan dyes in food samples by high-performance liquid chromatography-ultraviolet/mass spectrometry. J Chromatogr A 1188:124–131. https://doi.org/10.1016/j.chroma.2008.02.065
98. Fashi A, Salarian AA, Zamani A (2018) Solvent-stir bar microextraction system using pure tris-(2-ethylhexyl) phosphate as supported liquid membrane: a new and efficient design for the extraction of malondialdehyde from biological fluids. Talanta 182:299–305. https://doi.org/10.1016/j.talanta.2018.02.002
99. Georgiadis DE, Tsalbouris A, Kabir A et al (2019) Novel capsule phase microextraction in combination with high performance liquid chromatography with diode array detection for rapid monitoring of sulfonamide drugs in milk. J Sep Sci 42:1440–1450. https://doi.org/10.1002/jssc.201801283
100. Kermani M, Jafari MT, Saraji M (2021) Self-rotating stir mesh screen sorptive extraction for analyzing chlorpyrifos by ion mobility spectrometry. Anal Methods 13:2631–2644. https://doi.org/10.1039/d1ay00595b
101. Kabir A, Mesa R, Jurmain J, Furton KG (2017) Fabric phase sorptive extraction explained. Separations 4:21. https://doi.org/10.3390/separations4020021
102. Alampanos V, Kabir A, Furton K, Samanidou V (2021) Magnet integrated fabric phase sorptive extraction of selected endocrine disrupting chemicals from human urine followed by high-performance liquid chromatography—photodiode array analysis. J Chromatogr A 1654:462459. https://doi.org/10.1016/j.chroma.2021.462459
103. Mao X, Hu B, He M, Fan W (2012) Stir bar sorptive extraction approaches with a home-made portable electric stirrer for the analysis of polycyclic aromatic hydrocarbon compounds in environmental water. J Chromatogr A 1260:16–24. https://doi.org/10.1016/j.chroma.2012.08.062
104. Qin Z, Bragg L, Ouyang G et al (2009) Solid-phase microextraction under controlled agitation conditions for rapid on-site sampling of organic pollutants in water. J Chromatogr A 1216:6979–6985. https://doi.org/10.1016/j.chroma.2009.08.052
105. Roldán-Pijuán M, Lucena R, Cárdenas S, Valcárcel M (2014) Micro-solid phase extraction based on oxidized single-walled carbon nanohorns immobilized on a stir borosilicate disk: application to the preconcentration of the endocrine disruptor benzophenone-3. Microchem J 115:87–94. https://doi.org/10.1016/j.microc.2014.02.015
106. Casado-Carmona FA, Alcudia-León M del C, Lucena R, Cárdenas S (2019) Portable stir membrane device for on-site environmental sampling and extraction. J Chromatogr A 1606:360359. https://doi.org/10.1016/j.chroma.2019.07.013
107. Casado-Carmona FA, Lucena R, Cárdenas S (2021) Magnetic paper-based sorptive phase for enhanced mass transference in stir membrane environmental samplers. Talanta 228:122217. https://doi.org/10.1016/j.talanta.2021.122217
108. Casado-Carmona FA, Jiménez-Soto JM, Lucena R, Cárdenas S (2022) Portable stirring device for the on-site extraction of environmental waters using magnetic hydrophilic-lipophilic balance tape. Anal Chim Acta 1189:339186. https://doi.org/10.1016/j.aca.2021.339186
109. Casado-Carmona FA, Lucena R, Cárdenas S (2023) Lab in a bottle, open-source technologies for the design of affordable environmental samplers integrating on-site extraction. J Environ Chem Eng 11:109713. https://doi.org/10.1016/j.jece.2023.109713

110. Makkliang F, Kanatharana P, Thavarungkul P, Thammakhet-Buranachai C (2018) A miniaturized monolith-MWCNTs-COOH multi-stir-rod microextractor device for trace parabens determination in cosmetic and personal care products. Talanta 184:429–436. https://doi.org/10.1016/j.talanta.2018.03.024
111. Badiee H, Zanjanchi MA, Zamani A, Fashi A (2019) A four-hollow fibers geometry of revolving solvent bar microextraction setup for the enrichment of trace ammonia. Talanta 199:170–177. https://doi.org/10.1016/j.talanta.2019.02.028

Matrix Solid-Phase Disperion

Dorota Wianowska and Małgorzata Olszowy-Tomczyk

Abstract Matrix solid phase dispersion (MSPD) is an extremely simple, fast and effective technique of sample preparation, in which sample destruction, homogenization and extraction takes place simultaneously in a single step of the MSPD procedure. This technique has been known for nearly 30 years. At that time, new types of sorbents, abrasive materials and extraction fluids were introduced to the MSPD procedure, and the process itself began to be assisted by ultrasounds, vortexing or microwaves. The result of these improvements is far-reaching miniaturization accompanied by greater isolation efficiency per unit of time achieved using more ecological and economical analytical procedures. Due to its flexibility and versatility, the MSPD technique is currently being implemented in various research laboratories for the isolation of endo- and exogenous compounds, including hazardous or prohibited compounds, volatile and non-volatile, present in various concentrations not only in solid but also in semi-solid and viscous samples, which can be generally grouped into environmental, biological, pharmaceuticals, food and everyday products samples. This chapter outlines the various analytical challenges where MSPD is useful and the sorbents that are currently being used to meet these challenges, with particular emphasis on new research areas where the MSPD process has come into use.

Keywords Sample preparation · Sorptive extraction · Miniaturized sample preparation method · Solventless extraction · SSDM · MSPD

Abbreviations

β-CD	β-Cyclodextrin
C_{18}	Chemically bound octadecyl phase

D. Wianowska (✉) · M. Olszowy-Tomczyk
Department of Chromatography, Institute of Chemical Sciences, Faculty of Chemistry, Maria Curie-Skłodowska University in Lublin, Pl. Maria Curie-Skłodowska 3, 20-031 Lublin, Poland
e-mail: dorota.wianowska@mail.umcs.pl

M. Olszowy-Tomczyk
e-mail: malgorzata.olszowy-tomczyk@mail.umcs.pl

M. Á. Rodríguez-Delgado et al. (eds.), *Microextraction Techniques*, Integrated Analytical Systems, https://doi.org/10.1007/978-3-031-50527-0_6

CDs	Cyclodextrins
CLC	Chiral liquid chromatography
DESs	Deep eutectic solvents
DLLME	Dispersive liquid–liquid microextraction
EDCs	Endocrine disrupting chemicals
EOs	Essential oils
FID	Flame ionization detector
GC	Gas chromatography
GCB	Graphitized carbon black
HLLME	Homogeneous liquid–liquid microextraction
HPLC	High performance liquid chromatography
ILs	Ionic liquids
IL-VF-MSPD	Ionic liquid based vortex-forced matrix solid phase dispersion
IP-SPE	Ion pair—solid-phase extraction
LC	Liquid chromatography
LLE	Liquid–liquid extraction
LOD	Limit of detection
LOQ	Limit of quantification
MALDI	Matrix assisted laser desorption ionization
MA-MSPD	Microwave-assisted matrix solid phase extraction
MCs	Microcystins
MEEKC	Microemulsion electrokinetic chromatography
MEPS	Microextraction in packed syringe or microextraction by packed sorbent
MI	Molecularly imprinted
MIM	Molecularly imprinted microsphere
MI-MSPD	Molecularly imprinted matrix solid-phase dispersion
MIPs	Molecularly-imprinted polymers
MOF	Metal–organic framework
MS	Mass spectrometry
MSPD	Matrix solid-phase dispersion
NMR	Nuclear Magnetic Resonance
NP	Normal phase
PLE	Pressurized liquid extraction
PSA	Primary and secondary amine
Q	Quadrupole
QuECHERS	Quick Easy Cheap Effective Raged and Safe
RP	Reversed phase
RSD	Relative standard deviation
SBSE	Stir bar sorptive extraction
SF	Solvent flotation
SLE	Supported liquid extraction
SPDE	Solid phase dynamic extraction
SPE	Solid-phase extraction
SPME	Solid-phase microextraction

SSDM	Sea sand disruption method
S-SIL	Silica-supported ionic liquid
q-TOF	Time of flight
UAE	Ultrasonically assisted extraction
UA-MSPD	Ultrasound-assisted matrix solid phase dispersion
UV	Ultraviolet detection
UPLC	Ultra-performance liquid chromatography
VA-MSPD	Vortexed-assisted matrix solid phase dispersion

1 Introduction

The term miniaturization is the key word of the modern world, a determinant of trends and directions of activities in many areas of our lives, including analytical chemistry and the related sample preparation process [1]. Behind miniaturization are the analytical capabilities of modern systems, especially chromatographic ones, which allow for the determination of compounds at increasingly lower concentration levels in the minimum amount of sample needed for a single and accurate analysis. In addition, miniaturization is supported by the desire to meet the challenges of modern analytics and innovative research areas, and the need to overcome the problems of classic methods of sample preparation, and especially making them more economical and ecological methods. This is all the more important as sample preparation is still a critical step in any analytical process [2–5]. Sample preparation usually consists of several stages, so it is not surprising that it is still one of the most laborious and time-consuming stages of any analytical procedure. In addition, this stage is extremely prone to errors, which often cannot be corrected at the later stages of the analytical procedure, because it is the properly targeted stage of sample preparation that guarantees the method of analysis independent of any changes in the sample matrix as well as accurate and indisputable results. Therefore, the correct preparation of the sample is not only the key to the success of the analysis, but also improves it, contributing to the increase in the number of analyses and the reduction of both labour time and costs.

Extraction is one of the most commonly used methods to prepare a sample for analysis. It owes its popularity to the ability to achieve all the objectives of the sample preparation step, as it allows for complete isolation of the analyte from complex and complicated matrices, concentration of the analyte, removal of accompanying interfering substances and replacement of the matrix with a solvent compatible with the target analytical technique. In addition, various physicochemical properties of analytes and matrices do not limit the area of its application and, as a result, it is used to isolate volatile and non-volatile compounds from solid, liquid and gaseous matrices [5]. The current trend in the use of the extraction method is focused on miniaturization in the broad sense, miniaturization understood as the use of scaled-down extraction systems capable of processing very small sample volumes using (if any) significantly reduced volumes of organic solvents, and simplification of analytical procedures by combining several stages of sample preparation or analytical procedure into one, while eliminating, importantly, the loss/degradation of sample components [1, 5, 6].

In the last decade, microextraction techniques of sorption extraction gained popularity, effectively displacing the classical methods of solvent extraction. One of the first such techniques is solid-phase microextraction (SPME). One of the newest is the supported liquid extraction (SLE) technique, touted as the best kept secret in sample preparation. Currently the block of miniaturized sorption extraction techniques includes stir bar sorptive extraction (SBSE), solid phase dynamic extraction (SPDE), microextraction in packed syringe (MEPS), microextraction by packed sorbent (MEPS), the combination of liquid–liquid extraction and dispersive solid phase extraction known as Quick Easy Cheap Effective Raged and Safe (simply QuECHERS) [1]. An important item in the aforementioned block of techniques is the matrix solid-phase dispersion (MSPD) technique, known for over 30 years. This is one of the least equipment-demanding techniques for the isolation of compounds from solids, with efficiency equal to advanced extraction techniques such as pressure liquid extraction (PLE). Due to its simplicity, the MSPD procedure can be performed by anyone, which makes it one of the most attractive and more frequently used methods of extraction applied in various areas of research, often going beyond the framework of chemical analytics. The validity of these statements is confirmed by the number of review papers dedicated to the progress and applications of the MSPD technique that have been published in the recent period [4, 6–10].

2 Fundamentals

2.1 General Information About MSPD

The MSPD technique, the stages of which are shown in Fig. 1, was introduced by Baker in the 1990s. This is a simpler version of the solid phase extraction (SPE) process. It involves grinding the sample with a solid abrasive material to obtain a semi-dry and homogeneous material with a specific structure. The abrasive material is most often a sorbent, which not only releases the analyte from the matrix but also increases the selectivity of the extraction. However, the sorbent can be replaced with another solid material, such as sand, to obtain a cheaper version of the process with the same isolation efficiency [9]. Homogenization is carried out in a mortar with a pestle made of glass, agate or quartz, and the addition of a small amount of solvent increases the dispersion of the sample components in the space of the abrasive material. The mixture obtained by grinding is quantitatively transferred to a syringe barrel (SPE column) with sintered paper at the bottom, pressed to ensure the best possible contact surface with the eluent, and then eluted dropwise under reduced pressure. Finally, the obtained extract is subjected to an analytical procedure.

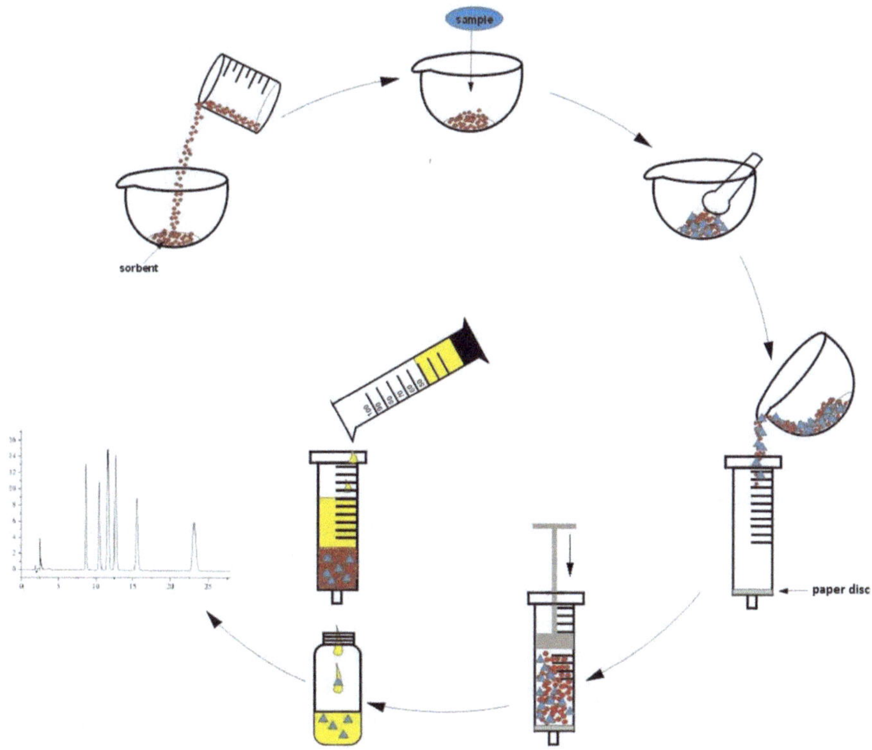

Fig. 1 Steps of the MSPD procedure

The efficiency of the MSPD process depends on several factors, which can be easily optimized by selecting the appropriate type of abrasive, specifying the ratio of the mass of the sample to the abrasive, mixing time, composition of the eluent and/ or its volume. Most often, due to the limited sorption capacity of the typically used sorbents, a four-fold excess of the sorbent mass in relation to the mass (amount) of the sample is used, with 0.5 g of the sample being typically used. Depending on the degree of hardness of the sample matrix, the homogenization time varies from 5 to 10 (15) min in the case of harder matrices. Taking into account that the elution stage is governed by the principles of frontal analysis and the first drops of the extract are the richest in compounds, only a small amount of extractant is needed for quantitative elution. As a result, MSPD allows you to reduce the consumption of organic solvent. Moreover, it promotes the concentration of compounds. An important feature of this extraction technique is that it does not require special equipment, and by combining sample breaking, extraction and purification in one step, it reduces sample preparation time. These attributes explain why MSPD is recognized as a very simple, cheap and quick sample preparation procedure that can be easily implemented in any research laboratory [9]. In addition, a wide range of different, more or less selective, ad(ab)sorptive abrasive materials suitable for use in the MSPD procedure makes it a technique with a great application potential.

To date, various MSPD materials have been introduced and employed such as e.g. silica- or carbon-based materials, nanoparticles, molecularly imprinted polymers, molecular sieves or ionic liquids [4, 6–10]. For example, Wianowska et al. which is discussed in more detail later in this chapter, applied sand as abrasive material for the extraction of volatile and non-volatile compounds from plants and herbs [11–14]. Sowa et al. showed that aniline deposited on silica gel particles successfully isolate triterpenes [15]. Yet, conventionally applied dispersion MSPD sorbents are silica gel, florisil and alumina. These are inorganic sorbents working, to use the typical chromatographic term, in the normal phase (NP) mode. For this reason, they are briefly referred to as normal phase materials (NP sorbents) capable of retaining polar analytes from less polar liquids by adsorption. Since this phenomenon is associated with access to functional groups present on the surface of the adsorbent, the extraction efficiency determined by the sorption capacity, a parameter characteristic of sorption processes, is often low and insufficient for effective concentration of the analytes. In order to increase the effectiveness of isolation, sorbents with a chemically bonded phase are used, which show absorptive properties in interactions with polar and non-polar components of the sample, while increasing the selectivity of the extraction process. One of the most commonly used sorbents of this type is silica gel with a chemically bound octadecyl phase (C_{18} sorbent). This material, in contrast to the previously mentioned group of inorganic sorbents, is used in the reversed phase (RP) system, enabling the extraction of non-polar (medium-polar) analytes.

As for the basic characteristics of the liquids used for elution of compounds from the MSPD blend, due to the destructive nature of the process, their choice mainly depends on the properties of the sorbent used and the target analytical technique. In general, non-polar eluting liquids are used when working with NP sorbents, while more polar liquids are used when working with RP sorbents. In the latter, eluents typically used include methanol, acetonitrile or acetone and mixtures of these solvents with water. Doping the organic solvent with water on the one hand allows to create a more selective eluting mixture, but is also a way to further reduce the consumption of organic, i.e. toxic liquids.

3 Novel Developments

The simplicity, effectiveness and versatility of MSPD makes it a technique worth working on, making it even better. In general, these efforts are focused on the use of new sorption or abrasive materials as alternatives to the commonly used silica-based materials in tandem with safer and environmentally friendly elution liquids and the development of more effective, faster and simpler MSPD procedures [16–36]. In the latter case, the efforts go in two directions. Firstly, to increase the effectiveness of isolation from difficult matrices (hard or swellable), the MSPD process is supported by the application of additional force or energy source [28, 34, 36–43]. Secondly, to increase the selectivity of the extraction process, especially when using an abrasive material with inert properties, the MSPD process is combined with other extraction techniques [25, 26, 44–46]. Examples of the use of various MSPD materials, under specific and characteristic for them conditions, together with the overall analytical performance of the analytical method are presented in Table 1.

3.1 New Sorption Materials

In the MSPD procedures, due to greater analytical possibilities, sorbents with a chemically bound phase are more willingly used. As mentioned, these sorbents, due to their significant sorption capacities, allow for more effective concentration of the analyte in the extract. In addition, owing to the access to a wide range of commercially available materials with different chemical properties of the functional groups forming the bonded phases, it is possible to select the one that will interact more selectively with the analyte. However, polar silica gel is most often used as a support of polar or non-polar functional groups. Thus, as a result of grinding the sample matrix with the RP sorbent, both polar and non-polar components of the sample interact with the support and the chemically bonded phase, and the expected high extraction selectivity is lost. Hence, the recently observed tendency to increase the sensitivity and selectivity of the analysis, especially of natural samples in the pre-analytical stage, is based on the use of very selective new materials. Providing a high selectivity of the extraction process, these materials have an extraordinary enrichment ability, which allows to reduce the consumption of solid and liquid reagents. Moreover, they make it possible to conduct g the MSPD process in a more miniaturized version. New MSPD materials, mostly adsorptive, with good chemical and mechanical properties are discussed below and illustrated in Fig. 2. However, it should be mentioned that this group also includes absorbents that have been known for a long time, but their usefulness in the MSPD process has been confirmed in recent years (e.g. polyvinylpolypyrrolidone) [16] and those that have recently been developed by combining well-defined polymers with inorganic substrates to create polymer-inorganic hybrids such as as SiO_2/polyvinylimidazole hybrid polymer [17] with much better absorption capacity.

One of a rapidly developing technique for the preparation of functional polymers having specific molecular recognition properties is molecular imprinting. Thus, molecularly imprinted polymers (MIPs) are new selective sorbents for the MSPD prcedures of organic compounds in complex natural matrices. Their selectivity mimics the interactions between natural receptors in antibody-antigen interactions. It is based on the concept of matching a three-dimensional structure of a sorbent to the structure of an analyte molecule. To achieve this matching, functionalized monomers are polymerized around a template analyte molecule, creating a highly cross-linked three-dimensional network polymer with affinity only for the target molecule used in the imprinting procedure.

Table 1 Examples of applications of the MSPD technique to different types of materials and analytes

Material (analyte)	MSPD conditions	Analysis type and its performance	Ref.
Rice (imidacloprid)	0.5 g of the sample was mixed with the MIP sorbent (1 g); blended for 8 min; rinsed with 20% aqueous methanol solution (5 mL) and eluted with methanol (8 mL)	LC-MS/MS LOD: 2.4 ng/g	[20]
Carrot and yacon (phosphorothioate organophosphorus pesticides)	0.2 g of the sample was mixed with the MIP sorbent (0.3 g) and 10% magnesium chloride (0.05 mL); methanol–water 1:2, v/v (0.3 mL) and 10% magnesium chloride solution (0.1 mL) were added, the whole was incubated at room temperature for 3 h, rinsed with methanol–water 1:9, v/v (5.0 mL) and eluted with acetonitrile-trifluoroacetic acid 99:1, v/v. (6.0 mL)	GC LOD: 0.012–0.026 ng/g	[21]
Schisandra Chinensis (Turcz.) Baill. Fructus (lignans)	25 mg of the powdered sample was mixed with TS-1 (50 mg); blended for 150 s and eluted with methanol (500 μL)	MEEKC LOQ: <2.77 μg/mL	[24]
Herba Lysimachiae (quercetin)	0.1 g of the sample was mixed with MIP (0.1 g); blended for 10 min; washed with 2% aqueous methanol (4 mL) and eluted with acetic acid–methanol (2:98, v/v) (3 mL)	HPLC-UV LOD: 0.25 μg/mL	[18]
Onion, apples (Golden Delicious), black tea (Yunnan) (quercetin)	0.2 g of the grounded sample was mixed with sand (0.8 g); blended for 5 min with methanol as dispersing solvent (1 mL) and eluted with methanol (10 mL)	HPLC-PDA LOQ: 0.1162 μg/mL	[12]
Arabidopsis thaliana L. Heynh., single leaf (gibberellins)	0.30–0.80 mg of the sample was mixed with the C_{18} sorbent (2 g), washed with methanol (10 mL) and acetonitrile (200 μL) and finally centrifugated (10 min)	UPLC-MS/MS LOD: 10.1–72.3 amol	[47]

(continued)

Table 1 (continued)

Material (analyte)	MSPD conditions	Analysis type and its performance	Ref.
Fructus Corni Torr. (5-HMF, iridoid glycosides)	20 mg of the powdered sample was mixed with silica gel (40 mg); blended for 3 min with [Domim]HSO_4 (6 mL) followed by vortexing (3 min) and centrifugation (10 min)	UHPLC LOD: 0.02–0.08 μg/mL	[28]
Hawthorn and black elder flowers, green tea and nettle leaves, yerba mate, St John's wort, green coffee beans (chlorogenic acids and their derivatives)	0.8 g of the sample was mixed with sand (0.8 g); blended for 10 min and eluted with ethanol/water (75/25%, v/v) (25 mL)	LC-MS	[11]
Chamomile, thyme, mint, sage, marjoram, savory, oregano (essentials oils components)	0.2 g of the grounded sample was mixed with C_{18} (0.8 g); blended for 10 min with 1,4-dioxan as dispersing solvent (1 mL) and eluted with hexane–ethyl acetate mixture (9:1, v/v) (10 mL)	GC-MS GC-FID	[14]
Scots pine and cypress needles (essentials oils components)	0.2 g of the cut needles sample was mixed with sand (4.8 g), blended for 10 min with 3 mL 1,4-dioxan as dispersing solvent (1 mL) and eluted with ethyl acetate (10 mL)	GC-MS GC-FID	[13]
Rice samples (insecticides: chlorfenapyr and abamectin)	2 g of the sample was mixed with basic alumina (4 g) and acetonitrile (10 mL) followed by microwave irrigation at 600 W in a microwave-assisted MSPD	LC-MS/MS LOD–0.8 ng/g	[43]
Orange samples (auxins: indole-3-acetic acid, indole-3-propionic acid, indole-3-butyric acid, 1-naphthaleneacetic acid)	0.1 g of the sample was mixed with MIP (0.1 g); rinsed with 5.0 mL of methanol–water (1:9, v/v) and eluted with dichloromethane–acetic acid (95:5, v/v) (3 mL))	HPLC-UV-VIS LOD: 1.0–2.4 ng/g	[19]
Egg yolk (Sudan dyes)	0.1 g of the sample was mixed with MIM (0.2 g); rinsed with 4.0 mL of methanol–water (1:1, v/v) and eluted with acetone–acetic acid (95:5, v/v) (3 mL)	HPLC 87.2 < Recovery < 103.5% (RSD < 6.1%)	[46]

(continued)

Table 1 (continued)

Material (analyte)	MSPD conditions	Analysis type and its performance	Ref.
Raw propolis (caffeic acid, ferulic acid, morin, luteolin, quercetin, apigenin, chrysin, kaempferide)	0.05 g of the sample was mixed with S-SIL containing 10% [C_6MIM]Cl (0.02 g), blended for 3–4 min, then defatted with *n*-hexane (20 mL) and eluted with methanol (15 mL)	HPLC-DAD LOD: 5.8–22.2 ng/mL LOQ: 19.2–74.0 ng/mL	[27]
Fortified propolis sample (pesticides: dichlorvos, diazinon, methyl parathion, malathion, cumaphos)	1 g of the sample was dissolved in 10 mL of *n*-hexane and an aliquot of 1 mL was mixed with C18 (1 g) or SiO_2–PVI polimer (1 g), then eluted with acetonitrile–dichloromethane (25:75 v/v) (8 mL)	GC-MS C18 recovery: 83–126% (RSD < 12%) SiO_2–PVI recovery: 81–122% (RSD $\leq$ 11%)	[48]
Olive fruit and oil samples (phospholipids)	1 g of the sample was mixed with TiO_2NP (2 g), blended for 10 min, pre-extracted with acetone (5 mL $\times$ 3) and the eluted with chloroform–methanol (1:2, v/v) (3 mL)	MALDI-TOF/MS Intra-day precision: 1.23 < RSD < 4.25% Inter-day precision: 1.77 < RSD < 4.47%	[35]
P. ginseng leaves (saponins: ginsenoside Rg2, Rg1, Re, Rd, Rb1)	25 mg of the powdered sample was mixed with MOF-808 (20 mg), blended for 60 s and eluted with 80% methanol–water solution (200 μL)	UHPLC–QTOF-MS Recovery: 87.04–103.78%, (RSD < 5%) LOD: 0.087–0.114 μg/mL LOQ: 0.292–0.379 μg/mL	[22]
Breast milk (ibuprofen enantiomers)	0.5 g of the sample was mixed with diatomaceous earth (0.30 g), Na_2SO_4 (0.30 g), PSA-bonded silica (0.26 g) and cyclodextrin (0.02 g), blended for 5 min and then vortexed (1 min) with the addition of methanol (2 mL) and centrifugated (15 min)	CLC-UV Recovery: 71.0–88.2%, (RSD < 9%) Quantification in the range of 0.15–6.0 μg/g	[31]

(continued)

Table 1 (continued)

Material (analyte)	MSPD conditions	Analysis type and its performance	Ref.
Muscle tissue (amitriptyline, atenolol, carbamazepine, chlorpropamide, chlorthalidone, diclofenac, diltiazem, enalapril, fluoxetine, flurazepam, furosemide, glibenclamide, nimesulide, propranolol, salbutamol)	0.5 g of the sample was mixed with diatomaceous earth (0.5 g) and Na_2SO_4 (0.5 g), blended for 5 min, then methanol was added (5.0 mL), followed by vortexing (1 min) and centrifugation (10 min)	LC-MS/MS LOQ: 5–1000 ng/g	[34]
Breast milk (parabens)	0.2 mL of the sample was mixed with 50 mg of poly(indole-thiophene) coated magnetic graphene oxide (MGO@PIT) and 550 mg of Na_2SO_4, then methanol (1.0 mL) was added, followed by vortexing for 2.0 min, finally 1-octanol (100 μL) was added as the extraction solvent	LC-UV (LOD: 25 ng/mL) LC-MS/MS (LOD: 0.5 ng/mL) Recoveries > 83% RSD for intra- and inter-day precisions were less than 7.5% and 11.3%, respectively Quantification in the range of 50–4000 ng/mL	[36]

MIPs preparation is not simple. The most commonly reported issues include incomplete analyte template removal, non-uniform distribution, and poor site availability. However, the increased stability and resistance to a wide range of pH values, temperatures and solvent types compensate for the high price of materials available on the market. However, these materials can be prepared in the laboratory. An example of the synthesis and use of a MIP sorbent in the analysis of polyphenols, with the limit of detection (LOD) method established at 0.25 μg/mL, is available in [18]. The basic features of this and other representative MSPD procedure are summarized in Table 1. Another interesting article of the use of molecularly imprinted matrix solid-phase dispersion (MI-MSPD) in the plants analysis, showing the full spectrum of the possibilities of these materials, is presented in [19] on the example of the analysis of phytohormones from the auxin group in orange samples. According to the quoted paper, the only parameter that requires optimization is the sample to sorbent mass ratio. Under the optimal conditions, LOD is in the range of 1.2–2.4 ng/g.

The MI-MSPD procedures have also been used in the LC analysis of insecticides [20] and the GC analysis of phosphorothioate pesticides [21] in food samples. In the latter case, owing to the use of new molecularly imprinted polymer nanomicrospheres, which were synthesized using a typical structural analogue of tolclophosmethyl as a template by surface-graft polymerization on nanosilica, the detection limit of the method was obtained in the range of 0.012–0.026 ng/g and the recoveries ranged from 85.4 to 105.6% with RSD $\leq$ 9.6%.

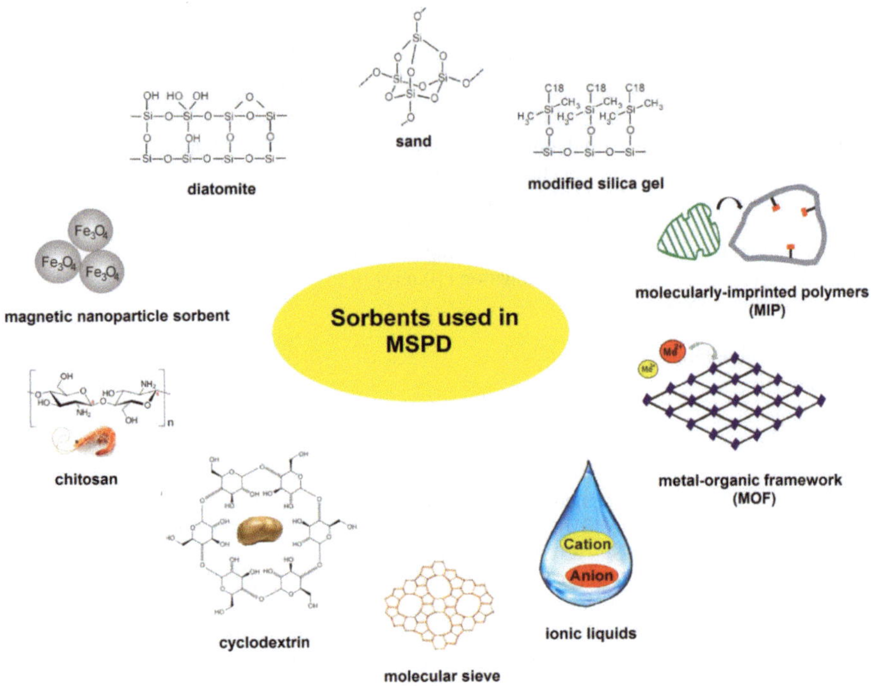

Fig. 2 Novel materials used in the MSPD procedure

In a different approach to the preparation of composite materials, owing to the use of the embedding method, not only the problems encountered in the synthesis of MIPs were eliminated, but also sorbents with exceptional enrichment capacity were obtained. In this context, special attention should be paid to sorbents in which a metal–organic framework (MOF) was used as a support for synthetic composite materials MOF-MIP.

The MOF materials are highly porous coordination polymers in which the three-dimensional structures of organic linkers with metal ions are formed through coordination bonds. There are few applications of these new sorbents in the context of microextraction of organic compounds. An exception is the article by Zhang et al. [22] which reports the synthesis of MOF (marked as MOF808) for accurate and sensitive analysis of ginsenosides in *Panax ginseng* C.A. Meyer root by means of MSPD and UPLC coupled to a quadrupole time-of-flight MS detector. In compliance with the article, under the optimal conditions, using only 20 mg of sorbent during 60 s grinding, analytical recoveries ranging from 87.04 to 103.78% are obtained, with RSD below 5% and LOD in the range from 0.087 to 0.114 μg/mL (see Table 1). According to the authors, the proposed MOF-assisted MSPD procedure, compared to the traditional extraction method and other published procedures, is characterized by higher extraction efficiency, simpler operation and provides a purer extract with less use of organic reagents.

Molecular sieves are another example of selective dispersion materials used in MSPD. These extremely selective materials are crystalline metal aluminosilicates composed of interconnected three-dimensional networks of tetrahedral oxides. Currently, more than 200 different molecular sieve frameworks have been described in the literature [9]. However, only two have been used in MSPD so far. These are sieves marked as SBA-15 and TS-1, the latter of which has a three-dimensional channel system with linear and zigzag locations. Both are attractive materials for MSPD due to their uniform and adjustable pores size, their large volumes, well-defined channels giving a high surface area to volume ratio, easy surface functionalization and hydrothermal stability. Their use allows for significant miniaturization of the sample preparation process for chromatographic analysis. For example, in [23], the SBA-15 molecular sieve was used to prepare a sample of orange fruit peels for the analysis of flavonoids using the UPLC-UV technique. Optimal extraction conditions boiled down to dispersing 25 mg of pre-ground sample in 25 mg of sorbent and then eluting the target compounds with only 0.5 mL of methanol, giving LOD of 0.02–0.03 μg/mL. In turn, in [24], the TS-1 mesoporous molecular sieve was used to prepare a sample of *Schisandrae Chinensis Fructus* for the analysis of lignans by microemulsion electrokinetic chromatography (MEEKC). The developed method, adapted to the preparation of a very small amount of sample (25 mg) with the use of an equally small amount of sorbent (50 mg) and low consumption of elution solvent, in accordance with the principles of green chemistry, showed good precision with the limit of quantification (LOQ) below 2.77 μg/mL. Compared to conventional MSPD procedures, the proposed methodology turned out to be extremely efficient, which was reflected in the modification of the MSPD name to the micro-MSPD version.

One of the newer research trends, already mentioned in the context of Sowa's research, is the use of sorbents obtained by modifying the silica gel surface with specialized liquids. These specialized liquids used in MSPD are mainly ionic liquids [25–28]. Nevertheless, the literature also describes the use of, e.g., poly(N-vinylimidazole) and deep eutectic solvents (DESs) for this purpose [29].

Ionic liquids (ILs) are non-molecular solvents that are organic and inorganic salts with the melting point below 100 °C. They are characterized by unique properties such as adjustable viscosity, miscibility with water and organic solvents, and low vapor pressure associated with high thermal stability. Moreover, they can be immobilized in the micropores of silica gel to obtain the silica-supported ionic liquid (S-SIL). The ILs ILs then lose their liquid state, but retain their beneficial properties. Since the S-SIL material has many micropores filled with ILs, S-SIL-based extraction improves the mass transfer rate and achieves a high level of recovery while reducing IL consumption. These features explain why S-SIL is now so often used as a dispersion adsorbent in the plant MSPD process. An example of the use of S-SIL materials in the MSPD process is the work [27] in which a methodology for the determination of phenolic compounds in a difficult material such as propolis was proposed. In the work, its authors showed that, compared to the classic ultrasonically assisted extraction (UAE) and extraction in the Soxhlet apparatus, their method allows for lower consumption of the sample and organic solvents and a shorter extraction time. In terms of the method performance, the limits of detection and quantification were in the range of 5.8–22.2 ng/mL and 19.2–74.0 ng/mL, respectively. The recoveries ranged from 65.51 to 92.32%, with RSDs lower than 8.95%.

As mentioned, the factor determining the success of the MSPD extraction is the selectivity and sorption capacity of the sorbents used in the process. In this respect, cyclodextrins (CDs) are an interesting material enabling the isolation of racemic mixtures, as well as configurational and constitutional isomers. The structure of these chiral materials has a characteristic truncated-cone shape. It is based on cyclic oligosaccharides built of D-glucose units linked by α(1,4)-glucosidic bonds. These materials, by selectively binding various molecules in their hydrophobic cavities, can form supramolecular host–guest complexes with high molecular recognition potential and excellent adsorption properties. One of the most commonly used CDs materials is β-CD containing seven glucose units in its structure. For example, in [30] this material was used for MSPD microextraction of various phenolic isomers from the honeysuckle flowers (*Lonicera japonica* Thunb.) before further analysis by UPLC-UV-Q-TOF/MS and NMR to determine and characterize the exact structure isolated compounds. In this method, 25 mg of samples were homogenized with 75 mg of β-CD using 0.5 mL of methanol–water mixture (80:20, v/v) as elution solvent, obtaining an LOD ranging from 1.62 to 3.33 ng/mL and recovery in the range of 87–105%. In [31], a mixture of β-CD and primary and secondary amine (PSA) sorbents was used for the isolation and quantification of ibuprofen enantiomers from human breast milk, combining a vortex-assisted MSPD and direct chiral liquid chromatography (CLC) with UV detection. The presence of the chiral β-CD sector was found to promote a variety of interactions resulting in good analytical performance o. In addition to the secondary interactions, hydrogen bonding or dipole–dipole interactions with the hydroxyl groups, the specific shape of β-CD and the appropriate size of the cavity, enabling the formation of inclusion complexes with enantiomers, determine the greater adsorption of the latter. Under optimal conditions (see Table 1), the proposed method provided good repeatability and accuracy, with RSDs of 6.4% and 8.3% for intra-day and inter-day precision, respectively, and recoveries in the range of 71.0–88.2%.

In this review of newer sorbents, it is also worth noting the possibilities of using chitin and chitosan as an alternative abrasive and adsorption materials in the miniaturized MSPD extraction process. Chitosan is produced as a result of deacetylation of one of the most common polysaccharides, i.e. chitin. It is mainly obtained from the hard outer skeleton of marine animals and insect cuticles. This material attracts attention with its special properties, which includ good biocompatibility, biodegradability and adsorption capacity guaranteed by a large surface area. In addition this material is characterized by, non-toxicity, renewability, and hydrophobicity [32]. It has many free hydroxyl and amino groups in its structure, allowing forming the hydrogen bonds and participation in electrostatic and ion-exchange interactions. For example, [33] describes the use of medium-molecular chitosan prior to the UPLC-Q-TOF/MS analysis of natural compounds (phenols) contained in plum fruits at various concentration levels. Optimized MSPD parameters were determined by choosing the amount of chitosan (25 mg), grinding time (60 s), and methanol:water (6:4, v:v) mixture as the eluting solvent. The method showed LOD in the range of 69.6–358.4 ng/g, and recoveries exceeded 80%. In [34], chitin and chitosan were used to extract various pharmaceuticals from fish samples using MSPD-LC–MS/MS. However, recoveries were low compared to other tested materials, with the best results being obtained with diatomaceous earth. Under optimal conditions, recoveries ranged from 58 to 128%, with RSD below 15% and LOQ values for all analytes ranging from 5 to 1000 ng/g. According to the authors, the reason for the unsatisfactory results was the irreversible sorption of the target compounds on both materials, because chitin and chitosan contain basic nitrogen centers, which gives these materials a greater ability to adsorb the analytes. To confirm the validity of the conclusion, the authors cited a higher extraction efficiency on chitin, which, compared to chitosan, contains less alkaline nitrogen centers.

To conclude the review of newer sorption materials that have been used in MSPD, it is worth mentioning nanoparticles. In general, two types of nanoparticles find application in the new MSPD procedures. The first type is titanium dioxide nanoparticles which, for example, Shen et al. [35] applied for the selective extraction, visualization and analysis of phospholipids from olive fruit and oil by a matrix assisted laser desorption ionization time-of-flight mass spectrometry (MALDI-TOF/MS). The advantage of these materials is chemical stability in a wide range of pH and good adsorption capacity. The second type is magnetic nanoparticles of graphene oxide, used e.g. in [36], which allow for a significant simplification of the MSPD procedure and the possibility of their reuse, which makes this procedure even more ecological (see Table 1).

Nevertheless, over the last decade, not only materials with adsorption or absorption capabilities have been developed. As mentioned earlier, attention was also paid to sand in an alternative approach to the MSPD process, called the sea sand disruption method (SSDM). The rationale for using this inert material was to further simplify the MSPD process, reduce its cost, and make it more environmentally friendly. Currently, there are many examples of analytical procedures indicating the usefulness of this approach in the effective isolation of both essential oils (EOs), non-volatile analytes, and those that are not necessarily easy to recover from plant matrices [12, 49–51]. For example in [13, 14] it was shown that the optimal SSDM conditions for extracting EOs from conifer needles are determined by the sand to plant mass ratio of 24:1 with a volume of 3 mL of 1,4-dioxane used as the eluting solvent. The use of lower mass ratios, including the 4:1 ratio most often used in the conventional MSPD processes with a sorbent, results not only in lower efficiency of EOs components, but also proves to be difficult in the practical application. The blending step requires much more physical effort. Moreover, the bed of the homogenized material is less permeable to the eluting solvent.

At this point, however, it is worth emphasizing that the SSDM approach is eagerly used to release natural compoundsfrom hard matrices [9, 49, 50]. Another example of the use of this technique is the isolation of thermally unstable compounds, which is more effective compared to the amounts released by high-temperature extraction techniques [51–54]. This issue will be briefly discussed later. Nevertheless in order to balance the advantages of the SSDM process, it should be noted that the use of sand as an abrasive material leads to a reduction in its selectivity, which may be a problem when using less selective detectors such as UV [9, 44]. Therefore, the currently observed trend is the development of more complex procedures in which the efficiency of homogenization of the SSDM process, which goes hand in hand with the ability to release compounds even from strong interactions with other components of the matrix, is combined with the selectivity of other extraction techniques, especially sorptive extraction, e.g. in processes such as ion pair—solid-phase extraction (IP-SPE) [44].

3.2 Assisted MSPD Extraction

As mentioned, the MSPD process can be adjusted to improve the extraction efficiency. This usually involves selecting the appropriate type of sorbent, the composition of the eluent and/or its volume, as well as determining the sample to sorbent mass ratio and the mixing time. Occasionally, modifiers such as chelators, salts, acids, bases, and co-sorbents are added during the blending to increase the MSPD efficiency. A new trend in MSPD of very hard/difficult materials is the use of assisted methods of dispersion by means of ultrasounds, microwaves or vortex mixing. This not only shortens the extraction time, but also reduces the consumption of organic solvents and the amount of sample needed to fully isolation of trace compounds.

In vortexed-assisted MSPD (VA-MSPD), vortex mixing is used to further improve the isolation of compounds from solid biological matrices and save time. The procedure is carried out in two different ways. In the first one, after destruction/dispersion of the sample with a sorbent in a mortar, the resulting homogeneous mixture is quantitatively transferred to a centrifuge tube and then a solvent is added to the sample to obtain a suspension. The content of the tube is then vortexed for several minutes to increase the contact surface (usually from 1 to 3 min, see Table 1) and then centrifuged to recover the extract ready for analysis [34, 37]. In the second, the step of blending the sample in a mortar is omitted. A pre-prepared sample, e.g. by freezing, is dispersed by 2 min vortexing in a mixture of solids, then the blend is transferred to a SPE column filled with a co-sorbent at the bottom and finally eluted with a solvent by gravity or after applying a low vacuum [38, 39]. An interesting way of combining the vortexed-assisted MSPD approach with a non-toxic ionic liquid was proposed by Du et al. in [28] in the procedure called ionic liquid-based vortex-forced matrix solid phase dispersion (IL-VF-MSPD). In the cited paper, 1-dodecyl-3-methyl-1H-imidazolium bisulfate was applied as the ionic liquid used to elute 5-hydroxymethylfurfurol and iridoid glycosides from 20 mg of *Fructus Corni*. Silica (20 mg) was used as a dispersant and vortex mixing (3 min) was applied to contact the solid phase with 6 mL of the IL which, after centrifugation, was injected into the UPLC system (5 μL) achieving a limit of quantification (LOQ) in the range of 0.02–0.08 μg/mL. The recoveries were in the range of 95.2–103% (RSD $<$ 5.0%).

In ultrasound-assisted MSPD (UA-MSPD), the sample homogenization or solvent elution step is amplified by ultrasounds. In the first case, a syringe cylinder filled with sorbent and sample, closed on both sides, is periodically sonicated in an ultrasonic bath [40]. In the second, a previously homogenized sample placed in a sealed vessel and flooded with the eluting solvent is subjected to ultrasound [41]. The use of the phenomenon of cavitation accompanying the passage of an acoustic wave through the extraction system is conducive in the first case to increasing the degree of sample homogenization and facilitating the release of compounds from the solid matrix, and in the second to increasing the mass transfer while reducing the volume of solvents used. Generally compared to the usual MSPD process, UA-MSPD shortens the extraction time and increases its efficiency even up to 25% [42].

The approach to enhance the MSPD efficiency with the use of microwaves is slightly different. In this strategy, microwaves are generally used after homogenization to facilitate elution of the compounds. For example, Zhang et al. [43] used microwave-assisted MSPD (MA-MSPD) coupled with LC-MS/MS to accurately determine two common and health-hazardous insecticides in rice samples. In this method, a 2 g sample of rice was homogenized with 4 g of basic alumina and, after addition of solvent, exposed to microwave radiation in a 600 W microwave oven to give a limit of detection (LOD) of 0.8 ng/g (the recoveries were in the range of 88.6–96.5% with RSD values of 2.6 and 8.1% for intra- and inter-day precision respectively).

The block of assisted MSPD techniques also includes a procedure in which the previously mentioned magnetic nanoparticles of graphene oxide are used. In this method, unlike those previously presented, it is possible to reuse of the sorbent. In addition, the extraction time is reduced by eliminating the need to fill the SPE column. The approach is known as magnetically assisted MSPD (MA-MSPD). It was introduced by Fotouhi et al. [36] for the determination of parabens in breast milk. In the developed method, a modified magnetic nanosorbent in the form of poly(indole-thiophene) magnetic graphene oxide in the amount of 50 mg was mixed with 200 μL of milk and 550 mg of Na_2SO_4. The mixture was transferred to a beaker containing 5.0 mL of distilled water, mechanically stirred for several minutes, and then the nanosorbent was separated from the suspension with a strong magnet and immersed in methanol (1.0 mL) to desorb the analytes from the sorbent, after which the sorbent was again removed from the eluate with a magnet, allowing it to be reused.

3.3 *Coupling MSPD with Other Isolation Techniques*

As mentioned in the Introduction, one of the areas of contemporary research activity is the development of efficient and sensitive methods that not only allow drawing quantitative conclusions from very small amounts of samples, but also lead to savings in reagents. And all this is due to the miniaturization of extraction systems. Above are presented those of the newly developed MSPD procedures in which the demands of selectivity, efficiency and miniaturization are intertwined into one due to the use of non-standard dispersion materials and those in which the efficiency of extraction is supported by the energy of microwaves, ultrasounds or the effect of vortex mixing. A characteristic feature of the newly developed MSPD procedures is also the combination of the MSPD process with other isolation techniques, in particular liquid–liquid microextraction techniques such as dispersion liquid–liquid microextraction (DLLME) and homogeneous liquid–liquid microextraction (HLLME). Another trend of changing MSPD selectivity, especially when using non-selective dispersion materials, is based on combining MSPD with backward extraction and/or extraction proceeding with the formation of neutral ion pairs of the analyte. These and other ways to change selectivity in the MSPD process will be outlined below.

DLLME and HLLME represent a new approach to liquid–liquid extraction (LLE) where the extraction efficiency depends on the contact area between the two liquid phases. In the former, the amount of organic solvent (extractant) is kept to a minimum, and in order to guarantee an appropriate contact surface, a slightly larger amount of another less toxic solvent is introduced that has an affinity for both the aqueous and organic phases. The liquid introduced into the extraction system is the so-called dispersant increasing the contact surface between the phases. The difference of the second approach consists in the use of alternative extraction solvents miscible with the water phase in an unlimited way. In HLLME, ILs or DESs are used as extractans. The general concept of ILs is presented above. So when it comes to DESs, they generally consist of two cheap components capable of self-association, often through hydrogen bonding interactions, to form an eutectic mixture with a melting point lower than the melting point of each individual component [45]. These eutectic solvents are soluble in water when they are made up of hydrophilic components, resulting in a homogeneous extraction system that guarantees the highest contact surface area between the phases. To convert this system into a two-phase system characteristic of LLE, an aprotic solvent is added to induce phase separation.

The combination of DLLME with the MSPD process was first presented by Yan et al. [46]. It is also worth emphasizing that these authors were the first to attempt to use molecularly imprinted microspheres (MIM) as an MSPD sorbent for the selective extraction and determination of Sudan dyes in egg yolks by the MIM-MSPD-DLLME-HPLC-UV method. MIM was synthesized by aqueous suspension polymerization using phenylamine-naphthol as a dummy template. In the developed method, briefly, 0.1 g portions of yolk were dispersed with 0.2 g MIM and eluted with 3 mL of acetone/acetic acid (95:5, v/v) which was used as the dispersing liquid in DLLME for further purification and enrichment of analytes prior to HPLC separation. The developed method combined the high selectivity of MIM, the excellent MSPD dispersion of complex solid samples and the high enrichment factor (over 18–20-fold) obtained by further purification using the DLLME technique.

Wang et al. [25] proposed combining diatomaceous earth with the surface-deposited IL (1-hexyl-3-methylimidazolium tetrafluoroborate [C_6MIM][BF_4]), used in amounts of 1.5 g and 0.12 mL, respectively, with HLLME for the analysis of illegal dyes (chrysoidins, safranins O, auramines O and rhodamines B) in spice samples by the IL-based MSPD-HLLME procedure. It should be clarified that [C_6MIM][BF_4]

acted as the extraction solvent. In turn, the ion-pairing agent, $[NH_4]PF_6$, was used in HLLME method for further purification of the extract. For this purpose, the target analytes were eluted with 6 mL of water. Then, 1.0 mL of 2.0 M of $[NH_4]PF_6$ was added to the eluate (the molar ratio of $[C_6MIM][BF_4]$ to $[NH_4]PF_6$ was 1:4) to facilitate the separation of the newly formed $[C_6MIM][PF_6]$ phase. Finally, acetonitrile was added to dilute the extract and an aliquot was analyzed by UPLC-UV with LOD of 6.7–26.8 μg/kg.

Another interesting and clever method combining the high selectivity of ILs deposited on the surface of the sorbent with excellent dispersion of the MSPD process of solid samples is the method of ionic liquid-matrix solid-phase dispersion-solvent flotation (IL-MSPD-SF) developed by Zhan et al. [26] for the determination of acetanilide herbicides in rice samples. The proposed method resulted in high recoveries (89.4–108.7%) with RSD < 7.1%. LOQs were in the range from 38.0 to 84.7 μg/kg.

A simple, economic, and eco-friendly method able to detect triazole fungicides in tomato samples using a DES-based MSPD extraction followed by liquid–liquid back-extraction was porposed by Gallo et al. [55]. The developed method enables the MSPD extraction with alumina as a dispersant sorbent by replacing the organic solvents with a DES as extraction solvent during the MSPD blend elution. Choline chloride-ethylene glycol in a molar ratio of 1:2 (n/n) and ethyl acetate was used as a deep eutectic-organic solvent. Back extraction of analytes from the DES solution into ethyl acetate allows sample concentration overcoming the limited DES low vapor pressure, improving the method sensitivity. The LOQ was in the range of 5–11 ng/g and recoveries varied from 61 to 116%.

4 Main Application

By providing a complete fractionation of the sample matrix components and the ability to selectively elution of compounds from the sample, MSPD is widely used to isolate a variety of endogenous and exogenous compounds from solid, semisolid or viscous materials. A brief overview of MSPD applications in various analytical areas is presented in the aforementioned Table 1. A broader insight into the possibilities and contemporary applications of MSPD, along with a statistical analysis of the results of searching for information in available databases (Scopus, Web of Science) can be found in the review articles [6–10]. An updated view of the percentage use of MSPD by research areas, reflecting the thematic specificity of journals publishing the articles devoted to MSPD, together with the involvement of MSPD in the preparation of different samples types is presented in Fig. 3 a and b, respectively. This comprehensive search was conducted using keywords ("matrix solid-phase dispersion", "matrix solid-phase disruption", "MSPD", "SSDM" and "plants", "foods", "animals", "environmental samples", "fishes", tissues", "soils", "pharmaceuticals', and "cosmetics". The search was limited to the English language. In addition, the abstracts were pre-screened before studying the whole documents.

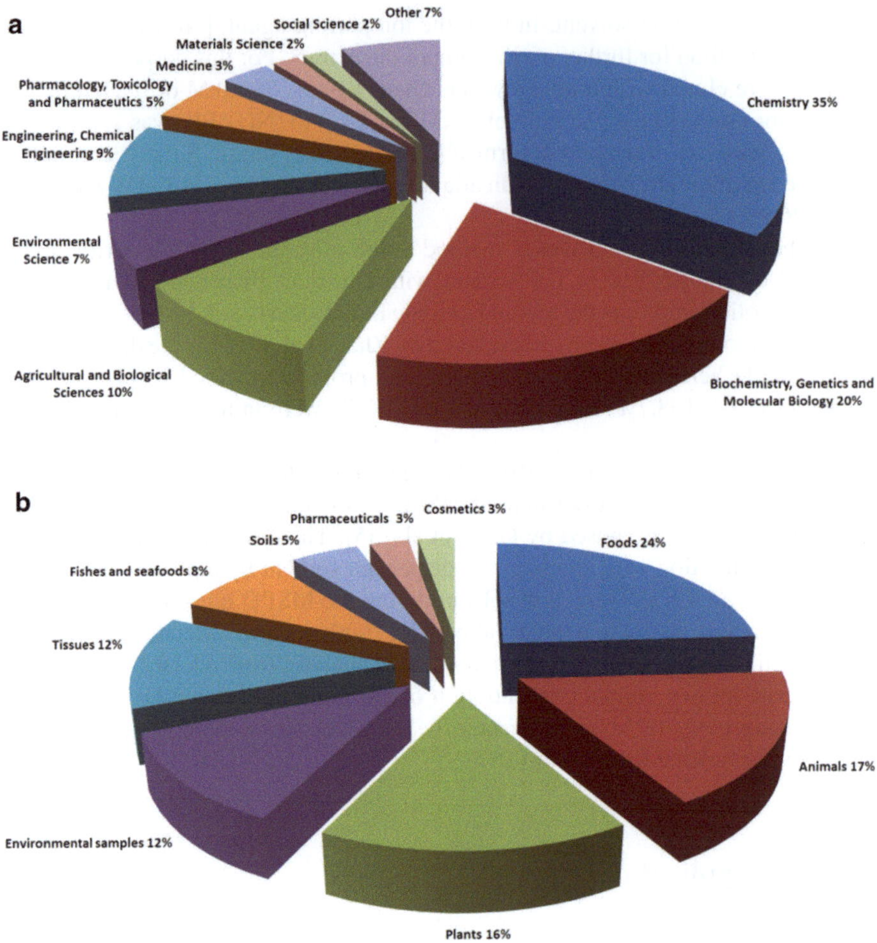

Fig. 3 Scopus search results depicting the percentage use of MSPD by research areas, reflecting the thematic specificity of journals publishing the manuscripts devoted to MSPD (**a**), together with the involvement of MSPD in the preparation of different samples types (**b**)

These data show that, apart from the typical field of MSPD application in chemical analysis of compounds, accounting for 35% of all applications, MSPD is widely used in the area of biochemical and molecular biology analyses (20% of applications). In addition, this method is also applied in the field of agricultural and environmental sciences, engineering, pharmacology, toxicology, pharmaceutics, and medicine. The percentage of MSPD use for the preparation of different sample types shown in Fig. 3b proves that the main area of MSPD applications relates to the preparation of processed food samples. Further places in the frequency of using this method are taken by the preparation of pre-unprocessed animal and plant tissue samples.

Undoubtedly, the method is the least frequently used for the preparation of pharmaceutical and cosmetics samples [31, 34, 36, 56]. The use of MSPD for the preparation of environmental samples for pollutant analysis occupies an intermediate place. However, it should be emphasized that the analysis of compounds whose presence is in admissible or whose concentrations are limited to very low levels, especially pesticides and drugs residues, is the main area of MSPD application in the analysis of food and animal tissues. In addition, numerous papers have recently been published on the use of MSPD for multi-residue screening analyses in plant and animal tissues [16, 42, 56–58]. Yet, in plants research, the use of MSPD to analyze the main and characteristic components of a given plant comes to the fore. One of the trends in the use of MSPD in plant research is the analysis of unstable compounds that either can be degraded/transformed using conventional extraction methods, especially those that are applied at elevated temperature, or are formed during these processes not being a result of cellular metabolism [11, 49, 50, 59]. There is also great interest in the use of the solvent-free MSPD method to analyze the composition of plant essential oils [13, 14].

The versatility and flexibility of adapting the MSPD technique to solving various analytical problems is presented below, highlighting two areas of application of this technique, i.e. in the analysis of secondary plant metabolites and the determination of hazardous (potentially hazardous) substances in various types of natural and artificial matrices.

4.1 Analysis of Secondary Plant Metabolites

The term "secondary plant metabolites" identifies those compounds that are not directly involved in the normal growth and development of the plant. As already noted, their analysis is one of the main application areas of MSPD [11, 13, 14, 19, 23, 24, 33, 40, 49, 50, 59–64]. This is understandable, because the importance of these compounds for human health and many industries attracts the attention of researchers from many different fields. Nevertheless, the popularity of this topic is also supported by the number of currently known secondary plant metabolites (over 50,000). Among them, the group of polyphenolic compounds is the most numerous. These compounds are also the most widespread compounds found in nature. Due to their moderate polarity and large molecular sizes, they are routinely analyzed in the reversed phase LC systems. Therefore, the MSPD procedures typically use the C_{18}-bounded silica sorbent and water-organic mixtures. Non-modified silica or Florisil, i.e. sorbents characteristic of the NP systems, are much less frequently used [40, 61, 62]. Among the new selective sorbents, MIPs, molecular sieves, titanium dioxide nanoparticles, cyclodextrins, chitosan or silica-supported ILs are most often used in these applications [18, 24, 27, 30, 33, 35]. In another approach, consistent with the principles of green chemistry, sand is used instead of the MSPD sorbent in order to increase the effectiveness of plant tissue disruption while reducing the cost of analysis [11–14, 49, 50].

As stated, a typical MSPD extraction of secondary plant metabolites proceeds in the reversed phase system. The best example of possibilities of the application of the RP systems is the paper by Deng et al. [47] where a modified MSPD procedure for profiling of plant hormones from the gibberellins group in a single leaf is presented. The modification concerned not only the reduction of the sample quantity (<1 mg) but also the amount of sorbent used (2 mg) in a single analysis and the way in which the MSPD process was realized. In this approach the grinding step, extraction and purification were performed in one microcentrifuge tube without any sample transfer step, resulting in an obvious decrease in the sample loss and an increase in the sensitivity (LOD was established on the attomol level).

In the context of the use of MSPD in the analysis of plant constituents, it is worth paying attention to those applications that relate to the extraction of essential oils (EOs). Due to the MSPD conditions, i.e. grinding in an open mortar, the suitability for EOs extraction can be considered as less attractive due to the high volatility of essential oil components. In addition, the analysis of the EOs composition is not easy and its result depends on the method of their isolation as shown in [53, 54, 65]. The Dawidowicz's research team was one of the first to demonstrate the usefulness of the MSPD process in the chromatographic analysis of EOs composition [13, 14]. Similarly, comparing the total amount and composition of EOs from various species of herbs and needles of coniferous trees with the amounts obtained by the steam distillation, recognized as the standard method of obtaining EOs and one of the most effective extraction techniques, i.e. pressure liquid extraction (PLE), they demonstrated that the efficiency of the MSPD process is equivalent to that obtained by both of the above-mentioned methods. Thus, MSPD is suitable for the isolation of these compounds, even if the C_{18} sorbent is replaced with sand, as mentioned earlier. In addition, they found that the MSPD method provides the most representative profile of all essential oil components because no heat is applied. Therefore, this environmentally friendly method was proposed by them as the main extraction procedure for the differentiation of essential oil components in plants for scientific and industrial purposes. In [66], the researchers proposed a different method of extracting essential oils using the MSPD technique with the solventless blending step, making the process even more environmentally friendly. The results presented in the cited article showed that when using the C_{18} sorbent in the MSPD process of volatile compounds, the use of a solvent at the grinding stage (the so-called dispersing liquid) is redundant, because the sorption capacity of the octadecyl brush is sufficient for the quantitative retention of isolated compounds. By studying various plants, the authors proved that the proposed method does not depend on the composition of essential oils and the volatility of individual components of the mixture under study. Then they showed that the extraction efficiency of the simplified MSPD method is equivalent to the conventional MSPD method and the PLE technique, which is a much more complex and technically advanced method of extracting plant components.

Concluding the review of the MSPD application in the analysis of secondary plant metabolites, it is worth bringing the previously mentioned topic of using sand in the MSPD procedure. Wianowska [12] successfully used it in the SSDM procedure for the analysis of quercetin in onions, apples and tea, revealing that only in the case of onions the SSDM results are not comparable with those obtained using PLE. This discrepancy became the basis for further studies, which showed that the instability of quercetin glycosidic derivatives under the PLE conditions and their degradation to quercetin aglycone is responsible for the overestimation of the amount of quercetin in onions by the PLE technique [5, 51]. The conclusion about the instability of phenolic compounds under the high-temperature extraction conditions leading to an overestimation of the extraction efficiency of some compounds, and thus to their incorrect quantitative estimation in plant materials, was independently confirmed by the results presented in [11, 49, 50, 67]. Two variants of the SSDM procedure were used in these experiments, with and without dispersant liquid, and both variants revealed comparable amounts of compounds. On their basis, a general conclusion was drawn that SSDM does not cause any transformations and/or degradation processes of secondary metabolites. Thus, the use of SSDM/MSPD in plant analysis not only allows to determine the actual concentration of individual compounds in plants, but also to determine which derivatives 'are native plant components and what is their concentration level. Moreover using the MSPD process carried out under the conditions where the relationship between the reciprocal of the analyte efficiency and the mass ratio of the sorbent to the plant is linear, it is possible to estimate the actual content of a compound in a plant sample [63]. To appreciate the importance of this simple method of assessing the actual content of a compound, it should be added that there are few materials certified for the content of organic compounds, and hundreds of thousands of different organic compounds are known.

4.2 *Analysis of Hazardous Substances*

Apart from the analysis of plant components, the second important area of application of the MSPD technique is the analysis of hazardous substances, not only exogenous but also endogenous. The role of MSPD in this research area cannot be overlooked, if only for the reason that the MSPD technique was introduced to facilitate this type of analysis.

One of the priority groups of hazardous substances that are increasingly appearing in various elements of the environment all over the world, and especially in its bloodstream, i.e. the aquatic environment, is a group of organic pollutants known as endocrine disrupting chemicals (EDCs). These substances are not only ubiquitous but also permanent. These include many families of compounds that can cause disorders in the human endocrine system even at low doses. As a result, they are toxic and raise concerns about the potential negative effects not only on humans but also on wildlife, especially as they undergo bioaccumulation. The threats posed by these pollutants make it necessary to constantly monitor them. Therefore, new methods of their sensitive and selective analysis are being developed. An example of one such method using the MSPD technique is provided in the article [68], which presents the MSPD procedure in tandem with LC-MS/MS for the simultaneous analysis for

45 contaminants, including antibiotics, non-steroidal anti-inflammatory drugs, β-blockers, antidepressants, antimicrobials and preservatives in sewage sludge. Vela-Soria et al. [69] developed an accurate, selective and sensitive MSPD-UHPLC-MS/MS method for the simultaneous determination of 10 EDCs, including parabens and benzophenone-UV filters, in human placental tissue samples with a LOQ in the range of 0.2–0.4 ng/g and a non-precision of 5.4–12.8%. The main advantage of both examples of the above-mentioned methods is the possibility of comprehensive determination of many compounds in complex matrices, and owing to the use of MSPD, sample preparation is easier and faster to perform compared to other commonly used methods.

Casado et al. [70] developed a method combining MSPD sample preparation (using a mixture of sorbents, a strong cation exchanger and PSA) with LC-MS/MS for a more selective and sensitive analysis of azole antifungal drugs (absolute recoveries ranged from 70 to 118%, and the LOQs of the method ranged from 5 to 8 ng/g). These substances, apart from non-steroidal anti-inflammatory drugs, are another group of compounds recognized as emerging environmental pollutants as a result of their widespread use and relatively high stability during biological treatments in sewage treatment plants or during chlorination and disinfection of treated sewage with UV rays, leading to their ineffective removal. In addition, another manifestation of the toxic effects of these pharmaceuticals is their inhibitory effect on certain enzymes.

As mentioned, an important area of research is the analysis of drug residues in food. In [56], MSPD with UPLC-MS/MS was used to extract and determine the residues of 10 steroid hormones in food matrices, achieving a LOD of 10 ng/kg with recovery of hormones estimated for chicken, pork, beef and sausage ranging from 77 to 99% with RSDs less than 10%.

An example of substances prohibited for use in food products in any concentration due to their allergenic and/or asthmatic effects are synthetic dyes. Nevertheless, despite the ban, it happens that these substances are present in food products. The usefulness of MSPD in the analysis of Sudan dyes has already been cited earlier [46]. Similarly, the use of MSPD for the isolation of four artificial colorants from chili spice samples was mentioned [25]. As the tested dyes are water-soluble polar compounds, they were eluted from the MSPD blend using water instead of an organic solvent, which allowed the authors of this procedure to call the method an organic solvent-free MSPD procedure. In turn, in [71], a miniaturized version of the MSPD was developed for the rapid and simultaneous determination of nine regulated water-soluble dyes in personal care and decorative products using the Florisil sorbent.

Analysis of agrochemicals such as herbicides, insecticides, pesticides and fungicides is another extremely important research area with significant MSPD activity. For years, these substances have been commonly and often, unfortunately, incorrectly used to protect crops. The consequence is a negative impact on the entire ecosystem. These compounds get into the food chain, contaminate soil and surface waters. The knowledge of the risks resulting from the excessive use of agrochemicals has led to changes in the applicable maximum residue levels (MRLs) and forced a revision of the analytical procedures used so far. Nevertheless, because the skillful use of these compounds increases the quantity and quality of crops, new plant protection products are still being developed. In [72], a simple method for the determination of metrafenone is presented, which is a new type of fungicide with the MRL level established in the European Union from 0.01 to 2.00 mg/kg. The key for clean-up of the proposed method was the use of an appropriate combination of dispersant and elution solvent, respectively a 1:1 mixture of alumina with silica gel and dichloromethane. As a result, the LOD was obtained at the level of 2 μg/kg with recovery >96% and RSD < 10%. In [73], the MSPD procedure with Florisil was proposed for the analysis of one of the most mobile, both in the aquatic and terrestrial environments, herbicides characterized by high toxicity to aquatic plants, i.e. penoxsulam. In the cited paper, the results obtained by the MSPD technique were compared with those obtained by classical LLE, showing their convergence, although considering the amount of time and effort involved, the authors found the MSPD extraction to be superior. Another interesting example of using MSPD in the analysis of hazardous compounds is the work developed by Medina-Dzul et al. [48] describing the MSPD-GC/MS analytical procedure for the simultaneous isolation and quantification of organophosphorus compounds in beeswax.

Summing up the review of the newly developed MSPD extraction methods, it is worth emphasizing its applicability in the analysis of microcystins (MCs) causing the worldwide problem of "blooms". MCs are a class of toxins produced by some freshwater cyanobacteria, mainly *Microcystis aeruginosa*, which are chemically covalently bound cyclic heptapeptides. More than 90 different MCs are known. Of these, the most common, toxic and the most studied isomeric form is MC marked as MC-LR. Qian et al. [60] developed a selective and sensitive method for the determination of MC-LR against other common MCs in vegetables, based on MSPD with a mixture of graphitized carbon black (GCB) and PSA, followed by HPLC-MS, setting the limits of detection of the proposed method at 13.0 μg/kg.

In conclusion, the versatility and flexibility of the MSPD process makes the technique applicable in the isolation of a wide variety of different compounds from hard or sticky and waxy matrices. However, the chemical nature of the typical analytes and thus the eluting solvents used in MSPD suggests that the technique is mainly used for the preparation of samples analyzed by liquid chromatography. Nevertheless, due to the availability of new selective sorbents, MSPD is increasingly used to prepare samples analyzed by gas chromatography [13, 14, 40, 41, 58, 64, 66]. MSPD's routine application area includes study of plants [11, 13, 14, 19, 23, 24, 33, 40, 49, 50, 59–64], meat [34, 37, 38, 56, 57, 69, 74], soils [73, 75], juices [19, 42], milk [58], oils [35] and breast milk [31, 36, 76]. New application of the MSPD method in very specific areas of analysis including forensic research are the examination of cosmetics [71], sediment [70, 72], human placenta [69, 74], molasses [77], bee products [27, 48] and even human hair [78].

5 Conclusions and Future Trends

One of the main challenges facing the analyst is to guarantee a sensitive and selective analysis of volatile and non-volatile compounds present at low concentrations levels in small amounts of complex matrices of solid, semi-solid and liquid samples. The access to modern extremely sensitive and selective chromatographic systems means that the responsibility for the quality of the analysis results rests at the sample preparation stage. At this stage, various more or less advanced extraction techniques are most often used. In the face of the growing need to reduce the cost of analyses and make them more environmentally friendly, the MSPD method enjoys the interest and attention of researchers from various fields of analytics.

The MSPD method is extremely simple, therefore it is fast and, what is important, cheap. It does not require special equipment and can be performed by anyone and anywhere. It allows you to carry out the entire sample preparation procedure in one stage of homogenization, extraction, purification and concentration, significantly reducing the time of sample preparation. Its effectiveness in isolating compounds is at least comparable to that of more sophisticated extraction techniques and therefore represents a simple and cheap alternative to them. However, by eliminating the need for high temperature to increase the efficiency of the extraction process, the MSPD reveals unique benefits and potential in the analysis of unstable compounds.

This chapter summarizes the recent achievements of the MSPD. These include the development of new adsorbents and compatible pro-environmental elution liquids as well as new more ecological and miniaturized procedures. These procedures allow to meet a variety of analytical challenges, which is a characteristic feature distinguishing the MSPD technique from other currently available sample preparation techniques. Since there is still room for improvement, this chapter identifies new tools to improve MSPD performance and selectivity by combining it with other extraction techniques, especially microextraction, and to further reduce sample preparation costs and organic solvent consumption by replacing the sorbent by sand and dissemination of the use of the solvent-free method even in the analysis of volatile compounds.

The authors have declared no conflict of interest.

References

1. Wianowska D, Gil M, Olszowy M (2020) Miniaturized methods of sample preparation. In: Hussain Ch (ed) Handbook on miniaturization in analytical chemistry: application of nanotechnology. Elsevier, pp 99–125. https://doi.org/10.1016/b978-0-12-819763-9.00005-2
2. Martín-Pozo L, de Alarcón-Gómez B, Rodríguez-Gómez R, García-Córcoles MT, Çipa M, Zafra-Gómez A (2019) Analytical methods for the determination of emerging contaminants in sewage sludge samples. A review. Talanta 192:508–533. https://doi.org/10.1016/J.TALANTA.2018.09.056

3. Wianowska D, Gil M (2019) Recent advances in extraction and analysis procedures of natural chlorogenic acids. Phytochem Rev 18:273–302. https://doi.org/10.1007/s11101-018-9592-y
4. Hoff RB, Pizzolato TM (2018) Combining extraction and purification steps in sample preparation for environmental matrices: a review of matrix solid phase dispersion (MSPD) and pressurized liquid extraction (PLE) applications. TrAC—Trends Anal Chem 109:83–96. https://doi.org/10.1016/J.TRAC.2018.10.002
5. Wianowska D, Gil M (2019) Critical approach to PLE technique application in the analysis of secondary metabolites in plants. TrAC—Trends Anal Chem 114:314–325. https://doi.org/10.1016/j.trac.2019.03.018
6. Shi M-Z, Yu Y-L, Zhu S-C, Yang J, Cao J (2022) Latest development of matrix solid phase dispersion extraction and microextraction for natural products from 2015–2021. Sep Purif Rev. https://doi.org/10.1080/15422119.2022.2094274
7. Tu H, Chen W (2018) A review on the recent progress in matrix solid phase dispersion. Molecules 23(11):2767. https://doi.org/10.3390/molecules23112767
8. Capriotti AL, Cavaliere C, Foglia P, Samperi R, Stampachiacchiere S, Ventura S, Laganà A (2015) Recent advances and developments in matrix solid-phase dispersion. TrAC—Trends Anal Chem 71:186–193. https://doi.org/10.1016/j.trac.2015.03.012
9. Wianowska D, Gil M (2019) New insights into the application of MSPD in various fields of analytical chemistry. TrAC—Trends Anal Chem 112:29–51. https://doi.org/10.1016/j.trac.2018.12.028
10. El-Deen AK (2023) An overview of recent advances and applications of matrix solid-phase dispersion. Sep Purif Rev. https://doi.org/10.1080/15422119.2023.2172734
11. Wianowska D, Typek R, Dawidowicz AL (2015) How to eliminate the formation of chlorogenic acids artefacts during plants analysis? Sea sand disruption method (SSDM) in the HPLC analysis of chlorogenic acids and their native derivatives in plants. Phytochemistry 117:489–499. https://doi.org/10.1016/j.phytochem.2015.07.006
12. Wianowska D (2015) Application of sea sand disruption method for HPLC determination of quercetin in plants. J Liq Chromatogr Relat Technol 38:1037–1043. https://doi.org/10.1080/10826076.2015.1012520
13. Dawidowicz AL, Czapczyńska NB, Wianowska D (2013) Relevance of the sea sand disruption method (ssdm) for the biometrical differentiation of the essential-oil composition from conifers. Chem Biodivers 10:241–250. https://doi.org/10.1002/cbdv.201200001
14. Dawidowicz AL, Wianowska D, Rado E (2011) Matrix solid-phase dispersion with sand in chromatographic analysis of essential oils in herbs. Phytochem Anal 22:51–58. https://doi.org/10.1002/pca.1250
15. Sowa I, Wójciak-Kosior M, Strzemski M, Sawicki J, Staniak M, Dresler S, Szwerc W, Mołdoch J, Latalski M (2018) Silica modified with polyaniline as a potential sorbent for matrix solid phase dispersion (MSPD) and dispersive solid phase extraction (d-SPE) of plant samples. Materials 11(4):467. https://doi.org/10.3390/ma11040467
16. Cao Y, Tang H, Chen D, Li L (2015) A novel method based on MSPD for simultaneous determination of 16 pesticide residues in tea by LC-MS/MS. J Chromatogr B Anal Technol Biomed Life Sci 998–999:72–79. https://doi.org/10.1016/j.jchromb.2015.06.013
17. Medina-Dzul K, Carrera-Figueiras C, Pérez-Padilla Y, Vilchis-Nestor RA, López-Téllez G, Sánchez M, Muñoz-Rodríguez D (2015) SiO_2/polyvinylimidazole hybrid polymer as a sorbent for extraction by matrix solid-phase dispersion (MSPD): synthesis, characterization, and evaluation. J Polym Res 22. https://doi.org/10.1007/s10965-015-0677-7
18. Hong Y, Chen L (2013) Extraction of quercetin from Herba Lysimachiae by molecularly imprinted-matrix solid phase dispersion. J Chromatogr B Anal Technol Biomed Life Sci 941:38–44. https://doi.org/10.1016/j.jchromb.2013.10.002
19. Yan H, Wang F, Wang H, Yang G (2012) Miniaturized molecularly imprinted matrix solid-phase dispersion coupled with high performance liquid chromatography for rapid determination of auxins in orange samples. J Chromatogr A 1256:1–8. https://doi.org/10.1016/j.chroma.2012.07.037

20. Chen L, Li B (2012) Determination of imidacloprid in rice by molecularly imprinted-matrix solid-phase dispersion with liquid chromatography tandem mass spectrometry. J Chromatogr B Anal Technol Biomed Life Sci 897:32–36. https://doi.org/10.1016/j.jchromb.2012.04.004
21. Zhou M, Hu N, Shu S, Wang M (2015) Molecularly imprinted nanomicrospheres as matrix solid-phase dispersant combined with gas chromatography for determination of four phosphorothioate pesticides in carrot and yacon. J Anal Methods Chem 2015. https://doi.org/10.1155/2015/385167
22. Zhang R, Xu N, Wang Y, Liu X, Wang S, Cao J (2020) Metal–organic framework assisted matrix solid-phase dispersion microextraction of saponins using response surface methodology. Electrophoresis 41(15):1354–1363. https://doi.org/10.1002/elps.202000042
23. Cao W, Hu SS, Ye LH, Cao J, Pang XQ, Xu JJ (2016) Trace matrix solid phase dispersion using a molecular sieve as the sorbent for the determination of flavonoids in fruit peels by ultra-performance liquid chromatography. Food Chem 190:474–480. https://doi.org/10.1016/j.foodchem.2015.05.133
24. Chu C, Wei M, Wang S, Zheng L, He Z, Cao J, Yan J (2017) Micro-matrix solid-phase dispersion coupled with MEEKC for quantitative analysis of lignans in Schisandrae Chinensis Fructus using molecular sieve TS-1 as a sorbent. J Chromatogr B Anal Technol Biomed Life Sci 1063:174–179. https://doi.org/10.1016/j.jchromb.2017.08.024
25. Wang Z, Zhang L, Li N, Lei L, Shao M, Yang X, Song Y, Yu A, Zhang H, Qiu F (2014) Ionic liquid-based matrix solid-phase dispersion coupled with homogeneous liquid-liquid microextraction of synthetic dyes in condiments. J Chromatogr A 1348:52–62. https://doi.org/10.1016/j.chroma.2014.04.086
26. Zhang L, Wang C, Li Z, Zhao C, Zhang H, Zhang D (2018) Extraction of acetanilides in rice using ionic liquid-based matrix solid phase dispersion-solvent flotation. Food Chem 245:1190–1195. https://doi.org/10.1016/j.foodchem.2017.11.029
27. Wang Z, Sun R, Wang Y, Li N, Lei L, Yang X, Yu A, Qiu F, Zhang H (2014) Determination of phenolic acids and flavonoids in raw propolis by silica-supported ionic liquid-based matrix solid phase dispersion extraction high performance liquid chromatography-diode array detection. J Chromatogr B Anal Technol Biomed Life Sci 969:205–212. https://doi.org/10.1016/j.jchromb.2014.08.022
28. Du K, Li J, Bai Y, An M, mei Gao X, xu Chang Y (2018) A green ionic liquid-based vortex-forced MSPD method for the simultaneous determination of 5-HMF and iridoid glycosides from Fructus Corni by ultra-high performance liquid chromatography. Food Chem 244:190–196. https://doi.org/10.1016/j.foodchem.2017.10.057
29. Gutiérrez-Solís MC, Muñoz-Rodríguez D, Medina-Peralta S, Carrera-Figueiras C, Ávila-Ortega A (2013) Matrix solid-phase dispersion extraction of organophosphorus pesticide using SiO_2-poly(N-vinylimidazole). IOP Conf Ser Mater Sci Eng 45:5–9. https://doi.org/10.1088/1757-899X/45/1/012022
30. Xu JJ, Cao J, Peng LQ, Cao W, Zhu QZ, Zhang QY (2016) Characterization and determination of isomers in plants using trace matrix solid phase dispersion via ultrahigh performance liquid chromatography coupled with an ultraviolet detector and quadrupole time-of-flight tandem mass spectrometry. J Chromatogr A 1436:64–72. https://doi.org/10.1016/j.chroma.2016.01.046
31. León-González ME, Rosales-Conrado N (2017) Determination of ibuprofen enantiomers in breast milk using vortex-assisted matrix solid-phase dispersion and direct chiral liquid chromatography. J Chromatogr A 1514:88–94. https://doi.org/10.1016/j.chroma.2017.07.072
32. Cao W, Hu S-S, Ye L-H, Cao J, Xu J-J, Pang X-Q (2015) Trace-chitosan-wrapped multi-walled carbon nanotubes as a new sorbent in dispersive micro solid-phase extraction to determine phenolic compounds. J Chromatogr A 1390:13–21. https://doi.org/10.1016/j.chroma.2015.02.060
33. Peng LQ, Li Q, xu Chang Y, An M, Yang R, Tan Z, Hao J, Cao J, Xu J, Hu SS (2016) Determination of natural phenols in olive fruits by chitosan assisted matrix solid-phase dispersion microextraction and ultrahigh performance liquid chromatography with quadrupole time-of-flight tandem mass spectrometry. J Chromatogr A 1456:68–76. https://doi.org/10.1016/j.chroma.2016.06.011

34. Hertzog GI, Soares KL, Caldas SS, Primel EG (2015) Study of vortex-assisted MSPD and LC-MS/MS using alternative solid supports for pharmaceutical extraction from marketed fish. Anal Bioanal Chem 407:4793–4803. https://doi.org/10.1007/s00216-015-8685-3
35. Shen Q, Dong W, Yang M, Baibado JT, Wang Y, Alqouqa I, Cheung HY (2013) Lipidomic study of olive fruit and oil using TiO_2 nanoparticle based matrix solid-phase dispersion and MALDI-TOF/MS. Food Res Int 54:2054–2061. https://doi.org/10.1016/j.foodres.2013.10.001
36. Fotouhi M, Seidi S, Shanehsaz M, Naseri MT (2017) Magnetically assisted matrix solid phase dispersion for extraction of parabens from breast milks. J Chromatogr A 1504:17–26. https://doi.org/10.1016/j.chroma.2017.05.009
37. Escarrone ALV, Caldas SS, Soares BM, Martins SE, Primel EG, Maia Nery LE (2014) A vortex-assisted MSPD method for triclosan extraction from fish tissues with determination by LC-MS/MS. Anal Methods 6:8306–8313. https://doi.org/10.1039/c4ay01518e
38. Chen JM, Yang CC, Chung WH, Ding WH (2016) Vortex-homogenized matrix solid-phase dispersion coupled with gas chromatography-electron-capture negative-ion mass spectrometry to determine halogenated phenolic compounds in seafood. RSC Adv 6:96510–96517. https://doi.org/10.1039/c6ra20680h
39. Chen YH, Chang CY, Ding WH (2016) Vortex-homogenized matrix solid-phase dispersion for the extraction of short chain chlorinated paraffins from indoor dust samples. J Chromatogr A 1472:129–133. https://doi.org/10.1016/j.chroma.2016.10.048
40. Albero B, Sánchez-Brunete C, Miguel E, Tadeo JL (2017) Application of matrix solid-phase dispersion followed by GC-MS/MS to the analysis of emerging contaminants in vegetables. Food Chem 217:660–667. https://doi.org/10.1016/j.foodchem.2016.09.017
41. Aznar R, Albero B, Sánchez-Brunete C, Miguel E, Martín-Girela I, Tadeo JL (2017) Simultaneous determination of multiclass emerging contaminants in aquatic plants by ultrasound-assisted matrix solid-phase dispersion and GC-MS. Environ Sci Pollut Res 24:7911–7920. https://doi.org/10.1007/s11356-016-6327-8
42. Barfi B, Asghari A, Rajabi M, Barfi A, Saeidi I (2013) Simplified miniaturized ultrasound-assisted matrix solid phase dispersion extraction and high performance liquid chromatographic determination of seven flavonoids in citrus fruit juice and human fluid samples: hesperetin and naringenin as biomarkers. J Chromatogr A 1311:30–40. https://doi.org/10.1016/j.chroma.2013.08.078
43. Zhang P, Ding J, Hou J, Zhao L, Chen Y, Ding L (2017) Dynamic microwave assisted extraction coupled with matrix solid phase dispersion for the determination of chlorfenapyr and abamectin in rice by LC-MS/MS. Microchem J 133:404–411. https://doi.org/10.1016/j.microc.2017.04.006
44. Wianowska D (2022) Combination of sea sand disruption method and ion-pair solid-phase extraction for effective isolation and purification of chlorogenic acid from plants prior to the HPLC determination. Molecules 27(12):5601. https://doi.org/10.3390/molecules27175601
45. Lomba L, Ribate P, Sangüesa E, Concha J, Garralaga P, Errazquin D, García CB, Giner B (2021) Deep eutectic solvents: are they safe? Appl Sci 11(21):10061. https://doi.org/10.3390/app112110061
46. Yan H, J.Qiao J, Wang H, Yang G, Row KH (2011) Molecularly imprinted solid-phase extraction combined with ultrasound-assisted dispersive liquid–liquid microextraction for the determination of four Sudan dyes in sausage samples. Analyst 136:2629–2634. https://doi.org/10.1039/c0an00951b
47. Deng T, Wu D, Duan C, Yan X, Du Y, Zou J, Guan Y (2017) Spatial profiling of Gibberellins in a single leaf based on microscale matrix solid-phase dispersion and precolumn derivatization coupled with ultraperformance liquid chromatography-tandem mass spectrometry. Anal Chem 89:9537–9543. https://doi.org/10.1021/acs.analchem.7b02589
48. Medina-Dzul K, Medina-Peralta S, Carrera-Figueiras C, Sánchez M, Muñoz-Rodríguez D (2017) Matrix solid-phase dispersion extraction of organophosphorous pesticides from beeswax. Int J Environ Anal Chem 97:831–840. https://doi.org/10.1080/03067319.2017.1363195

49. Wianowska D, Dawidowicz AL, Bernacik K, Typek R (2017) Determining the true content of quercetin and its derivatives in plants employing SSDM and LC-MS analysis. Eur Food Res Technol 243:27–40. https://doi.org/10.1007/s00217-016-2719-8
50. Typek R, Dawidowicz AL, Wianowska D, Bernacik K, Stankevič M, Gil M (2019) Formation of aqueous and alcoholic adducts of curcumin during its extraction. Food Chem 276:101–109. https://doi.org/10.1016/j.foodchem.2018.10.006
51. Wianowska D, Dawidowicz AL (2016) Effect of water content in extraction mixture on the pressurized liquid extraction efficiency—stability of quercetin 4′-glucoside during extraction from onions. J AOAC Int 99:744–749. https://doi.org/10.5740/jaoacint.16-0019
52. Teixeira DM, Patão RF, Coelho AV, da Costa CT (2006) Comparison between sample disruption methods and solid–liquid extraction (SLE) to extract phenolic compounds from Ficus carica leaves. J Chromatogr A 1103:22–28. https://doi.org/10.1016/J.CHROMA.2005.11.047
53. Wianowska D (2014) The influence of purge times on the yields of essential oil components extracted from plants by pressurized liquid extraction. J AOAC Int 97:1310–1316. https://doi.org/10.5740/jaoacint.13-318
54. Dawidowicz AL, Czapczyńska NB, Wianowska D (2012) The loss of essential oil components induced by the Purge Time in the Pressurized Liquid Extraction (PLE) procedure of Cupressus sempervirens. Talanta 94:140–145. https://doi.org/10.1016/j.talanta.2012.03.008
55. Gallo V, Della Posta S, Gentili A, Gherardi M, De Gara L, Fanali Ch (2023) Back-extraction applied to green matrix solid-phase dispersion for fungicides determination in tomatoes. Sep Sci Plus 2200140. https://doi.org/10.1002/sscp.202200140
56. Fan YB, Yin YM, Bin Jiang W, Chen YP, Yang JW, Wu J, Xie MX (2014) Simultaneous determination of ten steroid hormones in animal origin food by matrix solid-phase dispersion and liquid chromatography-electrospray tandem mass spectrometry. Food Chem 142:170–177. https://doi.org/10.1016/j.foodchem.2013.06.104
57. Li H, Sun N, Zhang J, Liang S, Sun H (2014) Development of a matrix solid phase dispersion-high performance liquid chromatography-tandem mass spectrometric method for multiresidue analysis of 25 synthetic colorants in meat products. Anal Methods 6:537–547. https://doi.org/10.1039/c3ay41664j
58. Cai H, Ji S, Zhang J, Shang G, Tao G, Peng C, Chen G, Hou R, Zhang L, Wan X (2017) Determination of 11 photoinitiators and their migration into tea and milk by gas chromatography-tandem mass spectrometry (MSPD-GC-MS/MS). Anal Methods 9:2957–2963. https://doi.org/10.1039/c7ay00156h
59. Wianowska D, Typek R, Dawidowicz AL (2015) Chlorogenic acid stability in pressurized liquid extraction conditions. J AOAC Int 98:415–421. https://doi.org/10.5740/jaoacint.14-200
60. Qian ZY, Li ZG, Ma J, ting Gong J, Xian QM (2017) Analysis of trace microcystins in vegetables using matrix solid-phase dispersion followed by high performance liquid chromatography triple-quadrupole mass spectrometry detection. Talanta 173:101–106. https://doi.org/10.1016/j.talanta.2017.05.079
61. Gao Y, Sun Y, Wang Y, Zhang J, Xu B, Zhang H, Song D (2013) A practical and rapid method for the simultaneous isolation, purification and quantification of geniposide from the fruit of Gardenia jasminoides Ellis by MSPD extraction and UFLC analysis. Anal Methods 5:4112–4118. https://doi.org/10.1039/c3ay40638e
62. Han Y, Wen J, Zhou T, Fan G (2015) Chemical fingerprinting of Gardenia jasminoides Ellis by HPLC-DAD-ESI-MS combined with chemometrics methods quality evaluation. Food Chem 188:648–657. https://doi.org/10.1016/j.foodchem.2015.05.039
63. Dawidowicz AL, Wianowska D (2009) Application of the MSPD technique for the HPLC analysis of rutin in Sambucus nigra L.: the linear correlation of the matrix solid-phase dispersion process. J Chromatogr Sci 47:914–918
64. Rashidipour M, Heydari R, Feizbakhsh A, Hashemi P (2015) Rapid monitoring of carvacrol in plants and herbal medicines using matrix solid-phase dispersion and gas chromatography flame ionisation detector. Nat Prod Res 29:621–627. https://doi.org/10.1080/14786419.2014.980247

65. Mardarowicz M, Wianowska D, Dawidowicz AL, Sawicki R (2004) Comparison of terpene composition in Engelmann Spruce (Picea engelmannii) using hydrodistillation, SPME and PLE. Zeitschrift Für Naturforsch C 59:641–648. https://doi.org/10.1515/znc-2004-9-1006
66. Wianowska D, Dawidowicz AL (2016) Can matrix solid phase dispersion (MSPD) be more simplified? Application of solventless MSPD sample preparation method for GC-MS and GC-FID analysis of plant essential oil components. Talanta 151:179–182. https://doi.org/10.1016/J.TALANTA.2016.01.019
67. Wianowska D (2014) Hydrolytical instability of hydroxyanthraquinone glycosides in pressurized liquid extraction. Anal Bioanal Chem 406:3219–3227. https://doi.org/10.1007/s00216-014-7744-5
68. Li M, Sun Q, Li Y, Lv M, Lin L, Wu Y, Ashfaq M, ping Yu C (2016) Simultaneous analysis of 45 pharmaceuticals and personal care products in sludge by matrix solid-phase dispersion and liquid chromatography tandem mass spectrometry. Anal Bioanal Chem 408:4953–4964. https://doi.org/10.1007/s00216-016-9590-0
69. Vela-Soria F, Rodríguez I, Ballesteros O, Zafra-G6mez A, Ballesteros L, Cela R, Naval6n A (2014) Simplified matrix solid phase dispersion procedure for the determination of parabens and benzophenone-ultraviolet filters in human placental tissue samples. J Chromatogr A 1371:39–47. https://doi.org/10.1016/j.chroma.2014.10.063
70. Casado J, Castro G, Rodríguez J, Ramil M, Cela R (2015) Selective extraction of antimycotic drugs from sludge samples using matrix solid-phase dispersion followed by on-line clean-up. Anal Bioanal Chem 407:907–917. https://doi.org/10.1007/s00216-014-8167-z
71. Guerra E, Celeiro M, Lamas JP, Llompart M, Garcia-Jares C (2015) Determination of dyes in cosmetic products by micro-matrix solid phase dispersion and liquid chromatography coupled to tandem mass spectrometry. J Chromatogr A 1415:27–37. https://doi.org/10.1016/j.chroma.2015.08.054
72. Li J, Li Y, Xu D, Zhang J, Wang Y, Luo C (2017) Determination of metrafenone in vegetables by matrix solid-phase dispersion and HPLC-UV method. Food Chem 214:77–81. https://doi.org/10.1016/j.foodchem.2016.07.061
73. Kaur P, Kaur K, Bhullar MS (2014) Quantification of penoxsulam in soil and rice samples by matrix solid phase extraction and liquid-liquid extraction followed by HPLC-UV method. Environ Monit Assess 186:7555–7563. https://doi.org/10.1007/s10661-014-3947-7
74. Villaverde-de-Sáa E, Rodil R, Quintana JB, Cela R (2016) Matrix solid-phase dispersion combined to liquid chromatography–tandem mass spectrometry for the determination of paraben preservatives in mollusks. J Chromatogr A 1459:57–66. https://doi.org/10.1016/j.chroma.2016.06.070
75. Iparraguirre A, Rodil R, Quintana JB, Bizkarguenaga E, Prieto A, Zuloaga O, Cela R, Fernández LA (2014) Matrix solid-phase dispersion of polybrominated diphenyl ethers and their hydroxylated and methoxylated analogues in lettuce, carrot and soil. J Chromatogr A 1360:57–65. https://doi.org/10.1016/j.chroma.2014.07.079
76. Samanidou VF, Frysali MA, Papadoyannis IN (2014) Matrix solid phase dispersion for the extraction of bisphenol a from human breast milk prior to HPLC analysis. J Liq Chromatogr Relat Technol 37:247–258. https://doi.org/10.1080/10826076.2012.745133
77. Kamal YT, Alam P, Alqasoumi SI, Foudah AI, Alqarni MH, Yusufoglu HS (2018) Investigation of antioxidant compounds in commercial pomegranate molasses products using matrix-solid phase dispersion extraction coupled with HPLC. Saudi Pharm J 26:839–844. https://doi.org/10.1016/j.jsps.2018.03.015
78. Argente-García A, Moliner-Martínez Y, Campíns-Falcó P, Verdú-Andrés J, Herráez-Hernández J (2016) Determination of amphetamines in hair by integrating sample disruption, clean-up and solid phase derivatization. J Chromatogr A 1447:47–56. https://doi.org/10.1016/j.chroma.2016.04.036

Solvent-Based Microextractions

Single-Drop Microextraction

Francisco Pena-Pereira, Inmaculada de la Calle, Vanesa Romero, Isela Lavilla, and Carlos Bendicho

Abstract Since its discovery in 1995 by Dasgupta et al., and further implementation by Jeannot et al. in 1996, single-drop microextraction (SDME) has undergone a dramatic increase, as demonstrated by the significant number of developments and the plethora of applications that have extended this technique to almost every area, being nowadays one of the most popular miniaturised extraction techniques. The diversity of analytes, possibilities of combining SDME to detection techniques, the continuous improvements in extractant phases and the simplicity of operation, account for the spread acceptance of SDME. In this chapter, the different approaches available under the concept of extraction in a drop, extractant phases, both conventional and novel ones, as well as couplings of SDME with different detectors will be addressed. Relevant applications of SDME will be provided emphasizing analytical characteristics such as detection limits, precision and enrichment factors. Apart from the well-established modes of SDME such as direct immersion single-drop microextraction (DI-SDME), directly suspended drop microextraction (DSDME), headspace single-drop microextraction (HS-SDME), continuous flow microextraction (CFME) and liquid–liquid-liquid microextraction (LLLME), novel sophisticated approaches have arisen in the last years, such as drop-to-drop solvent microextraction (DDSME) or bubble-in-drop microextraction (BID), which provide new avenues for the continuous improvement of this technique.

F. Pena-Pereira · I. de la Calle · V. Romero · I. Lavilla · C. Bendicho (✉)
Centro de Investigación Mariña, Departamento de Química Analítica y Alimentaria, Universidade de Vigo, Vigo QA2, 36310, Spain
e-mail: bendicho@uvigo.es

F. Pena-Pereira
e-mail: fjpena@uvigo.es

I. de la Calle
e-mail: incalle@uvigo.es

V. Romero
e-mail: vromero@uvigo.es

I. Lavilla
e-mail: isela@uvigo.es

M. Á. Rodríguez-Delgado et al. (eds.), *Microextraction Techniques*, Integrated Analytical Systems, https://doi.org/10.1007/978-3-031-50527-0_7

Keywords Single-drop microextraction · Liquid-phase microextraction · Continuous flow · Direct immersion · Directly suspended · Headspace

Abbreviations

AFS	Atomic fluorescence spectrometry
Ag@Au np	Silver-gold core–shell nanoprisms
APDC	Ammonium pyrrolidinedithiocarbamate
APIs	Active pharmaceutical ingredients
ATR-FTIR	Attenuated total reflectance-Fourier transform infrared spectroscopy
Au-np/TR	Gold nanoprisms/Tollen's reagent
BID	Bubble-in-drop
BPHA	N-benzoyl-N-phenylhydroxylamine
BTEXs	Benzene, toluene, ethylbenzene and xylenes
CCU-CF-SDME	Column clean-up continuous flow single-drop microextraction
CDs	Carbon dots
CE	Capillary electrophoresis
CE-MS	Capillary electrophoresis mass spectrometry
CE-UV	Capillary electrophoresis ultra-violet detector
CFME	Continuous flow microextraction
CV	Cold vapour
CZE	Capillary zone electrophoresis
DDSME	Drop-to-drop solvent microextraction
DES	Deep eutectic solvent
DI-SDME	Direct immersion single-drop microextraction
DMF	Dimethylformamide
DMSO	Dimethyl sulfoxide
DRS-FTIR	Diffuse reflectance-Fourier transform infrared spectroscopy
DSDME	Directly suspended droplet microextraction
DS-LLLME	Directly suspended liquid–liquid-liquid microextraction
ECD	Electron capture detector
EDXRF	Energy dispersive X-ray fluorescence
EF	Enrichment factor
ESI	Electrospray ionization
ETAAS	Electrothermal atomic absorption spectrometry
ETV	Electrothermal vaporisation
ETV-ICP-MS	Electrothermal vaporisation inductively coupled plasma mass spectrometry
FAAS	Flame atomic absorption spectrometry
FID	Flame ionization detector

GC	Gas chromatography
GC-ECD	Gas chromatography electron capture detector
GC-FID	Gas chromatography flame ionization detector
GC–MS/MS	Gas chromatography-tandem mass spectrometry
GC–MS	Gas chromatography mass spectrometry
GF	Graphite furnace
GF-AAS	Graphite furnace atomic absorption spectroscopy
HBA	Hydrogen bond acceptor
HBD	Hydrogen bond donor
HG	Hydride generation
HPLC	High performance liquid chromatography
HPLC–DAD	High performance liquid chromatography photodiode array detector
HPLC–UV	High performance liquid chromatography ultra-violet detector
HS-BID	Headspace bubble-in-drop
HS-SDME	Headspace single-drop microextraction
HS-SDME-SP	Headspace single-drop microextraction spectro-pipette
HTL	Homocysteine thiolactone
ICP-MS	Inductively coupled plasma mass spectrometry
ICP-OES	Inductively coupled plasma optical emission spectrometry
ILs	Ionic liquids
IMS	Ion-mobility spectrometry
IR	Infrared
LC	Liquid chromatography
LC–MS/MS	Liquid chromatography tandem mass spectrometry
LGL	Liquid–gas-liquid
LIBS	Laser-induced breakdown spectroscopy
LIS	Lab-in-syringe
LLE	Liquid–liquid extraction
LLL	Liquid–liquid-liquid
LLLME	Liquid–liquid-liquid microextraction
LLL-SDME	Liquid–liquid-liquid single-drop microextraction
LOD	Limit of detection
LPME	Liquid phase microextraction
MALDI	Matrix-assisted laser desorption/ionization
MALDI-MS	Matrix-assisted laser desorption/ionization mass spectrometry
MALDI-TOF-MS	Matrix-assisted laser desorption/ionization time-of-flight mass spectrometry
MBGs	Magnetic bucky gels
mCNTs	Magnetic-carbon nanotubes
MIL	Magnetic ionic liquid
MS	Mass spectrometry

MTME-SD	Magnetic tip microextraction to a single drop
MTP-SDME	Magnetic three phase single-drop microextraction
NCs	Nanoclusters
N-GQDs	Nitrogen-doped graphene quantum dots
NPs	Nanoparticles
NRs	Nanorods
NSAIDs	Non-steroidal anti-inflammatory drugs
NSRs	Nanostars
NTs/G-quadruplex/PPIX	Nanosheets G-quadruplex/protoporphyrin IX
PAHs	Polycyclic aromatic hydrocarbons
PAN	1-(2-Pyridylazo)-2-naphtol
Pa-SDME	Parallel single-drop microextraction
PBDEs	Polybrominated diphenyl ethers
PMAA	Poly(metacrylic acid)
pMBA	Para-mercaptobenzoic acid
PS-MS	Paper spray mass spectrometry
PTLM	Photothermal lens microscopy
PTV	Programmable temperature vaporisation
QDs	Quantum dots
qPCR	Quantitative polymerase chain reaction
REACH	Registration, Evaluation, Authorisation and Restriction of Chemicals
RSD	Relative standard deviation
RTGC	Room temperature gas chromatography
RTGC-IMS	Room temperature gas chromatography ion-mobility spectrometry
SDILNDμE	Single-drop ionic liquid non-dispersive microextraction
SDME	Single-drop microextraction
SDME-GC	Single-drop microextraction gas chromatography
SDME-LVSEP	Single-drop microextraction large-volume sample stacking electroosmotic flow pump
SDS	Sodium dodecyl sulphate
SENLIBS	Surface-enhanced laser-induced breakdown spectroscopy
SERS	Surface-enhanced Raman scattering
SFODME	Solidified floating organic drop microextraction
SI	Sequential injection
SUPRADES	Supramolecular deep eutectic solvent
SUPRAS	Supramolecular solvent
TOF	Time-of-fligth
TXRF	Total reflection X-ray fluorescence
UHPLC-MS/MS	Ultra-high performance liquid chromatography tandem mass spectrometry

1 Introduction

Sample preparation represents a key step of the analytical process in most analytical methodologies. Particularly, extraction techniques such as solid-phase extraction and solvent extraction have been typically employed for achieving the extraction and enrichment of target analytes, an efficient sample clean-up and/or to obtain an extract compatible with the analytical instrumentation. However, these conventional extraction techniques are not free from drawbacks, including the achievement of reduced enrichment factors, a large consumption of solvents and, thus, generation of wastes, apart from the tendency to form undesirable emulsions that impair the separation of phases in solvent extraction. The main limitations of these classical techniques led to the development of miniaturised extraction counterparts, commonly termed as microextraction techniques. Since their introduction in the mid-1990s as a consequence of very remarkable disruptive works [1–4], substantial efforts have been made for the development of convenient and complementary microextraction approaches. Thus, a wide range of single drop microextraction (SDME) modes can nowadays be selected for (virtually) solving any analytical problem.

The chapter provides an overview of the inception and evolution of SDME, highlighting the main achievements and applications of the technique, paying special attention to fundamental thermodynamic and kinetic aspects. The chapter focuses on liquid-phase microextraction (LPME) approaches involving microdrops of extractant phases in a nearly spherical configuration during the extraction process and, therefore, related LPME approaches that do not fulfil this criterion, such as dispersive liquid–liquid microextraction or LPME approaches involving supported liquid membranes, among others, can be found in Chaps. 8, 9 and 10 of this book. Recent review articles dealing with these contents can also be found elsewhere [5, 6].

2 Fundamentals

In this section, the most prominent SDME approaches are presented and theoretical aspects of two-phase and three-phase SDME systems are provided.

2.1 SDME Approaches

A number of SDME approaches, involving both two-phase and three-phase systems, have been reported in the literature. A schematic representation of SDME approaches described below is shown in Fig. 1.

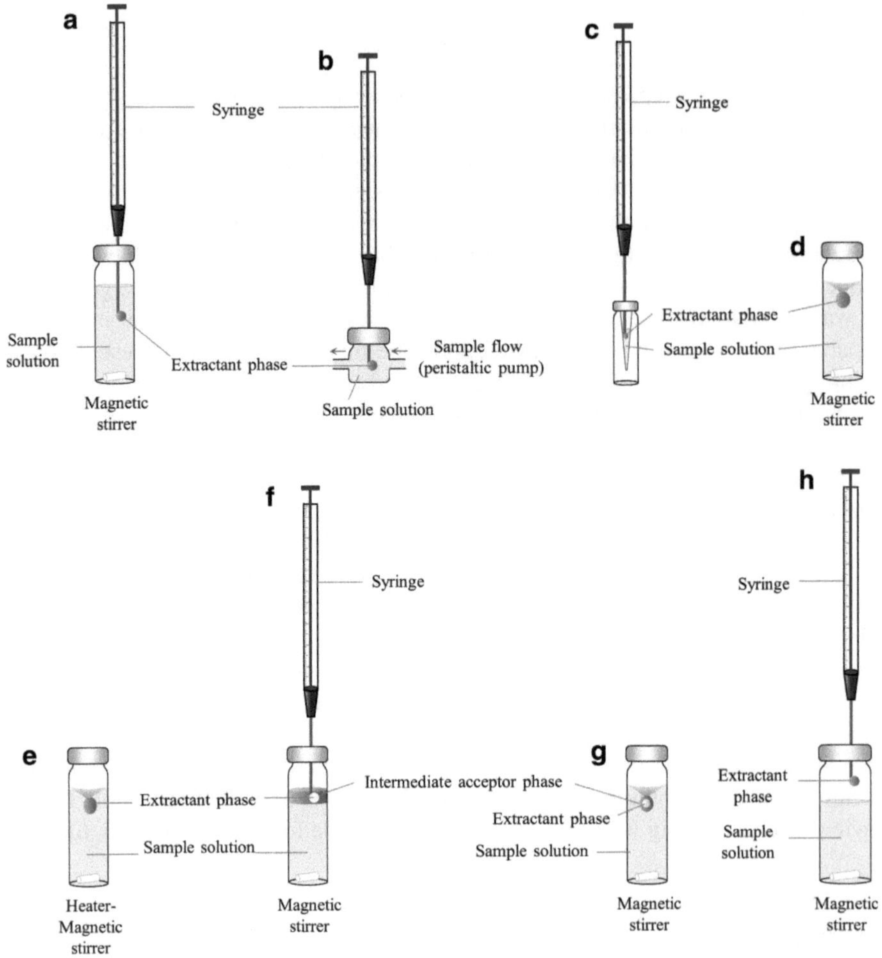

Fig. 1 Schematic representation of SDME approaches. DI-SDME **a**, CFME **b**, DDSDME **c**, DSDME **d**, SFODME **e**, LLLME **f**, DSLLLME **g** and HS-SDME **h**

2.1.1 Two-Phase SDME Approaches

Biphasic systems can be considered as purely miniaturised solvent extraction approaches, involving the direct exposure of a microdrop of immiscible extractant phase to an aqueous sample. Biphasic SDME has been mainly used with the aim of extracting target analytes (or analyte derivatives) displaying moderate to high hydrophobicity. The basic requirements of solvents to be used in two-phase SDME include immiscibility with the sample and highly reduced solubility on the aqueous phase. The first attempts to miniaturise the conventional solvent extraction reported on the exposure of a microdrop of immiscible organic solvent hanging from the tip of a capillary to an aqueous sample [4, 7], even though the use of a microsyringe was soon found to be highly convenient to facilitate the reproducible exposure of a microvolume of extractant phase, followed by the retraction and injection of the enriched extractant phase for analysis [8]. From these first studies, different biphasic SDME approaches have been reported in the literature. In direct (or immersed) SDME (DI-SDME, Fig. 1a), the microdrop of extractant phase is immersed in the sample (typically 1–2 cm below the surface of the aqueous sample solution) hanging from the tip of a microsyringe. Convection is commonly favoured to enhance the extraction kinetics, mainly by magnetic stirring or sample pumping. In the second case, the SDME mode is commonly termed as continuous flow microextraction (CFME, Fig. 1b) [9]. Furthermore, a highly miniaturised approach, particularly suitable for the enrichment of target compounds present in limited sample volumes (*e.g.*, clinical samples) named as drop-to-drop solvent microextraction (DDSME, Fig. 1c) has been reported [10], in analogy to a previous approach described by Dasgupta et al. in the mid-90s [7]. In spite of the convenience of using a microsyringe to integrate unitary steps, the performance of these biphasic SDME approaches is severely limited by the instability of the microdrop under certain experimental conditions. Particularly, extended extraction times, high agitation or relatively high temperatures, which generally lead to enhanced extraction, affected the balance of forces on the microdrop, leading to drop dislodgement. Thus, alternative biphasic SDME approaches that do not make use of the syringe as a holder but to collect the enriched microdrop at the end of the process have been reported to overcome the above limitations, namely directly suspended droplet microextraction (DSDME) and solidification floating organic drop microextraction (SFODME). In DSDME (Fig. 1d), a microvolume of an immiscible extractant phase showing lower density than water is directly injected at the top of a highly stirred aqueous sample, thus forming a self-stable drop that maintains a nearly spherical configuration during the extraction process. The fact that the extractant phase is freely suspended in the sample allows the use of very high stirring rates, which favours the extraction kinetics. On the other hand, however, the collection of the enriched microdrop at the end of the process becomes more complicated with regard to SDME modes that make use of the syringe as a holder of the microdrop. Different options have been considered to facilitate the collection stage immediately prior to analysis, including the design of collection devices with narrow-neck tubes [11] or the use of syringes for improved extraction with subsequent collection of the enriched acceptor phase at its tip [12]. In addition, the introduction of an analogous biphasic SDME approach, namely SFODME (Fig. 1e) [13], has also simplified the collection step. SFODME exploits the physicochemical properties of certain solvents used as extractant phases to induce their physical separation after the enrichment process. In particular, apart from being immiscible with water and showing lower density that water, the solvents used as extractant phases in SFODME must fulfil the additional requirement of showing a melting point close to room temperature (*ca.* 10–30 °C). Thus, the process of analyte extraction from the sample solution is carried out in an analogous way to DSDME but at a temperature higher than the melting point of the extractant phase to maintain it in its liquid phase during the extraction process (commonly requiring a thermostatic bath). The collection of the enriched drop is performed at lower temperature than the melting point by locating the vial in an ice bath, that leads to the solidification of the drop, thus facilitating the collection of the enriched solid-microdrop by using a spatula or tweezers. Then, the enriched solvent melts immediately before analysis. The numerous requirements that must be met by potential extractant phases significantly limit their choice to a reduced set of solvents whose properties might not comply with the principle of "like dissolves like".

2.1.2 Three-Phase SDME Approaches

Three-phase systems have been reported for enhanced selectivity. Thus, liquid–liquid-liquid microextraction systems (LLLME) and headspace SDME (HS-SDME) have been reported in the literature for the enrichment of ionisable and volatile (or semivolatile) compounds, respectively. These three-phase SDME approaches involve liquid–liquid-liquid and liquid–gas-liquid mass transfer processes, respectively.

In LLLME, a reduced volume of an immiscible solvent showing lower density than water is layered over a stirred aqueous sample and, in turn, a microdrop of extractant phase (by default of aqueous nature) is immersed into the intermediate acceptor phase. The exposure of the extractant phase can be carried out by using a syringe as holder during the microextraction process (Fig. 1f) or by delivering the extractant phase microdrop in the intermediate acceptor phase (Fig. 1g), in the so called directly suspended three phase liquid phase microextraction (DS-LLLME). In the latter case, the syringe is not used as a holder during the extraction process but to collect the enriched microdrop for analysis. Simultaneous extraction (from the sample to the intermediate acceptor phase, equivalent to DI-SDME) and back-extraction (from the intermediate acceptor phase to the extractant phase microdrop) of target analytes occur in LLLME and DS-LLLME. Careful adjustment of the pH of both the aqueous sample and the extractant phase bearing in mind the pKa values of target analytes is mandatory for achieving an efficient extraction of target analytes. In particular, ionisable compounds must be in their neutral form to be extracted by the intermediate organic layer, whereas ionisation of the analyte must occur at the interface organic layer-aqueous receiving phase to favour the back-extraction into the aqueous extractant phase. By way of example, the LLLME of amines present in an aqueous sample required the pH adjustment of the sample solution and the aqueous microdrop at alkaline and acidic values, respectively [14]. Furthermore, the use of chelating agents in the extractant phase has expanded the applicability of LLLME for the extraction of metal ions and organometallic compounds [15, 16].

An alternative three-phase SDME mode, named as HS-SDME, is widely used for the enrichment of volatile compounds present in condensed samples [17]. HS-SDME is based on the extraction of analytes (or analyte derivatives) by a microdrop of extractant phase exposed to the headspace above a sample, in a closed vial (Fig. 1h). Volatile and semi-volatile compounds are thus transferred from the sample solution to the gaseous phase above it and subsequently extracted by the extractant phase microdrop when the "like dissolves like" principle is met. Apart from showing appropriate partition coefficients to ensure an efficient retention of target molecules, the extractant phases should show a reduced vapour pressure and relatively high boiling point. Obviously, the immiscibility of the solvent with the sample is not relevant in HS-SDME since these phases are not in contact with each other during the extraction process. Magnetic agitation is typically employed for efficiently transfer the volatile to the gaseous phase, even though increased temperatures and/or increase of the ionic strength can also be required to improve the process.

In comparison with biphasic systems, the selectivity of three-phase SDME approaches is highly increased bearing in mind that the extractant phase is physically separated (by an intermediate immiscible liquid phase or a gaseous phase) from the sample solution. Thus, potential interferences associated with certain compounds present in the sample can be avoided or minimised. For instance, the extraction of non-ionisable compounds by the extractant phase microdrop is avoided in liquid–liquid-liquid (LLL) approaches, even if they are extracted by the intermediate solvent phase. Analogously, non-volatile compounds present in aqueous samples cannot be transferred to the headspace and therefore their extraction by a microdrop exposed to the gaseous phase is not produced.

2.2 *Theoretical Aspects of SDME*

2.2.1 Thermodynamics of SDME

In SDME modes involving two phases, *i.e.*, aqueous sample and microdrop of extractant phase, the distribution constant, K_{ds}, can be expressed as the ratio of activities of the analytes in the drop of extractant phase and the sample. Bearing in mind that analytes are commonly present at trace and ultratrace levels, activities can be approximated by the corresponding concentrations:

$$K_{ds} = \frac{a_d}{a_s} \approx \frac{C_d}{C_s} \tag{1}$$

where a_d and a_s are the activities of the analytes in the extractant drop phase and sample, respectively, and C_d and C_s are the corresponding concentrations of the analytes in the extractant drop phase and sample, respectively.

Under equilibrium conditions, the amount of analyte present in the extractant phase, n_d, can be expressed as shown in Eq. (2), regardless the number of phases involved in the SDME process:

$$n_d = \frac{K_{ds} V_d C_0 V_s}{K_{ds} V_d + K_{hs} V_h + V_s} \tag{2}$$

where C_0 is the initial concentration of the analyte in the sample, V_s, V_h and V_d are the sample, headspace and drop volumes, respectively, and K_{ds} and K_{hs} are the drop/sample and headspace/sample distribution constants, respectively. It should be noted that the term $K_{hs}V_h$ present in the denominator of Eq. (2) is omitted when two-phase SDME modes are used. It is worth mentioning that the Eq. (2) can be simplified as shown in Eq. (3) when $V_s \gg V_d$ (and $V_s \gg V_h$, if applicable):

$$n_d = K_{ds}V_dC_0 \tag{3}$$

According to this expression, the amount of analyte extracted is independent on the sample volume when the above conditions are fulfilled. This aspect is of particular importance since it would (ideally) allow integrating sampling and sample preparation without significant modifications on the extracted amount of analytes when very large sample volumes are involved.

The expression for the concentration of the analyte in the extractant phase under equilibrium conditions in two-phase SDME approaches, in which a single equilibrium is involved, can be directly deduced from Eq. (2), being equivalent to the one valid for conventional liquid–liquid extraction (LLE):

$$C_d = \frac{K_{ds}C_0}{1 + K_{ds}\left(\frac{V_d}{V_s}\right)} \tag{4}$$

Two additional expressions can be deduced for the concentration of the analyte in the extractant phase under equilibrium conditions of three-phase SDME approaches. Two equilibria are involved in LLL and liquid–gas-liquid (LGL) microextraction processes, respectively.

In LLLME approaches, the concentration of the analyte in the extractant phase under equilibrium conditions can be expressed as:

$$C_d = \frac{K_{ds}C_0}{1 + K_{as}\left(\frac{V_a}{V_s}\right) + K_{ds}\left(\frac{V_d}{V_s}\right)} \tag{5}$$

where K_{as} is the intermediate acceptor phase/sample distribution constant and V_a is the corresponding volume.

Analogously, in the headspace mode, the concentration of the analyte in the extractant phase under equilibrium conditions can be estimated as:

$$C_d = \frac{K_{ds}C_0}{1 + K_{hs}\left(\frac{V_h}{V_s}\right) + K_{ds}\left(\frac{V_d}{V_s}\right)} \tag{6}$$

2.2.2 Kinetics of SDME

Two-Phase SDME Approaches

It has been experimentally verified that a first-order kinetic model fits well with the experimental data of concentration of the analyte in the extractant phase *vs* time, t, with both two and three-phase SDME approaches [4, 17, 18]:

$$C_d = C_d^{eq}\left(1 - e^{-kt}\right) \tag{7}$$

where C_d^{eq} corresponds to the concentration of the analyte in the extractant phase under equilibrium conditions and k is the observed rate constant.

The complexity of the observed rate constant, k, depends on a high extent on the number of phases involved. The Eq. (8) shows the expression reported in the literature for two-phase SDME modes [4]:

$$k = \frac{A_{ds}}{V_d}\overline{\beta}_d(1 + K_{ds}\frac{V_d}{V_s}) \tag{8}$$

where A_{ds} is the interfacial area and $\overline{\beta}_d$ is the overall mass transfer coefficient with respect to the extractant phase, which can be expressed as follows assuming rapid transfer across the liquid–liquid interface:

$$\frac{1}{\overline{\beta}_d} = \frac{1}{\beta_d} + \frac{K_{ds}}{\beta_s} \tag{9}$$

where β_d and β_s are the individual mass transfer coefficients for the extractant phase and sample, respectively.

The above expression can be written according to the film theory, which assumes that steady-state diffusion occurs from one phase (sample) to another (extractant phase) through stagnant layers (named as Nernst diffusion films) of thicknesses δ_d and δ_s adjacent to the interface in the extractant phase and sample, respectively, as follows:

$$\frac{1}{\overline{\beta}_d} = \frac{\delta_d}{D_d} + \frac{K_{ds}\delta_s}{D_s} \tag{10}$$

where D_d and D_s are the corresponding diffusion coefficients of the analyte.

The impact of experimental parameters on the extraction kinetics can be deduced from Eq. (10). Accordingly, rapid extraction can be attained when maximizing A_i, β_d and β_s, while minimizing V_s for a given combination of target analyte and extractant phase, which in turn defines K_{ds}. An efficient agitation reduces the thicknesses of stagnant layers and increases the mass transfer coefficients and, thus, the extraction kinetics. In fact, a log–log relationship between $\overline{\beta}_d$ and the stirring rate has been verified in two phases SDME approaches [4].

Three-Phase SDME Approaches

An expression for the time dependence of the analyte concentration extracted by a microdrop in a three-phase LLLME approach, has been given by the following equation [18]:

$$C_d = C_0\left(\frac{V_s}{V_d}\right)\left\{\frac{k_1k_3}{\lambda_1\lambda_3} + \frac{k_1k_3}{\lambda_2(\lambda_2-\lambda_3)}e^{-\lambda_2 t} + \frac{k_1k_3}{\lambda_3(\lambda_2-\lambda_3)}e^{-\lambda_3 t}\right\} \tag{11}$$

where λ_2 and λ_3 correspond to:

$$\lambda_2 = \frac{1}{2}\left\{(k_1+k_2+k_3+k_4) + \left[(k_1+k_2+k_3+k_4)^2 - 4(k_1k_3+k_2k_4+k_1k_4)\right]^{1/2}\right\} \tag{12}$$

$$\lambda_3 = \frac{1}{2}\left\{(k_1+k_2+k_3+k_4) - \left[(k_1+k_2+k_3+k_4)^2 - 4(k_1k_3+k_2k_4+k_1k_4)\right]^{1/2}\right\} \tag{13}$$

In addition, the first-order rate constant k depends on the individual rate constants as follows when the steady-state approximation is assumed [14]:

$$k \approx \frac{k_1k_3}{k_2+k_3} \tag{14}$$

Under these conditions, k can be expressed as:

$$k = \frac{A_{da}A_{as}K_{as}\overline{\beta}_{as}\overline{\beta}_{da}}{V_s\left(A_{da}K_{da}\overline{\beta}_{as} + A_{as}\overline{\beta}_{da}\right)} \tag{15}$$

It can be inferred from this expression that the extraction kinetics in three-phase LLLME approaches is enhanced by an increase on the intermediate acceptor phase/sample distribution ratio, K_{as}, the interfacial areas involved, A_{da} and A_{as}, and the mass transfer coefficients, $\overline{\beta}_{as}$ and $\overline{\beta}_{da}$, as well as a decrease on the sample volume, V_s. Efficient agitation of the sample favours the mass transfer across the sample-intermediate acceptor phase and, in turn, induces convection in the intermediate immiscible phase. Thus, the thicknesses of the four Nernst diffusion films involved in the process are decreased and the mass transfer coefficients increased.

A kinetic model for HS-SDME has also been reported by assuming "steady-state approximations" [19]. Accordingly, Eq. (7) has been deduced as a valid expression for the time dependence of the analyte concentration in the microdrop exposed to the headspace above the sample, where the rate constant k is expressed as:

$$k = \frac{A_{dh}A_{hs}\overline{\beta}_{dh}\overline{\beta}_{hs}}{V_d\left(A_{dh}\overline{\beta}_{dh}K_{dh} + A_{hs}\overline{\beta}_{hs}\right)}\left(K_{ds}\frac{V_d}{V_s} + 1\right) \tag{16}$$

$\overline{\beta}_{dh}$ and $\overline{\beta}_{hs}$ can be expressed as a function of each mass transfer coefficient and the corresponding distribution constants:

$$\overline{\beta}_{dh} = \frac{\beta_{dh}\beta_d}{\beta_{dh} + K_{dh}\beta_d} \tag{17}$$

$$\overline{\beta}_{hs} = \frac{\beta_s\beta_{hs}}{\beta_s + K_{hs}\beta_{hs}} \tag{18}$$

where β_s, β_{hs}, β_{dh} and β_d are the mass transfer coefficients for the aqueous sample, headspace by water sample, extractant phase by headspace and extractant phase, respectively.

On a general basis, mass transfer into the microdrop represents a slow step due to its purely diffusive nature, as reported in the literature [17]. In addition, mass transfer in the aqueous sample is also slow and represents another limiting step in the extraction process. Thus, the above Eqs. (17) and (18) can be simplified to β_d and β_s/K_{hs}, respectively, assuming that the mass transfer in the condensed phases are the limiting steps. Accordingly, Eq. (16) can be simplified as:

$$k = \frac{A_{dh}A_{hs}\beta_d\beta_s}{V_d K_{hs}\left(A_{dh}\beta_d K_{dh} + A_{hs}\left(\frac{\beta_s}{K_{hs}}\right)\right)}\left(K_{ds}\frac{V_d}{V_s} + 1\right) \tag{19}$$

This equation is particularly valid for highly volatile analytes. Regarding less volatile compounds, the Eq. (19) can be simplified further since K_{hs} is very small and, therefore, it can be assumed that $A_{hs}\beta_s/K_{hs} \gg A_{dh}\beta_d K_{dh}$. Under these conditions, the expression for the rate constant can be simplified to:

$$k \approx \frac{A_{dh}\beta_d}{V_d}\left(K_{ds}\frac{V_d}{V_s} + 1\right) \tag{20}$$

It can be deduced from this equation that diffusion into the extractant phase microdrop is the rate limiting step in the extraction of less volatile analytes.

As discussed above, the rate-limiting steps of the process can correspond to the mass transfer in both condensed phases, namely the extractant phase and the aqueous sample. Agitation of the sample favours the mass transfer and induces convection in the headspace. Diffusion coefficients in the gaseous phase are four orders of magnitude higher than in condensed phases and, therefore, mass transfer in the headspace has been typically considered a fast process. Recent contributions, however, have revealed that interfacial gas-phase constraints are non-negligible, affecting both the evaporation and uptake, and can be significantly minimised under reduced sampling pressures [20].

3 Novel Developments

The evolution of SDME modes in chronological order is presented in Sect. 3.1, describing advantages, limitations and requirements, as well as recent developments. In addition, relevant information on the extractant phases used in SDME is described in Sect. 3.2, paying special attention to neoteric solvents. Furthermore, coupling of SDME to different analytical instruments, including their improvements via automation, is discussed in Sect. 3.3.

3.1 Evolution of SDME Modes: Main Achievements

The first SDME-related works were reported in 1995. In these contributions, aqueous droplets containing colorimetric reagents showed much potential for trapping NH_3, SO_2 [3] and Cl_2 [2] gases from air samples. The miniaturisation of conventional LLE occurred one year later, with two works based on the exposure of a microdrop hanging from the tip of a capillary tube to aqueous phases [4, 7]. Thus, the direct immersion mode was exploited for extraction of 4-methylacetophenone in a small drop of n-octane (immiscible in sample aqueous solution) [4], whereas an organic drop placed in a flowing aqueous sample enabled the determination of sodium dodecyl sulphate [7].

Capillary tubes initially used as holders of the extractant phase were then replaced by chromatography syringes, which simplified the process by integration of unitary steps [8]. In fact, the microsyringe was found to be suitable for both holding the solvent drop during extraction and facilitating the drop introduction into the instrument.

General advantages of SDME include the use of a very small volume of organic solvent, and the absence of sample carryover or memory effects due to the solvent renovation for each extraction. However, the small volume of solvent drops prevents the possibility to perform measurement replicates. Different SDME modes were sequentially introduced to solve several inconveniences found during their application, such as extended extraction times and high temperatures, and drop instability (or even drop detachment) in two phase-SDME modes, especially at high stirring rates.

A three-phase SDME mode, namely LLLME, was introduced in 1998 [14]. LLLME involves two consecutive extractions, typically from aqueous sample to an organic donor phase and then, to an aqueous acceptor drop. LLLME was firstly applied using a Teflon ring to place the octane phase between the two aqueous phases, and evolved a year later to potentially improve the enrichment factors by reducing the volume of the extractant phase [14]. Usually, the extractant phase microdrop is immersed on the second phase (*e.g.*, 100 μL of organic solvent) which floats over the aqueous sample. This strategy, adequate for ionisable compounds, allows achieving an efficient sample clean-up and high enrichment factors.

CFME was developed in 2000 [9], presumably based on a previous work [7]. CFME involved the use of a glass extraction chamber, a peristaltic pump for the delivery of aqueous sample at a constant rate into the chamber (0.2–2 mL), and exposure of the solvent drop to the sample into the extraction chamber. Thus, the flowing sample solution continuously interacts with the solvent microdrop. A sample flow rate of 0.2–1.0 mL/min and an extraction time of 10–15 min used to be appropriate. In practice, the fundamental difference between DI-SDME and CFME lies in the way in which the convective-diffusive transport is favoured (magnetic agitation *vs.* sample pumping). The inconveniences of CFME were therefore similar to those of DI-SDME, which together with the requirement of unconventional equipment has meant that the contributions involving CFME have been scarce (*ca.* 20 publications).

Another three-phase SDME mode valid for the enrichment of volatiles that allows the achievement of efficient clean-up is HS-SDME, firstly introduced in 2001 [17] on the basis of the seminal works of Dasgupta and co-workers [2, 3]. Unlike other SDME modes, the physical separation of the extractant phase and sample solution occurring in HS-SDME enables its application to complex matrices. In comparison with two-phase SDME modes such as DI-SDME or CFME, HS-SDME enables the use of higher stirring rates with negligible risks of drop dislodgement and reduced sample matrix interferences. In order to extend the applicability of HS-SDME to slightly volatile and non-volatile compounds, a derivatisation strategy can be applied to form volatile derivatives that can be transferred from the aqueous sample to the headspace and trapped into the microdrop.

Inspired by a previous work where an organic drop of chloroform (1.3 μL) was placed inside a flowing aqueous drop (25–45 μL) [7], another two-phase SDME mode termed as DDSME was developed in 2006 for the extraction of target analytes present in clinical samples such as blood or saliva [10]. This SDME mode enabled a certain clean-up of the sample and transfer of the analytes to another (organic) phase compatible with the analytical instrumentation, even though the reduced sample-to-extractant phase volume ratio and the instability of the extractant phase microdrop under agitation conditions severely limited the achievable enrichment factors with reasonable extraction times. The same year, an SDME mode named as DSDME was developed [21] to overcome the limitations of previously reported two-phase SDME modes in which the syringe was used as extractant phase holder during the extraction process. DSDME is based on the use of a suspended solvent droplet in the microliter range suspended in the centre of the sample solution. However, the collection of the organic solvent after the extraction process represents the main difficulty of this SDME approach, since the solvent acquires the form of a thin layer or it is dispersed in the sample solution when stirring is stopped. Going further, a SDME mode analogous to DSDME, named as SFODME, was firstly reported in 2007 to facilitate the droplet collection after extraction [22]. SFODME requires, as in DSDME, an immiscible solvent with low volatility, low water solubility and less density than water. Additionally, solvents applicable in SFODME show melting points near room temperature (*e.g.*, 1-undecanol and 2-dodecanol). During extraction, commonly performed in a thermostatic bath, the sample vial is kept at a temperature above the melting point of the solvent. Thus, the solvent is in liquid state in SFODME and behaves analogously to solvents used in DSDME. After extraction, the vial is cooled in an ice bath to facilitate the collection of the droplet, which solidifies, and further melts before analysis.

In 2008, inspired by both DSDME and LLLME, another three-phase SDME approach termed as DS-LLLME was introduced. In DS-LLLME, the extractant drop (usually aqueous) is suspended in the organic solvent and the latter one on the aqueous sample solution [23]. This SDME mode, however, has not attracted widespread interest.

The introduction of the different SDME modes mentioned above was followed by additional developments toward improved extractability with reduced analysis time. In this vein, the formation of an undesirable air bubble in a drop of organic solvent during microextraction processes has been repeatedly reported [24] and, a decade ago, this drawback of SDME has been demonstrated to be advantageous under controlled conditions [25]. The intentional incorporation of air bubbles into the solvent drop results in an increasing surface area, which favourably affects the extraction kinetics, as discussed in Sect. 2.2.2, and can allow obtaining higher enrichment factors. However, the reproducibility can be compromised and, additionally, large air bubbles lead to the instability of the drop. The application of vacuum for reducing the interfacial gas-phase constraints observed in HS-SDME [20] or, more recently, the use of a gas bubble flow of N_2 to favor mass transfer of volatile analytes from the sample solution to the headspace [26], have contributed to improve experimental conditions for the application of SDME approaches.

3.2 Extractant Phases in SDME

Since its first application, different extractant phases have been used in SDME comprising organic solvents, ionic liquids (ILs), deep eutectic solvents (DES), supramolecular solvents (SUPRASs), nanomaterials and aqueous drops.

Recent examples of extractant phases in SDME are included in Table 1. Different considerations should be taken into account when selecting the most suitable extractant phase, such as the nature of the sample or the target analyte/s, the microextraction modality, as well as the analytical instrumentation used. For two-phase SDME modalities, such as DI-SDME and CFME, the selected extractant phase must be immiscible with the sample and have low water solubility, high boiling point and low vapour pressure. In the case of DSDME, in addition to the aforementioned conditions, the extractant phase should have lower density than the liquid sample to be directly suspended on it. Another microextraction modality included within two-phase systems is SFODME, where the extraction solvent should have a melting point near to room temperature to facilitate its collection after solidification induced by cooling. In the case of three-phase SDME modalities, when the system is constituted by LGL phases, *i.e.*, HS-SDME, the extractant phase should have a high boiling point and low vapour pressure to minimize any evaporation during the extraction process retaining the extractant phase size and shape. Another widely used modality is LLLME, where the critical selection relies on the intermediate liquid phase, which must be immiscible with the sample and have low solubility in water, and lower density than the sample to be adequately suspended ensuring the contact between phases. Besides, to minimize evaporation, the selected solvent should have high boiling point and low vapour pressure. Furthermore, the acceptor phase not only must be a solvent related to the target analyte according to the "like dissolves like", but this solvent should be immiscible with the intermediate phase.

As mentioned before, the selection of the most appropriate extractant phase for each SDME mode mainly depends on the physicochemical properties of the extractant phase (boiling point, vapour pressure, density, viscosity, water solubility, etc.), which will determine the formation of the drop, the extraction efficiency, as well as the compatibility with the analytical instrumentation.

Within the different type of extractant phases, organic solvents are still the most used in SDME in its different modalities (Table 1). A wide variety of organic solvents with different physicochemical properties are commercially available and ready to use without needing previous synthesis or purification procedures, offering a highly convenient option for both two-phase and three-phase systems. Moreover, considering that conventional LLE approaches usually involve the use of organic solvents, SDME makes feasible the miniaturisation of these classical methods, as long as the involved organic solvent fulfil the mentioned requirements as extractant phase, minimising the amount of solvent usage. To be selected as extractant phase, the organic solvent must fulfil a series of characteristics to avoid drop dissolution in the sample matrix or solvent evaporation during the extraction process. However, not every organic solvent meets these characteristics and can be incompatible with some analytical techniques. In addition, several organic solvents used in SDME are currently restricted or even banned by Registration, Evaluation, Authorisation and Restriction of Chemicals (REACH) regulation in the European Union, *e.g.*, benzene, carbon tetrachloride, chloroform and toluene. In the last years, different solvent selection guides have been established with the aim of substituting those highly toxic organic solvents with less harmful ones [27–29].

ILs have been studied as greener substitutes of conventional organic solvents for extraction and preconcentration of different target analytes. ILs are organic salts constituted by an organic cation and an organic or inorganic anion that are liquid at temperatures below 100 °C. ILs offer unique physicochemical properties such as thermal stability, negligible vapour pressure, low volatility, tunable polarity and viscosity, miscibility with organic solvents and water, and structural designability. Since the first description of the use of ILs as extractants in SDME in 2003 [30], more than 65 works related to the application of ILs using the different SDME modes have been reported. Imidazolium-based ILs have been the most studied extractant phases in SDME. In addition, magnetic ILs (MILs) have emerged as improved ILs offering an inherent magnetism, which allows their manipulation using an external magnetic field, increasing the stability of the droplet during extraction [31]. Concurrently, MILs possess the typical physicochemical properties of ILs, including extremely low volatility, modulable viscosity and solubility, high ionic conductivity and high solvation properties. In the last years, different classes of MILs have been tested in

Table 1 Extractant phases in SDME techniques

SDME approach	Extractant phase	Examples of extractant phases
Two phase SDME approaches		
DI-SDME	Organic solvent	Toluene, cyclohexane, toluene/butyl acetate, toluene/hexanol, toluene/cyclohexane, octanol, decane, dichloromethane, pentanol, n-octanol, carbon tetrachloride, chloroform, amylacetate, ethanol:acetonitrile, 1-decanol, 1-undecanol, 1-dodecanol
	ILs	1-hexyl-3-methylimidazolium hexafluorophosphate
	MILs	Trihexyltetradecylphosphonium tetrachloromanganate, trihexyl(tetradecyl)phosphonium tris(hexafluoroacetylaceto)nickelate(II)
	SUPRASs	Decanoic acid and tetrahydrofuran in water, tetrabutyl ammonium hydroxide in water
	NPs-containing organic solvent	Ag-dodecanethiol NPs in toluene, Ag-citrate NPs in toluene
	DES	Hexanoic acid/thymol (1:1)
CFME	Organic solvent	Dichloromethane/ethyl acetate, carbon tetrachloride, cyclohexane
DSDME	Organic solvent	Hexane, heptane:iso-octane, toluene, xylene, 1-butanol, 1-octanol, undecanol, 2-octanone, n-butyl acetate, hexyl acetate
	SUPRASs	1-decanol in ethanol
SFODME	Organic solvent	1-undecanol, 1-dodecanol, 2-dodecacol, 1-(2-thiazolylazo)-2-naphtol, n-hexadecane; 1-chlorooctadecane, 1,10-dichlorooctadecane, 1-bromohexadecane
	SUPRASs	1-dodecanol in tetrahydrofuran; decanoic acid in water; tetrabutyl ammonium hydroxide in water
Three phase SDME approaches		
LLLME	Intermediate phase	Chloroform, dodecane, chloroform/methanol, 1-octanol
	Acceptor phase	Phosphoric acid solution, tetramethylbenzidine/hydrogen peroxide solution, sodium hydroxide solution, Au NSRs/Ag(I)
HS-SDME	Organic solvent	1-undecanol, dimethylformamide, ethyl acetate: 1,4-butanediol; toluene; n-octane; 1-butanol; methylbenzoate; n-butyl acetate; amyl acetate
	ILs	1-butyl-3-methylimidazolium bis(trifluoromethylsulfonyl)imide, 1-hexyl-3-methylimidazolium tetrafluoroborate

(continued)

Table 1 (continued)

SDME approach	Extractant phase	Examples of extractant phases
	MILs	1-octyl-3-methylimidazolium tetrachloroferrate, 1-ethyl-3-methylimidazolium tetraisothiocyanatocobaltate(II), trihexyl(tetradecyl)phosphonium tris(hexafluoroacetylaceto) manganate(II), trihexyl(tetradecyl)phosphonium tetrachloromanganate(II)
	Aqueous drop	HCl aq. solution, 5,5'-dithiobis-(2-nitrobenzoic) acid solution, Ag(I)/Fe(III)/*o*-phen/SDS in acetate buffer solution, phosphoric acid aq. solution; boric acid aq. solution, acetylacetone/ammonium acetate in acetic acid aq. solution, fluorescein/dimethylformamide aq. solution
	NPs-containing drop	Ag modified N-GQDs, Au/Ag NPs; Au/Cu NPs, Au nanoprisms/Tollen's reagent, Au NPs, CDs, Ag-citrate NCs/glutathione, Ag-PMAA NCs
	DESs	menthol:thymol (1:1), tetrabutylammonium chloride:lactic acid (1:2), choline chloride:4-chlorophenol (1:2), choline chloride:oxalic acid (1:2), tetrabutylammonium bromide:dodecanol (1:2), choline chloride:chlorophenol (1:2) - mCNTs

SDME and related microextraction techniques. The paramagnetic component of the MIL can be a metal complex (anionic or cationic) or a radical. Most applications are related to the use of MILs based on paramagnetic anions, mainly tetrachloroferrate(III) ($[FeCl_4]^-$) and tetrachloromanganate(II) ($[MnCl_4]^{2-}$) [32]. However, in spite of being considered as green solvents, it should be noted that some ILs are reported as toxic, *e.g.*, those containing fluorinated anions [32]. Furthermore, ILs synthesis usually includes a series of steps that use volatile organic solvents and thermal treatments [33]. Also, hydrophobic ILs can be adsorbed onto the sediments becoming persistent contaminants in the environment [34].

DESs are based on the combination of a hydrogen bond acceptor (HBA), such as a quaternary salt, and a hydrogen bond donor (HBD). At a certain temperature (defined as eutectic point temperature), the different components can form a homogeneous liquid phase mixture, which presents a lower freezing point than those of the individual components. DESs have similar characteristics to ILs, such as low volatility associated with low vapour pressure, chemical and thermal stability, and high tunability. The chemical formation process, along with the source of their starting materials are the main differences between DESs and ILs. DESs can be obtained by simple and low-cost synthetic procedures, which are based on heating a mixture of their individual components. Choline chloride is the most used quaternary salt due to its low cost and biodegradability. Amides, carboxylic acids, and alcohols are commonly used as HBD [35]. A disadvantage of DESs is their high viscosity, which could lead to practical problems, including difficult handling with a micro-syringe and slow mass transfer in dissolutions resulting in longer extraction times. Besides, some DESs are not compatible with commonly used analytical techniques, such as gas or liquid chromatography, requiring dissolution in a volatile solvent or re-extraction into another solvent. Recent developments in DESs preparation are focused in developing new synthetic strategies with the aim of lowering their viscosity, making DESs more suitable for chromatographic applications. For instance, Rodinkov et al. [36] prepared a DES based on a equimolar mixture of thymol:menthol, which provides a lower viscous DES compatible for direct chromatographic analysis. The thymol:menthol DES has been applied for the extraction of toxic volatile organic compounds in atmospheric air using HS-SDME.

Surfactant-based solutions have been also used as alternative extractant phases to organic solvents in SDME. The addition of surfactants to aqueous solutions or hydro-organic media results in non-polar sites due to the generation of organised structures, like micelles, mixed micelles, vesicles or microemulsions, forming the so-called SUPRASs, also defined as nanostructured liquids immiscible with water [37]. SUPRASs are generated through self-assembly processes by dispersing a coacervating agent and an alkanoic acid in a continuous phase (water). First, surfactants coacervate three-dimensionally and then they aggregate getting immiscible with water phase. SUPRASs have unique properties, such as self-assembly synthetic procedure offering convenient and easy preparation, solvent tunability by changing hydrophobic or hydrophilic groups of the amphiphile, multi-functionality with high number of available binding sites, high surface area, low volatility, and non-flammability being a greener alternative to conventional organic solvents. The presence of distinct polarity areas in SUPRASs constituents provides good solvation capabilities for a wide range of organic and inorganic compounds. Different types of analytes are potentially extractable through different interactions, such as electrostatic and π-cation interactions, hydrogen bonds and Van der Waals forces [38].

Aqueous-based extractant phases have been also applied in SDME, mainly in three-phase systems, *i.e.*, LLL for LLLME and LGL for HS-SDME. Different aqueous solutions have been employed as acceptor phases in LLLME mode, including phosphoric acid solution, sodium hydroxide solution and colloidal liquid dispersions of nanoparticles (NPs). In the case of LGL systems, aqueous drops containing specific reagents that can selectively react with the target analyte have been reported for colorimetric and fluorescence analysis [39–41]. Furthermore, it is worth to mention that aqueous drops are easily compatible with fibre optic probes enabling fast accurate measurements for the analysis of organic and inorganic analytes. For instance, Skok et al. [39] reported an aqueous drop containing 5,50-dithiobis-(2-nitrobenzoic) acid as an optical probe combined with HS-SDME for the rapid detection of sulphite in food. Furthermore, in the last years several colloidal aqueous dispersions of nanomaterials have been employed as aqueous-based extractant phases playing the role of both acceptor phase and optical probe [42]. Analyte extraction in the aqueous drop promotes changes in the physicochemical properties of the nanomaterial, *e.g.*, fluorescence intensity, colour intensity or hue, etc., which can be used for sensing and quantification. Most applications involve the use of noble metals NPs (Ag, Au and bimetallic Ag-Au) as colorimetric probes combined with HS-SDME. Besides, fluorescence probes based on graphene quantum dots, carbon dots and metal nanoclusters have been also reported in literature for HS-SDME mode. Also, aqueous drops can be applied as extractant phases in reversed DI-SDME for the extraction of polar analytes from water immiscible samples. Although water is considered the greenest option when selecting an extractant phase for SDME and related techniques, the use of high concentrations of other components of the aqueous-based drop, *e.g.*, acids, bases or nanomaterials, can make these methodologies less environmentally friendly in accordance with the metrics for assessing the greenness of the analytical procedure [43].

3.3 SDME Combined with Different Analytical Techniques

SDME has been combined with virtually all analytical techniques, but its combination with chromatographic techniques is particularly fruitful. In this regard, ca. 70% of original articles involve the combination of SDME with gas chromatography (GC) or liquid chromatography (LC), as discussed in Sect. 3.4. In fact, original SDME was based on the use of an organic solvent microdrop, a chromatography microsyringe and, whenever needed, in-drop derivatisation, leading to significant advantages in sample treatment prior to injection into the chromatograph. Nevertheless, the combination of SDME with spectrometric techniques such as electrothermal atomic absorption spectrometry (ETAAS), microvolume UV–Vis spectrophotometry or spectrofluorometry, and mass spectrometry (MS) has also been significantly reported.

The combination of SDME with an analytical technique is relatively easy when nature and volume of extractant phase are directly compatible with the characteristics and the sample introduction system of the concerned technique. Thus, for example, SDME is especially suitable for GC and normal phase LC if a non-polar organic solvent is employed in the drop. On the contrary, reverse-phase LC and capillary electrophoresis are more difficult to use with certain organic solvents. Although, in some cases, it is feasible to evaporate the enriched drop solvent and re-dissolve it with another suitable solvent, this procedure is not without problems. Likewise, carrying out dilution of the enriched drop eliminates some advantages of SDME. Apart from introducing additional unitary steps, a significant decrease of the achieved preconcentration factor is produced. Nowadays, conventional organic solvents, considered hazardous and polluting, are gradually being replaced in SDME by a priori less toxic solvents such as ILs or DESs. However, their compatibility with certain analytical techniques is far from ideal. For example, most of these solvents are compatible with LC but their volatility and viscosity characteristics certainly hinder their use with GC. In fact, some strategies have been developed to favour compatibility with this latter technique. In addition, as mentioned above, appropriate volumes must be introduced into the corresponding instrument. Nevertheless, different strategies have been proposed to minimise the loss of the preconcentration achieved in the SDME procedure when the volume required for analysis is larger than the drop volume, *e.g.*, conventional nebulizers or cuvettes [44]. For example, electrothermal vaporisation (ETV) can be used for drop introduction in inductively coupled plasma-mass spectrometry (ICP-MS), thus avoiding nebulisation. Miniaturised optical instruments, accessories and optical probes allow solving the coupling of SDME with UV–vis or fluorescence detection in the drop. Other less conventional lab-made solutions will be discussed below. In general, the key to extend the use of SDME lies in being able to make measurements at the microliter level. The following is a brief description of the combination of SDME with various analytical techniques.

3.3.1 Combination of SDME with Separation Techniques

As mentioned above, separation techniques are the analytical techniques most commonly used with SDME, in particular GC. In fact, the first publications on SDME utilised GC [4], and many of SDME modes were also reported from this first time in combination with this separation technique. Probably, this can be attributed to the ideal compatibility of most organic solvents used as extractants (*e.g.*, toluene, hexane, etc.) with the separation system and the different detectors, including electron capture detector (ECD), flame ionisation detector (FID) or MS systems, which have become the most popular in combination with SDME. In addition, the coupling of SDME with capillary electrophoresis (CE) has allowed the use of CE to be extended for the determination of some bioanalyte traces and to perform chiral analysis, which would not be possible without the preconcentration and clean-up provided by a microextraction technique. In addition, SDME has the advantage of facilitating simultaneous extraction and derivatisation (in-drop or in-syringe derivatisation), which undoubtedly simplifies and shortens analysis procedures for GC, LC or CE.

Since injection in GC requires solvents with low boiling point and viscosity, solvents such as ILs or DESs are not directly compatible with the instrument. These solvents can be accumulated in the chromatograph (injection port liner and column) leading to serious problems. Notwithstanding the above, many and varied strategies have been proposed to solve this compatibility issue. Thus, for example, a removable lab-made interface has been designed for this purpose (Fig. 2). The interface allows ILs to be retained while analytes pass into the column. However, the removable unit has a reduced average life of five injections, after which it must be replaced [45]. Other strategy based on a commercially-available thermal desorption system has been proposed. Desorption of the analytes is carried out at 240 °C for 5 min [46]. The exposure of the drop into the injection port allows the volatilization of the analytes without the requirement to inject IL, thus avoiding compatibility issues [47]. Programmable temperature vaporisation (PTV) injectors have been also used for direct injection of the IL drop [48]. Although with some exceptions, GC also shows incompatibility with DESs. Dilution of the drop in an appropriate solvent

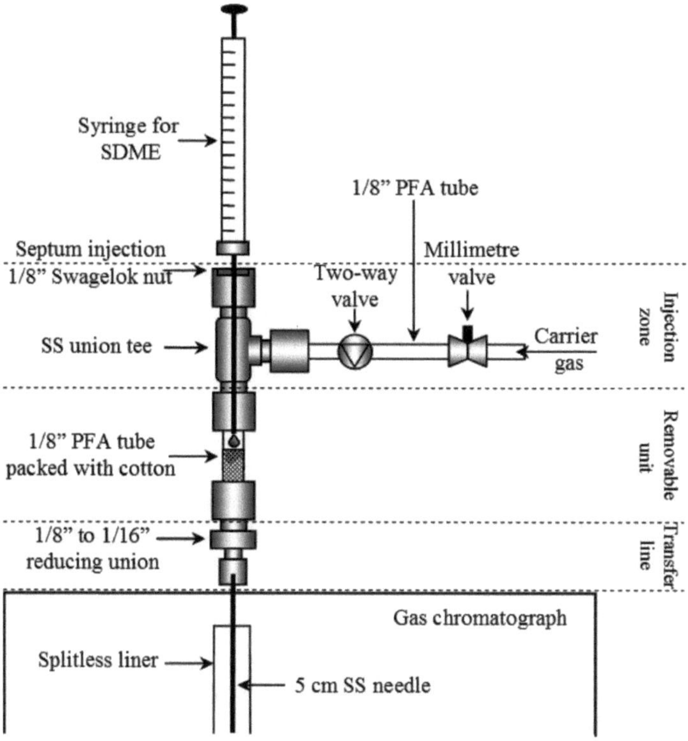

Fig. 2 Schematic diagram of the interface developed for the introduction of ionic liquids for GC–MS analysis. Reprinted from with [45] permission from the American Chemical Society

prior to the injection is generally the strategy used to protect the instrument from possible damages associated with DESs introduction [49, 50].

For combining SDME with LC, the extractant phase must be compatible with the mobile phase. Certain solvents can be directly used in the chromatograph (*e.g.*, n-hexane, o-dibutyl phthalate, cyclohexane, iso-octane, chloroform, toluene, xylene), even though the solvent is usually exchanged by evaporation and redissolution in the mobile phase [51]. However, as mentioned above, this strategy is not free of difficulties. When reversed-phase LC or CE are used for the determination of ionisable analytes, three-phase SDME approaches are often employed, taking advantage of the acid–base properties of analytes. Thus, LLLME is usual in both reverse-phase LC and CE since the enriched aqueous drop is compatible with the instrumentation. In addition, the HS-SDME mode involving aqueous drops is an important solution when volatile o semi-volatile analytes are being determined. ILs have also been used as extractants with LC and CE because they can be directly injected in the column or the capillary. For example, 1-hexyl-3-methylimidazolium hexafluorophosphate $[C_6MIM][PF_6]$ has been injected directly for high-performance liquid chromatography (HPLC) analysis using tetrahydrofuran and methanol at pH 4 as mobile phase

[52]. In the case of CE, the capillary tip can be directly used to suspend the IL droplet [53]. Furthermore, extractants used in SFODME mode (*e.g.*, 1-undecanol) are also compatible with these techniques [54].

Automation is desirable to avoid some of the problems of SDME (*i.e.*, drop instability) while improving precision and sample throughput. In general, autosamplers and syringe pump systems (lab-in-syringe, LIS) have been used toward the automation of the SDME-GC system. In this way, partial or total automation can be achieved. A fully automated IL-based HS-SDME coupled to gas chromatography-tandem mass spectrometry (GC–MS/MS) using an autosampler was developed to determine musk fragrances in environmental water samples [55]. An automated headspace BID microextraction (automated HS-BID) has been combined with gas chromatography-mass spectrometry (GC–MS) for determining nitro musks fragrances in environmental water samples [56]. Currently, commercial versatile autosamplers enable SDME automation in a relatively simple way, even though these systems are characterised by their expensiveness. Then, lab-made autosampler robots have been reported, especially in combination with HPLC. For instance, a lab-made cartesian robot equipped with a three-way solenoid microvalve and a six-port switching valve has been proposed for the determination of triazines in coconut water samples. An Arduino Mega board was used for synchronisation of robot action, valves and analytical instrument [57]. Recently, SDME in a 96-well plate mode has been proposed as an automated format with a high-throughput analysis [58].

The combination of SDME with CE can be performed off-line, on-line and in-line. In the on-line mode, the capillary inlet tip is commonly used for hanging the drop [59]. In the in-line mode, the different steps are carried out on the CE system through automatic control. For example, in LLLME, the following stages are included: hydrodynamic injection of the aqueous extractant phase; immersion of the capillary inlet into the organic phase and drop formation; extraction of the analyte and hydrodynamic injection of the enriched drop into the capillary for analysis [60].

3.3.2 Combination of SDME with Atomic Spectrometry

The combination of SDME and atomic spectrometry has been carried out mainly with ETAAS and ETV-ICP-MS because the extractant phase volume is of the same order of magnitude as the volume required for analysis. For the same reason, other atomic techniques commonly used in analytical laboratories such as flame atomic absorption spectrometry (FAAS), atomic fluorescence spectrometry (AFS), inductively coupled plasma optical emission spectrometry (ICP-OES) or ICP-MS, which use nebulizers for sample introduction, are less suitable for microextraction purposes. Thus, the graphite furnace (GF) and ETV units are suitable for the introduction of a few microlitres, fitting well with the drop volume used in microextraction. In fact, the ETV solves the problems associated with the introduction of organic solvents into the plasma. Aqueous solutions, organic solvents and ILs have been used as extractant phases in SDME in combination with both ETAAS and ETV-ICP-MS. Aqueous solutions of Pd(II) or Au(III) have been exploited in HS-SDME to retain volatiles generated in situ (both chemically and photochemically) from elements such as As, Se, Sb, Hg or Te [61, 62]. Hydrogen is originated as a byproduct, which leads to the reduction of noble metal ions present in the drop. The formed Pd(0) or Au(0) are able to decompose and retain the volatiles while acting as a matrix modifier in ETAAS, leading to very low limits of detection (LODs). This strategy has also been reported for simultaneous determination of As, Sb, Bi, Pb, Sn and Hg by ETV-ICP-MS [63]. When the sample is an oil, an aqueous drop can be used in the DI-SDME mode. By this way, Cd has been extracted from vegetable oil using an aqueous drop of 0.1 mol L^{-1} HNO_3 [64].

Organic solvents and ILs are commonly used in the DI-SDME mode for extracting metals in the form of complexes and ion pairs. However, certain aspects must be taken into consideration when these solvents are injected into the GF or ETV. For example, analyte losses can occur when volatile metal chlorides are formed as a result of the use of chlorine-containing organic solvents as extractant phases in SDME. In addition, extended temperature programs can be required with these solvents when compared with conventional alternatives. This is especially noteworthy in the case of ILs due to their thermal stability, high viscosity and immiscibility with aqueous solutions (*e.g.*, matrix modifiers), which can lead to difficulties in both ETAAS and ETV-ICP-MS. Dilution of the enriched IL drop with ethanol and nitric acid has been considered to reduce the viscosity of these neoteric solvents, making them compatible with the GF [65]. Other atomizers such as tungsten coils (W-coil) with a higher heating rate have also been proposed as possible solutions to these problems [66]. The combination of SFODME with ETAAS and ETV-ICP-MS has been also reported. Potential issues associated with solvent solidification in ETAAS have been minimised by means of hot injection (at *ca.* 80 °C) [67] or dilution of the enriched 1-undecanol drop with ethanol [68]. Losses of some volatile elements have been reported in ETV-ICP-MS due to the high drying temperatures required by this type of solvents. The use of mixed solvents (by addition of a solvent showing lower boiling point) has also been described. Thus, the mixture of 1-dodecanol (b.p. 259 °C) with p-xylene (b.p. 138 °C) allowed to set the drying temperature at 200 °C [69].

In general, although automated systems have been proposed, manual drop injection is still more commonly used. A certain degree of automation has been achieved with sequential injection (SI) systems. For this purpose, home-made flow-through extraction cells and commercial multiposition valves have been used [70, 71]. A fully automated HS-SDME system using a LIS system coupled with ETAAS has been proposed for Hg determination. In this case, Hg vapour is generated inside a microsyringe under reduced pressure conditions (up to 0.14 atm) without analyte losses [72].

GC can also be used after HS-SDME for sample introduction into the plasma. However, this coupling presents greater difficulties, and it is therefore less used in combination with microextraction. For example, SDME coupling with GC-ICP-MS has been used to speciation of volatile organometallic compounds [73].

As mentioned above, although the combination of FAAS with different preconcentration techniques is very popular in analytical laboratories to overcome its lack of sensitivity, FAAS has been rarely used with SDME. This is presumably attributed to the higher volumes required by the nebulisation system compared to those used in SDME. For this reason, the drop is usually diluted prior to measurement. For example, SFODME has been employed in combination with a FI-FAAS system for Cu(II) determination in water by using a 1,5-diphenylcarbazide-containing 1-undecanol microdrop as extractant. Dilution of the solidified drop to 300 μL with ethanol was required for this purpose [74]. Speciation of Cr(III) and Cr(VI) using DSDME required a home-made micro-sample introduction system and dilution of the extraction solvent (1-octanol) with an ethanol/0.1% (v/v) HNO_3 solution [75].

Likewise, AFS has been used after SDME with dilution and/or acid treatment of the drop, so that the typical advantages of SDME are not fully exploited. For example, HS-SDME using IL as extractant phase has been combined with AFS for the determination of organomercurials. In this case, the enriched drop was mixed with an acidic potassium permanganate solution to oxidise organomercurial species to inorganic mercury [76]. In addition, a method based on the combination of SFODME with hydride generation (HG)-AFS for Se(IV) determination has been reported, involving complex formation with ammonium pyrrolidinedithiocarbamate (APDC) and its extraction by a microvolume of 1-undecanol. Back-extraction was then performed using 300 μL of a 5 M HNO_3 solution for HG-AFS analysis [77]. 1-undecanol has also been used as extractant phase in SFODME for determining Hg by cold vapour (CV)-AFS. The method involved the extraction of a hydrophobic complex of the metallic ion (mercury diethyldithiocarbamate) followed by the dilution of the enriched drop with 2 mL of ethanol for analysis [78].

X-ray fluorescence-based techniques are a priori compatible with SDME because only a few microlitres are required for measurement. In fact, DI-SDME has been combined with energy dispersive X-ray fluorescence spectrometry (EDXRF) [79] and total reflection X-ray fluorescence (TXRF) [80] as a strategy to improve the LODs. However, the deposition of the enriched extractant phase for analysis represents the main problem to be addressed with these techniques. Particularly, it is necessary to avoid spillage on the sample carrier surface, for which different alternatives have been proposed. In the case of EDXRF, this issue has been solved by applying consecutive deposition/drying of the organic phase in small portions on a Whatman filter. Regarding TXRF, it required solvent exchange by replacing benzene by a significantly more polar solvent, specifically an aqueous solution containing an internal standard [80]. In this case, the enriched aqueous solvent was deposited on the sample carrier made of quartz, which was previously coated with a hydrophobised silicone solution.

Laser-Induced Breakdown Spectroscopy (LIBS) has also been proposed for measurement after SDME in spite of the problems shown by the technique when dealing with liquid phases. Two possible strategies have been evaluated. On the one hand, the analysis of the droplet suspended in the syringe tip. On the other hand, the analysis of the dried droplet on a substrate (surface-enhanced LIBS technique, SENLIBS). While the first option was found to be not suitable for analytical purposes due to lack of sensitivity and precision, SENLIBS was considered a promising alternative in combination with microextraction [81]. SDME with a toluene drop has been used for extraction of Cr, Mn, Ni, Cu and Zn APDC chelates followed SENLIBS measurement after drying the drop on an aluminium substrate [82].

3.3.3 Combination of SDME with UV–Visible and Fluorescence Spectrometry

As already noted, the incompatibility of the drop volume with conventional UV–vis spectrophotometry or fluorospectrometry has led to a later development of the combination of these techniques with SDME. In fact, drop dilution and/or use of microvolume cuvettes are still considered as a solution even though it implies a loss of sensitivity. Notwithstanding the above, special optical systems such as liquid droplets, liquid films/droplets and falling drops were initially proposed in combination with SDME without requiring dilution [83]. In general, the miniaturisation of sample compartment, radiation sources and detectors have been key to achieve SDME-compatible systems. At present, microvolume spectrophotometers and spectrofluorimeters, optical probes and smartphones are available for this purpose, greatly facilitating the combination of SDME with molecular spectroscopy.

Commercial confined drop-based UV–vis spectrometers and fluorospectrometers greatly facilitate the measurement of the enriched extractant phases from SDME. These systems are equipped with a system for depositing the drop between two pedestals to form a measurement column within the optical path. The characteristics of the extractant phase must be compatible with this system in order to form the column, while avoiding chemical attack of the pedestals and minimise losses by evaporation during measurement. In addition to aqueous solutions, many organic solvents (e.g., N,N'dimethylformamide, toluene, xylene, methyl isobutyl ketone), ILs and DESs can be used for measurement with these instruments [83]. In general, these systems have a lower path length in comparison with conventional spectrophotometers, which is undoubtedly a drawback when it comes to take advantage of the full potential of SDME. However, it may be compensated by the high preconcentration factors that can be reached in some cases. A large number of applications have been proposed involving SDME approaches involving both miniaturised UV–vis spectrometry and fluorospectrometry. In-drop derivatisation with conventional reagents for forming coloured or fluorescent compounds were initially purposed. Some examples include the colorimetric determination of nitrite by HS-SDME using a Griess reagent-containing aqueous drop for simultaneous extraction and derivatisation [84] or the fluorimetric determination of formaldehyde in textile samples using in-drop Hantzsch reaction [85]. The fluorescence enhancement or quenching of the droplet has also been exploited for sensing, *e.g.*, for determining bromide after in situ generation of volatile bromine and its trapping by a fluorescein-containing aqueous drop [41]. Within the different extractant phases, those containing nanoparticulated materials have greatly extended the use of these couplings. Thus, for example, the fluorescence quenching of quantum dots (QDs) when exposed to H_2Se was used for Se(IV) determination [86]. Remarkably, NPs have made it possible extending the application of SDME to the amplification of signals of DNA and microRNA analysis [87]. A confined drop-based UV–vis spectrometer has also been used for turbidity measurement after HS-SDME and precipitation of the analyte in the drop [88]. This strategy eliminates typical interferences in turbidimetry, such as absorbance, turbidity and/or fluorescence from the matrix and scattering and background fluorescence associated with optical windows.

Although to a lesser extent, optical systems that avoid the necessity of transferring the drop to the measurement instrument have also been used with SDME. Thus, an optical probe was proposed as both microdrop holder and measurement cell, then enabling the continuous monitoring of the extraction. As a proof of concept, sulphites were determined in this work by using the optical probe to expose an aqueous microdrop containing Fe(III) and 1,10-phenanthroline (adjusted at pH 5.6) to the headspace above the sample. In particular, volatile SO_2 reduces Fe(III) to Fe(II) in the drop, leading to the formation of a red coloured complex with 1,10-phenanthroline [89]. Optical probes can be also used with DI-SDME and LLLME modes [90]. A micro-pipette system was designed for integrating different unitary steps, fulfilling the functions of extractant handler, microdroplet holder and microcuvette for in situ absorbance measurement [40]. LIS automation in combination with miniaturised fibre optic systems is another interesting option that allows to eliminate drop transfer for analysis (Fig. 3) [91, 92].

The camera of mobile phones is nowadays an analytical detector of great interest and has also been used in combination with SDME in order to capture images of the drop. In this sense, the low sensitivity and selectivity reached by the camera can be solved by the preconcentration and clean up capacity of the microextraction strategy. Thus, for example, HS-SDME combined with mobile phone has been used for formaldehyde determination using a box illuminated by a white LED to obtain reproducible digitisation conditions [93]. In addition, pH-induced aggregation of

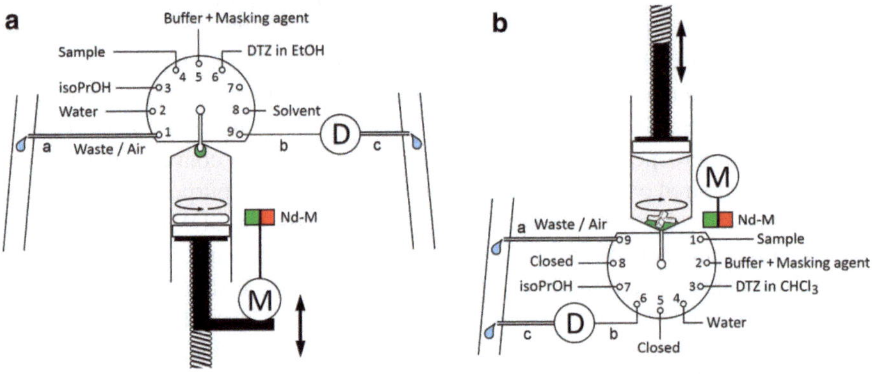

Fig. 3 **a** Schematic diagram of the DI-SDME into a floating drop **b** Schematic diagram of the in-drop stirring assisted DI-SDME (tubing dimensions of a: 5 cm, 0.8 mm i.d., b: 5 cm, 0.5 mm i.d., c: 5 cm, 0.5 mm i.d., all made from PEEK capillary. DTZ: dithizone, isoPrOH: isopropanol, M: DC motor, Nd-M neodymium magnets). Reprinted with permission from Elsevier [92]

biomass carbon dots (CDs) has been exploited for ammonia detection using a smartphone with colorimetric and fluorescent readout after HS-SDME [94]. This combination has also been recently used for determining hydrogen sulphide in biosamples with silver–gold core–shell nanoprisms (Ag@Au np) through the inhibition of absorbance of the Ag@Au np induced by the analyte [95].

3.3.4 Combination of SDME with Mass Spectrometry

Direct coupling of SDME with MS has also been reported. This coupling has allowed the applications of SDME to be extended to the determination of biomolecules, including proteins, peptides, carbohydrates, lipids, metabolites, bacteria, etc., in complex matrices. Matrix-assisted laser desorption/ionization (MALDI)-MS is generally used for this purpose. The volume required for measurement by MALDI-MS fits well with that used in SDME. Furthermore, the drop can be deposited directly on the target plate preventing further drop manipulation.

The extractant used in SDME must be miscible with the MALDI matrix solution, have a certain volatility and ensure the formation of homogeneous crystals with the matrix. In addition to conventional organic solvents such as octanol, toluene, chloroform, octane, etc., ILs have also been used as extractants, in many cases in combination with NPs. For instance, a toluene drop containing gold or silver NPs and tetraalkylammonium bromide has been proposed for DI-SDME of peptides. The isoelectric point of the peptides and the surface charge of the gold NPs were key to achieving the separation [96, 97]. A drop of IL (1-butyl-3-methylimidazolium hexafluorophosphate) containing platinum NPs was used for extraction of Gram-negative bacteria and subsequent detection using a time-of-flight (TOF) detector (MALDI-TOF-MS). In this case, the positively-charged head group of IL attracts membrane proteins of bacteria [98]. In general, NPs, in special silver NPs, have received a high degree of interest for ionisation and preconcentration in MS combined with SDME [99].

The combination of DI-SDME with paper spray mass spectrometry (PS-MS) has been recently proposed as an improved strategy [100]. In this case, the drop is added onto the tip of a paper triangle and a high voltage is applied to generate the ionisation.

Automation is also desirable in the coupling SDME-MS. In this sense, droplet-in-droplet and droplet-on-droplet microfluidic microextraction systems have recently been developed using a liquid-handling robot to work in the nanolitre-scale [101].

3.3.5 Combination of SDME with Electrochemical Detection Techniques

Electrochemical detection has been rarely used with SDME in spite of its interest in eliminating matrix interferences. This is probably due to the fact that solvents used in the drop are not always compatible with this type of detection. In this sense, the combination of SDME with potentiometric detection has been proposed for the determination of caffeine in saliva by means of a SI system. It was carried out by solvent change (evaporation and redissolution in an aqueous phase). As mentioned above, this strategy leads to the loss of the preconcentration factors achieved with SDME [102]. Voltammetric detection has also been used for the determination of ascorbic acid after SDME using a magnetic IL dissolved in ethanol as extractant phase. After recovery of the IL by a magnet, it was diluted in ethanol and transferred onto the surface of a carbon paste electrode (CPE) modified with TiO_2 NPs [103]. Conductimetry has also been used with HS-SDME integrated in a lab-made automatic flow-batch system for ammonium determination. This mode of SDME eliminates the solvent problem [104].

3.3.6 Combination of SDME with Other Techniques

Infrared (IR) spectroscopy has been used in a simple way after SDME. In general, the extraction of hydrophobic complexes of analytes by an organic drop eliminates two major problems of IR spectroscopy, namely poor sensitivity and interference from water. After extraction, the drop is directly deposited on the corresponding substrate and dried for the measurement. Thus, for example, Cr(VI) [105] and vanadate [106] complexes with hydroxy-N1,N2-diphenylbenzamidine have been extracted in a dichloromethane drop, whereas a dichloroethane drop has been used for extraction of a Mo(VI) complex with N1-hydroxy-N1,N2-diphenylbenzamidine [107]. Diffuse reflectance Fourier transform infrared spectroscopy (DRS-FTIR) has been used in these cases after SDME. To this end, the enriched drop was deposited over the KBr substrate and dried before filling the sample holder. Furthermore, perchlorate has been extracted as ion-pair with cetyltrimethylammonium in methyl isobutyl ketone. In this case, the drop is directly placed and dried on the zinc selenide crystal substrate of the attenuated total reflectance Fourier transform infrared (ATR-FTIR) spectrometer for analysis [108]. Recently, modified silver NPs have been implemented in the drop to enhance the signal intensity for tartrazine [109] and quaternary ammonium cationic surfactants [110].

Ion-mobility spectrometry (IMS) is a low-cost and fast analytical technique that is usually used after extraction methods to increase selectivity. In this regard, IMS has been combined with SDME [111–113]. The main drawbacks of this coupling come from the injection of organic solvents in the IMS system, which lead to high blanks. To solve these problems, IMS has been combined with IL-based HS-SDME with room temperature gas chromatography (RTGC) for on-site IMS analysis [111]. The design of a special injection unit for the retention of the IL in a glass wool before it reaches the IMS unit was necessary for this purpose [112]. More recently, LLLME has been proposed for the determination of psychotropic drugs in urine using IMS with electrospray ionisation (ESI). This coupling solves the problem caused by the need to mix the aqueous solution with methanol or acetonitrile to increase the ionisation efficiency in ESI [114].

Chemiluminescence has been also used in combination with SDME. Thus, for example, LLLME has been used for Sb(III) and total Sb determination through the inhibition of the chemiluminescence from the CdSe QDs/H_2O_2 system in the presence of antimony species using a tube luminometer for measurement [115].

Moreover, SDME has recently been combined with surface-enhanced Raman scattering (SERS) to increase the sensitivity of the technique. For instance, a drop of toluene was used as extractant phase for the preconcentration of p-mercaptobenzoic acid (p-MBA) using DI-SDME. After microextraction, repeated cycles of deposition/drying of the droplet onto the SERS substrate (*i.e.*, gold nanohole array substrate) were carried out. A sensitivity improvement of three orders of magnitude with a relative standard deviation (RSD) of 21% was achieved under these conditions. However, the procedure has not yet been proved to be suitable for application in real samples [116].

In addition, a procedure of HS-SDME using a gold NPs-containing aqueous drop has been proposed for the determination of captopril by a microchip coupled with photothermal lens microscopy (PTLM). This technique uses coaxial excitation and probe lasers focused onto the sample under an optical microscopy. This results in a local heating that produces a concave lens effect which responds to analyte concentration and thermo-optical properties of the medium. In the procedure proposed by Abbasi-Ahsd et al., the interaction of the thiol groups of captopril with the AuNPs present in the drop result in the reduction of the PTLM signal of AuNPs, leading to significantly increased linearity and sensitivity [117].

MILs-based SDME has been used for DNA extraction to direct quantitative polymerase chain reaction (qPCR) amplification, reaching high enrichment factors with an extraction time of 2 min [118]. This combination solves some of the reported problems related to the conventional methods used in DNA purification, such as the use of reagents that can inhibit qPCR in traditional methods or the low extraction efficiencies values achieved when magnetics beads are employed.

4 Performance Overview and Recent Applications

The overall performance of SDME approaches and recent selected applications are discussed in this section.

Figure 4a shows the evolution of the publications related to SDME from its first developments in the mid-1990s. The contributions involving SDME sharply increased since 2003 and peaked in 2009. Thereafter, the number of works gradually decreased and, since 2014, the scientific output involving SDME has remained almost constant at an average of 25 publications per year. In total, more than 700 articles have been published related to SDME. Figure 4b shows the annual evolution of publications involving each of the SDME techniques. Additionally, the pie chart (Fig. 4b) provides information on the proportion of the different SDME modes during this period (1995–2022). HS-SDME and DI-SDME are used in almost two thirds of the total, with similar figures, higher for HS-SDME than for DI-SDME, followed by other modes such as CFME, LLLME, DDSME, DSDME and SFODME. Regarding the evolution of each SDME mode per year, DI-SDME became the most popular SDME mode from 1995 to 2010, whereas HS-SDME was the most widely used approach during the period 2011–2022. The decline in the frequency of publication of DI-SDME contributions can be attributed to the increasing popularity of other two-phase microextraction techniques such as DLLME. Additionally, SFODME received more attention than other SDME modes previously developed such as CFME, LLLME, DSDME and DSDME in the period 2011–2022.

Ring graphs were also designed to show a general perspective of the type of analytes determined (Fig. 4c), the analytical technique coupled to SDME (Fig. 4d) and the type of samples analysed (Fig. 4e). As shown in Fig. 4c, organic compounds were the most evaluated analytes (71%), even though metals, organometallic

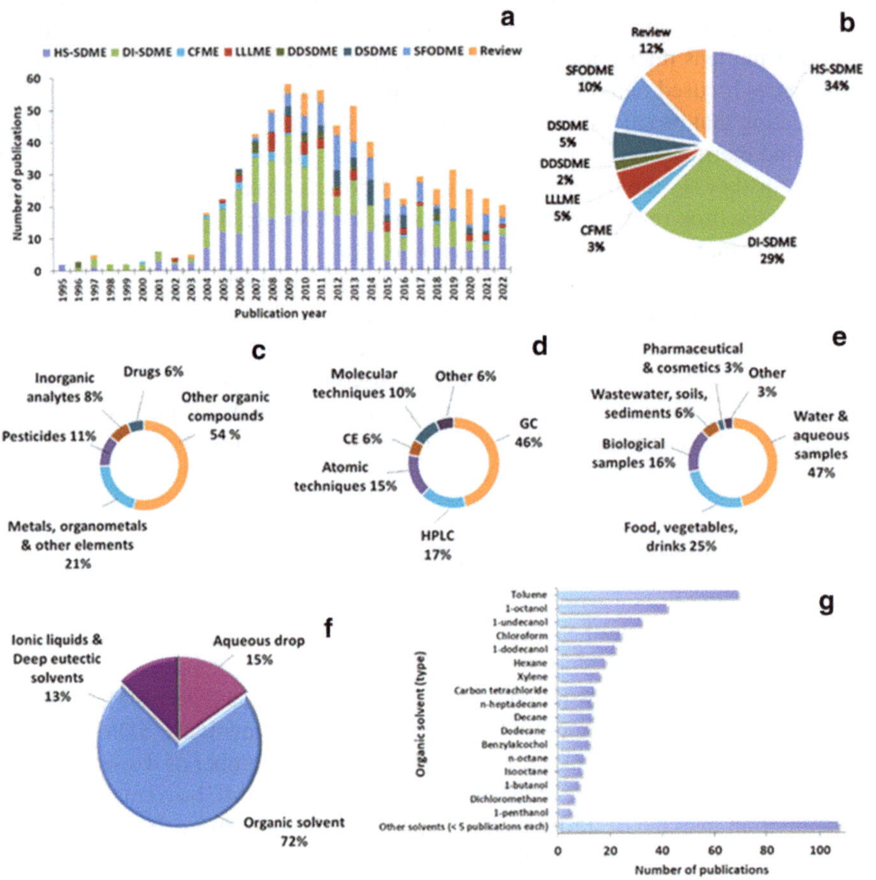

Fig. 4 Information of the publications related to SDME. **a** Vertical bar chart of the number of publications per year devoted to SDME. **b** Pie chart of the SDME modalities. **c** Ring graph related to the type of analyte. **d** Ring graph devoted to the techniques coupled to SDME. **e** Ring graph focused to the type of sample. **f** Pie chart of the type of drop. **g** Horizontal bar chart of the organic solvents most used

compounds and inorganic anions were also determined. As shown in Fig. 4d, the analytical techniques more commonly associated with SDME were separation techniques (69%). GC is used in most of contributions (43%) coupled to different detectors, including, in order of usage, MS, FID and ECD detectors. HPLC and CE have also been significantly used (17% and 6%, respectively). For contributions involving HPLC, the UV–vis detector was most commonly applied. The most used atomic techniques for the determination of metals were ETAAS and ETV-ICP-MS.

In relation with the type of samples analysed (Fig. 4e), aqueous samples (*e.g.*, river, tap, well, lake, drinking waters) were the most studied samples (47%), probably due to the low concentration levels of target analytes, which requires high enrichment factors to carry out the determination, apart from the relatively simple matrix of these samples. Nevertheless, more complex samples such as foods, drinks, biological samples (*e.g.*, blood, urine, saliva, hair) and environmental samples, such as soils or sediments, have been analysed.

As discussed in Sect. 3.2, extractant phases of different nature have been used in SDME, including organic solvents, aqueous drops, ILs and DESs. Organic solvents are by far the mostly used extractant phases, being used in more than 70% of the total number of publications (Fig. 4f). Notwithstanding this, aqueous microdrops containing metal ions, metallic nanomaterials, and surfactants have been significantly employed (15%) with a frequency similar to that of ILs and DESs (13%). Aqueous microdrops were mainly applied with three-phase SDME approaches, including both LLLME and HS-SDME.

Finally, the most used organic solvents in SDME can be seen in Fig. 4g. Specifically, toluene, 1-octanol, 1-undecanol, chloroform and 1-dodecanol were the five most commonly used organic solvents. It is worth mentioning that solvent selection depended to a high extent on the SDME mode used. For instance, toluene was used in DI-SDME, DDSDME, DSDME, HS-SDME, LLLME; 1-octanol was employed by HS-SDME, DSDME, DI-SDME, DDSME and 1-undecanol was mainly used in SFODME.

As it can be seen, SDME has been applied to a wide range of samples and analytes using different modalities/approaches (mainly DI, HS, LLLME, BID).

However, it could also be interesting to focus on the most recent applications to envisage future advances. In Table 2, selected applications have been included as an example of the most recent publications. Information related to the drop composition, LOD, EF, precision and the analytical technique were included for the different applications. As we can see, in the last five years, diverse drop types were applied, such as organic solvents, ILs, DESs, MILs, NADES, noble metals and NPs. Furthermore, SDME was coupled to almost all types of analytical techniques. High EF, up to 1600, were achieved for certain publications related, for instance with pesticides. Precision evaluated as RSD(%) was lower than 10% for most applications.

Applications to a huge variety of analytes have been carried out over the last years, including pesticides, alcohols, PAHs, parabens, terpenes, heavy metals, ammonium, hydrogen sulphide, DNA, etc. While most applications have been carried out using

Table 2 Selected applications of SDME for the determination of organic compounds, metal species, anions and biomolecules published in the last five years

SDME approach	Analyte	Solvent/Drop composition	Sample	Analytical technique	LOD	EF	Precision (RSD, %)	Ref.
SDME	Ranitidine	CH_2Cl_2	Waters	LC–MS/MS	9.4×10^{-9} mol/L	45	4.3	[119]
DI-SDME	Cu(II)	Decane	Tap water	GF-AAS	0.025 μg/L	45.2	6.8	[120]
BID	Carbamate pesticides	Toluene and butyl acetate	Waters	GC–MS	0.02–0.04 μg/L	536–1097	7	[121]
HS-SDME	Triazole pesticides	DESs	Fruit juice, vegetables	GC-FID	0.82–1 μg/L	273–346	3.9–6.2	[122]
HS-SDME-SP	As(III), As(V)	Ag(I) + Fe(III) + o-phen + SDS	Waters	UV–vis Spectrometry	0.1×10^{-9} mol/L	–	2	[40]
HS-SDME	Phenols, aliphatic alcohols	DES	Air, gas mixtures	GC-FID	0.001–0.01 mg/L	–	5	[36]
HS-SDME (gas-flow assisted)	BTEXs	1-undecanol	Waters	GC-FID	0.52–0.63 μg/L	–	3.2–5.1	[26]
HS-SDME	PAHs	DES	Waters	GC–MS	0.003–0.012 μg/L	–	3.9–7.2	[50]
Pa-SDME	Several organics	MIL	Waters	HPLC–DAD	1.5–3 μg/L	10–14	1–21	[58]
HS-SDME	Formaldehyde	N-GQDs-Ag	Milk	Luminescence	1.7×10^{-4}% w/v	–	0.7	[123]
LLLME-SDME	HTL	Several	Human urine	CZE	25×10^{-9} mol/L	340	4	[124]
DI-SDME	Vanadium	$[C_6MIM][PF_6]$	Water	Digital colorimetry	0.6 μg/L	50	4.8	[125]
HS-SDME	Methanesulphonates	$[BMIM]TF_2N$	APIs	HPLC–UV	15 ng/mL	–	1–2.8	[126]
DI-SDME	NSAIDs	3-phase	Human urine	In-line CE-UV	2–10 ng/mL	960–1610	0.6–1	[127]
HS-SDME	Organic pollutants	SUPRADESs	Waters	HPLC–UV	0.1–14.6 μg/L	2.1–961.1	2.3–9.9	[128]
HS-SDME	Methanol	DMF	Wine	GC-FID	1 μg/L	–	1.9–4.8	[129]

(continued)

Table 2 (continued)

SDME approach	Analyte	Solvent/Drop composition	Sample	Analytical technique	LOD	EF	Precision (RSD, %)	Ref.
DI-SDME	Acrylamide	n-octanol	Food samples	GC-ECD	0.6 μg/L	–	<6	[130]
HS-SDME	Formaldehyde	Au-np/TR	Food samples	Smartphone and UV–vis Spectr	30×10^{-9} mol/L	–	<4.6	[131]
HS-SDME	Ammonium	Boric acid	Waters	On-drop conductometry	2.8 μM		3	[104]
DI-SDME (Lab-in-syringe)	fluoroquinolones	NADES	Waters	HPLC-fluorescence	6–9 ng/L	35–45	0.2–5.9	[132]
HS-SDME	2-phenoxyethanol	Octane	Fish	GC–MS	0.2 μg/mL		<5	[133]
HS-SDME	ammonium	phosphoric acid	Water and food	UV–vis microspect	0.14 mg/kg	38–137	4.8–6.3	[134]
HS-SDME	Ethyl carbamate	Ethyl acetate	Wine	GC–MS	1.5 ng/mL	–	2.2–4.8	[135]
HS-SDME	CdTe QDs Te(IV)	$AuCl_4^-$	Waters	GF-AAS	0.13 μg/L Te(IV) 0.03 Total Te	14–101	<4.2	[62]
MTME-SD	Chloramphenicol	NTs/G-quadruplex/ PPIX	Milk, urine	Fluorescence	0.00856 fg/mL	–	<5.5	[136]
HS-SDME	NH_3, pH	CDs	Water	Smartphone	34 μM	–	–	[94]
HS-SDME	Methanol, ethanol	DMSO	Industrials oils	GC-FID	0.02 (m), 0.03 (e) μg/g	–	1.8–5.2	[137]
MTP-SDME	DNA, microRNA	Three-phase	Serum samples	UV–vis microspectr	0.15–0.34 aM	27	<5	[87]
HS-SDME	H_2S	Ag@Au-np	Egg, milk	Smartphone	65 nM	–	<4.8	[95]
HS-SDME (WCS)	H_2S, I_2, Br_2, Hg	Au@AgNPs and AuNRs	Waters	Smartphone	0.46 μM H_2S	–	4.4	[138]
SDME	Tartrazine	Toluene with AgNPs	Foodstuffs	DRS-FTIR	2.44 ng/mL	–	1.76	[109]

(continued)

Table 2 (continued)

SDME approach	Analyte	Solvent/Drop composition	Sample	Analytical technique	LOD	EF	Precision (RSD, %)	Ref.
Dynamic-SDME	Parabens	dichloromethane	Waters	HPLC–UV UHPLC-MS/MS	0.3 μg/L	–	5.54–17.94	[139]
On-line SDME	Amide herbicides	CCl_4	Rice samples	GC–MS	0.3–1.5 μg/kg	–	1.9	[140]
Pa-SDME	Pesticides	MIL	Human urine	HPLC–DAD	7.5 μg/L	4–15	3–17	[141]
SDME	muscimol, psilocin	Octanol	Urine	CE	0.004 mg/L (muscimol) 0.016 mg/L (PSC)	170	0.3–8.5	[142]
Three-phase SDME	HTL	$CHCl_3$	Urine	CE	0.04 μM		6.4–10.2	[143]
HS-SDME	Se(IV)	AuNPs	Seawater, tap water	Digital colorimetry	12 μg/L (LOQ)	–	4–5	[144]
HS-SDME	Terpenes	DESs	Spices	GC–MS	0.47–86.4 μg/g (LOQs)		<20	[145]
DI-SDME and CFME	Mn(II)	PAN/[C_4MIM][PF_6]	Tonic drinks, seafood	GF-AAS	3.9 and 7.9 ng/L	18 and 15	6.2 and 6.9	[146]
DI-SDME	PBDEs	Chlorobenzene	Waters	GC–MS/MS	6 ng/L	8–60	7.5–25	[147]
DI-SDME	p-MBA	Toluene	Water	SERS	10^{-8} mol/L	1000	21	[116]
SDILNDμE	As, Cd, Ni, Pb	[C_4MIM] [PF_6] IL	Eye makeup products	ETAAS	0.049–0.262 μg/L	48–57	<4.2	[65]
SDME-LVSEP	Herbicides	Pentanol	Soil	CE-MS	2–4 nM	600–1300	10–11	[148]
CCU-CF-SDME	PAHs	Cyclohexane	Waters	GC–MS	0.0012–0.0101 μg/L		0.5–6.4	[149]
HS-SDME	Hg(II)	AuNPs	Waters	Colorimetry	1 nM	–	–	[150]

(continued)

Table 2 (continued)

SDME approach	Analyte	Solvent/Drop composition	Sample	Analytical technique	LOD	EF	Precision (RSD, %)	Ref.
DI-SDME (LIS) (two modes)	Pb(II)	Dithizone	Water	UV–vis spectr	23–75 nmol/L	18	4.2–4.8	[92]
LLLME	Patulin	Ethyl acetate	Apple juice	LC–MS/MS	0.5 μg/L	–	3.6	[151]
HS-SDME	Total Br	AuNPs	E-waste polymers	UV–vis spectr	8.8 ng/mL	–	4.8	[152]
HS-SDME	formaldehyde	Acetylacetone	Textile, wastewater	Smartphone	0.1 mg/L	–	2.7–5.1	[93]
SDME	Sb(III)	BPHA-[C_4MIM][PF_6]	Water	GF-AAS	0.01 μg/L	112	4.2	[153]
HS-SDME	BTEX	DESs (MBGs)	Water, urine	GC-FID	0.05–0.9 ng/mL	520–587	<10%	[49]
SDME	Perchlorate	Ion-pair	Soil, water, milk, urine	ATR-FTIR	0.003 ng/mL	–	1.9–2.7	[108]
HS-SDME	Chlorobenzenes	MIL	Waters	GC–MS	4–8 ng/L	–	3–18	[154]
HS-SDME	Thiomersal	AuNRs	Pharmaceuticals	UV–vis spectr	0.5 ng/mL (Hg)	–	8.4	[155]

chromatographic separations coupled to mass spectrometry (e.g. LC–MS/MS, GC–MS), other detectors such as DAD, FID, UV have also been reported. Some spectrometric techniques such as ETAAS, SERS, UV–vis spectrophotometry, FTIR as well as digital colorimetry have also been applied.

5 Conclusions and Future Trends

SDME has established itself as one of the most powerful LPME techniques for analyte enrichment and sample clean-up. However, this technique has faced from the early developments, several drawbacks that have limited its use such as risk of drop detachment when supported on the tip of syringes, limited drop volume, extractant solvent volatility and impaired precision. New progress in this technique has occurred so as to alleviate these shortcomings such as some attempts to automate extraction procedures, implementation of novel extractants with enhanced preconcentration ability, better thermal stability, high viscosity, low volatility and adjustable miscibility as well as new extraction modes. Apart from conventional organic solvents used as extractant phases in first developments, others phases with more appealing properties such as ILs, DESs, SUPRASs, nanomaterials, etc. have emerged in the last years. Advances in the combination of SDME with a broad variety of detection techniques should also be highlighted. In order to increase sample throughput and improve precision, several automated procedures have been reported in the literature using lab-made autosamplers, lab-in-syringe platforms and well plate systems. Undoubtedly, robotics is called to play an important role in the advances to be expected in next years concerning SDME automation. Apart from the above properties, green chemistry guidelines should also be taken into account for the selection of non-toxic and environmentally-friendly extractants.

Acknowledgements The authors thank the MCIN/ AEI / 10.13039/501100011033 / FEDER, UE (Project PID2022-136337OB-I00) for financial support.

The authors have declared no conflict of interest.

References

1. Arthur CL, Pawliszyn J (1990) Solid phase microextraction with thermal desorption using fused silica optical fibers. Anal Chem 62:2145–2148. https://doi.org/10.1021/ac00218a019
2. Liu H, Dasgupta PK (1995) A renewable liquid droplet as a sampler and a windowless optical cell. Automated sensor for gaseous chlorine. Anal Chem 67:4221–4228. https://doi.org/10.1021/ac00119a003
3. Liu S, Dasgupta PK (1995) Liquid droplet. A renewable gas sampling interface. Anal Chem 67:2042–2049. https://doi.org/10.1021/ac00109a023
4. Jeannot MA, Cantwell FF (1996) Solvent microextraction into a single drop. Anal Chem 68:2236–2240. https://doi.org/10.1021/ac960042z

5. Sajid M (2022) Dispersive liquid-liquid microextraction: evolution in design, application areas, and green aspects. TrAC—Trends Anal Chem 152:116636. https://doi.org/10.1016/j.trac.2022.116636
6. Pedersen-Bjergaard S (2021) Analytical microextraction with supported liquid membranes. In: Lucena R, Cárdenas S (eds) Analytical sample preparation with nano- and other high-performance materials. Elsevier, pp 97–109
7. Liu H, Dasgupta PK (1996) Analytical chemistry in a drop. Solvent extraction in a microdrop. Anal Chem 68:1817–1821. https://doi.org/10.1021/ac960145h
8. He Y, Lee HK (1997) Liquid-phase microextraction in a single drop of organic solvent by using a conventional microsyringe. Anal Chem 69:4634–4640. https://doi.org/10.1021/ac970242q
9. Liu W, Lee HK (2000) Continuous-flow microextraction exceeding 1000-fold concentration of dilute analytes. Anal Chem 72:4462–4467. https://doi.org/10.1021/ac000240x
10. Wu H-F, Yen J-H, Chin C-C (2006) Combining drop-to-drop solvent microextraction with gas chromatography/mass spectrometry using electronic ionization and self-ion/molecule reaction method to determine methoxyacetophenone isomers in one drop of water. Anal Chem 78:1707–1712. https://doi.org/10.1021/ac052076p
11. Čabala R, Bursová M (2012) Bell-shaped extraction device assisted liquid-liquid microextraction technique and its optimization using response-surface methodology. J Chromatogr A 1230:24–29. https://doi.org/10.1016/j.chroma.2012.01.069
12. Xia Y, Cheng M, Guo F et al (2012) In-syringe demulsified dispersive liquid-liquid microextraction and high performance liquid chromatography-mass spectrometry for the determination of trace fungicides in environmental water samples. Anal Chim Acta 724:47–53. https://doi.org/10.1016/j.aca.2012.02.001
13. Dadfarnia S, Salmanzadeh AM, Shabani AMH (2008) A novel separation/preconcentration system based on solidification of floating organic drop microextraction for determination of lead by graphite furnace atomic absorption spectrometry. Anal Chim Acta 623:163–167. https://doi.org/10.1016/j.aca.2008.06.033
14. Ma M, Cantwell FF (1999) Solvent microextraction with simultaneous back-extraction for sample cleanup and preconcentration: preconcentration into a single microdrop. Anal Chem 71:388–393. https://doi.org/10.1021/ac9805899
15. Chamsaz M, Arbab-Zavar MH, Akhondzadeh J (2008) Triple-phase single-drop microextraction of silver and its determination using graphite-furnace atomic-absorption spectrometry. Anal Sci 24:799–801. https://doi.org/10.2116/analsci.24.799
16. Fan Z, Liu X (2008) Determination of methylmercury and phenylmercury in water samples by liquid-liquid-liquid microextraction coupled with capillary electrophoresis. J Chromatogr A 1180:187–192. https://doi.org/10.1016/j.chroma.2007.12.010
17. Theis AL, Waldack AJ, Hansen SM, Jeannot MA (2001) Headspace solvent microextraction. Anal Chem 73:5651–5654. https://doi.org/10.1021/ac015569c
18. Ma M, Cantwell FF (2000) Solvent microextraction with simultaneous back-extraction for sample cleanup and preconcentration: quantitative extraction. Anal Chem 70:3912–3919. https://doi.org/10.1021/ac980174n
19. Schnobrich CR, Jeannot MA (2008) Steady-state kinetic model for headspace solvent microextraction. J Chromatogr A 1215:30–36. https://doi.org/10.1016/j.chroma.2008.11.011
20. Psillakis E, Koutela N, Colussi AJ (2019) Vacuum-assisted headspace single-drop microextraction: eliminating interfacial gas-phase limitations. Anal Chim Acta 1092:9–16. https://doi.org/10.1016/j.aca.2019.09.056
21. Yangcheng L, Quan L, Guangsheng L, Youyuan D (2006) Directly suspended droplet microextraction. Anal Chim Acta 566:259–264
22. Khalili Zanjani MR, Yamini Y, Shariati S, Jönsson JA (2007) A new liquid-phase microextraction method based on solidification of floating organic drop. Anal Chim Acta 585:286–293. https://doi.org/10.1016/j.aca.2006.12.049
23. Sarafraz-Yazdi A, Mofazzeli F, Es'haghi Z (2007) Directly suspended droplet three liquid phase microextraction of diclofenac prior to LC. Chromatographia 67:49–53. https://doi.org/10.1365/s10337-007-0449-6

24. Jeannot MA, Cantwell FF (1997) Mass transfer characteristics of solvent extraction into a single drop at the tip of a syringe needle. Anal Chem 69:235–239. https://doi.org/10.1021/ac960814r
25. Williams DBG, George MJ, Meyer R, Marjanovic L (2011) Bubbles in solvent microextraction: the influence of intentionally introduced bubbles on extraction efficiency. Anal Chem 83:6713–6716. https://doi.org/10.1021/ac201323z
26. Bagherzadeh S, Faraji H, Shahbaazi HR, Azizinejad F (2022) Gas flow-assisted headspace-single drop microextraction to determine benzene, toluene, ethylbenzene and xylene in aqueous samples. Microchem J 183:108001. https://doi.org/10.1016/j.microc.2022.108001
27. Armenta S, Esteve-Turrillas FA, Garrigues S, de la Guardia M (2022) Alternative green solvents in sample preparation. Green Anal Chem 1:100007. https://doi.org/10.1016/j.greeac.2022.100007
28. Kokosa JM (2019) Selecting an extraction solvent for a greener liquid phase microextraction (LPME) mode-based analytical method. Trends Anal Chem 118:238–247. https://doi.org/10.1016/j.trac.2019.05.012
29. Tobiszewski M, Namieśnik J (2017) Greener organic solvents in analytical chemistry. Curr Opin Green Sustain Chem 5:1–4. https://doi.org/10.1016/j.cogsc.2017.03.002
30. Liu J-F, Jiang G-B, Chi Y-G et al (2003) Use of ionic liquids for liquid-phase microextraction of polycyclic aromatic hydrocarbons. Anal Chem 75:5870–5876. https://doi.org/10.1021/ac034506m
31. González-Martín R, Lodoso-Ruiz E, Trujillo-Rodríguez MJ, Pino V (2022) Magnetic ionic liquids in analytical microextraction: a tutorial review. J Chromatogr A 1685:463577. https://doi.org/10.1016/j.chroma.2022.463577
32. Bystrzanowska M, Pena-Pereira F, Marcinkowski Ł, Tobiszewski M (2019) How green are ionic liquids? a multicriteria decision analysis approach. Ecotoxicol Environ Saf 174:455–458. https://doi.org/10.1016/j.ecoenv.2019.03.014
33. de Jesus SS, Maciel Filho R (2022) Are ionic liquids eco-friendly? Renew Sustain Energy Rev 157:112039. https://doi.org/10.1016/j.rser.2021.112039
34. Maculewicz J, Świacka K, Stepnowski P et al (2022) Ionic liquids as potentially hazardous pollutants: evidences of their presence in the environment and recent analytical developments. J Hazard Mater 437:129353. https://doi.org/10.1016/j.jhazmat.2022.129353
35. Santos LB, Assis RS, Barreto JA et al (2022) Deep eutectic solvents in liquid-phase microextraction: contribution to green chemistry. TrAC—Trends Anal Chem 146:116478. https://doi.org/10.1016/j.trac.2021.116478
36. Rodinkov O, Znamenskaya E, Spivakovsky V et al (2022) Deep eutectic solvents-based headspace single-drop microextraction for the chromatographic determination of phenols and aliphatic alcohols in atmospheric air. Microchem J 182:107854. https://doi.org/10.1016/j.microc.2022.107854
37. Moradi M, Yamini Y, Feizi N (2021) Development and challenges of supramolecular solvents in liquid-based microextraction methods. TrAC—Trends Anal Chem 138:116231. https://doi.org/10.1016/j.trac.2021.116231
38. Ballesteros-Gómez A, Sicilia MD, Rubio S (2010) Supramolecular solvents in the extraction of organic compounds: a review. Anal Chim Acta 677:108–130. https://doi.org/10.1016/j.aca.2010.07.027
39. Skok A, Vishnikin A, Bazel Y (2022) A new approach for sulfite determination by headspace liquid-phase microextraction with an optical probe. Anal Methods 14:3299–3306. https://doi.org/10.1039/d2ay00943a
40. Leng G, Lin L, Worsfold PJ et al (2021) A simple and rapid head space-single drop microextraction-'spectro-pipette' (HS-SDME-SP) method for the on-site measurement of arsenic species in natural waters. Microchem J 168:106441. https://doi.org/10.1016/j.microc.2021.106441
41. García-Figueroa A, Pena-Pereira F, Lavilla I, Bendicho C (2017) Headspace single-drop microextraction coupled with microvolume fluorospectrometry for highly sensitive determination of bromide. Talanta 170:9–14. https://doi.org/10.1016/j.talanta.2017.03.090

42. Kailasa SK, Koduru JR, Park TJ et al (2021) Applications of single-drop microextraction in analytical chemistry: a review. Trends Environ Anal Chem 29:e00113. https://doi.org/10.1016/j.teac.2020.e00113
43. Sajid M, Płotka-Wasylka J (2022) Green analytical chemistry metrics: a review. Talanta 238:123046. https://doi.org/10.1016/J.TALANTA.2021.123046
44. Pena-pereira F, Lavilla I, Bendicho C (2016) Single-drop microextraction and related techniques. In: Valcárcel M, Cárdenas S, Lucena R (eds) Analytical microextraction techniques. Bentham Books, pp 327–379
45. Aguilera-herrador E, Lucena R, Cárdenas S, Valcárcel M (2008) Direct coupling of ionic liquid based single-drop microextraction and GC/MS. Anal Chem 80:793–800. https://doi.org/10.1021/ac071555g
46. Chisvert A, Román IP, Vidal L, Canals A (2009) Simple and commercial readily-available approach for the direct use of ionic liquid-based single-drop microextraction prior to gas chromatography determination of chlorobenzenes in real water samples as model analytical application. J Chromatogr A 1216:1290–1295. https://doi.org/10.1016/j.chroma.2008.12.078
47. Zhang J, Lee HK (2010) Headspace ionic liquid-based microdrop liquid-phase microextraction followed by microdrop thermal desorption-gas chromatographic analysis. Talanta 81:537–542. https://doi.org/10.1016/j.talanta.2009.12.039
48. Mokhtar SU, Chin S, Vijayaraghavan R et al (2015) Direct ionic liquid extractant injection for volatile chemical analysis: a gas chromatography sampling technique. Green Chem 17:573–581. https://doi.org/10.1039/C4GC01364F
49. Yousefi SM, Shemirani F, Ghorbanian SA (2018) Enhanced headspace single drop microextraction method using deep eutectic solvent based magnetic bucky gels: application to the determination of volatile aromatic hydrocarbons in water and urine samples. J Sep Sci 41:966–974. https://doi.org/10.1002/jssc.201700807
50. Mehravar A, Feizbakhsh A, Sarafi AHM et al (2020) Deep eutectic solvent-based headspace single-drop microextraction of polycyclic aromatic hydrocarbons in aqueous samples. J Chromatogr A 1632:461618. https://doi.org/10.1016/j.chroma.2020.461618
51. Xu L, Basheer C, Lee HK (2007) Developments in single-drop microextraction. J Chromatogr A 1152:184–192. https://doi.org/10.1016/j.chroma.2006.10.073
52. Pena-Pereira F, Lavilla I, Bendicho C et al (2009) Speciation of mercury by ionic liquid-based single-drop microextraction combined with high-performance liquid chromatography-photodiode array detection. Talanta 78:537–541. https://doi.org/10.1016/j.talanta.2008.12.003
53. Wang Q, Qiu H, Li J et al (2010) On-line coupling of ionic liquid-based single-drop microextraction with capillary electrophoresis for sensitive detection of phenols. J Chromatogr A 1217:5434–5439. https://doi.org/10.1016/j.chroma.2010.06.059
54. Sobhi HR, Yamini Y, Esrafili A, Abadi RHHB (2008) Suitable conditions for liquid-phase microextraction using solidification of a floating drop for extraction of fat-soluble vitamins established using an orthogonal array experimental design. J Chromatogr A 1196–1197:28–32. https://doi.org/10.1016/j.chroma.2008.05.005
55. Vallecillos L, Pocurull E, Borrull F (2012) Fully automated ionic liquid-based headspace single drop microextraction coupled to GC–MS/MS to determine musk fragrances in environmental water samples. Talanta 99:824–832. https://doi.org/10.1016/j.talanta.2012.07.036
56. Guo L, Binte Nawi N, Lee HK (2016) Fully automated headspace bubble-in-drop microextraction. Anal Chem 88:8409–8414. https://doi.org/10.1021/acs.analchem.6b01543
57. Medina DAV, Rodríguez Cabal LF, Titato GM et al (2019) Automated online coupling of robot-assisted single drop microextraction and liquid chromatography. J Chromatogr A 1595:66–72. https://doi.org/10.1016/j.chroma.2019.02.036
58. Mafra G, Vieira AA, Merib J et al (2019) Single drop microextraction in a 96-well plate format: a step toward automated and high-throughput analysis. Anal Chim Acta 1063:159–166. https://doi.org/10.1016/j.aca.2019.02.013

59. Choi K, Su JK, Yoo GJ et al (2009) Single drop microextraction using commercial capillary electrophoresis instruments. Anal Chem 81:225–230. https://doi.org/10.1021/ac801886b
60. Springer VH, Lista AG (2015) In-line coupled single drop liquid–liquid–liquid microextraction with capillary electrophoresis for determining fluoroquinolones in water samples. Electrophoresis 36:1572–1579. https://doi.org/10.1002/elps.201400602
61. Pena-Pereira F, Lavilla I, Bendicho C (2009) Miniaturized preconcentration methods based on liquid-liquid extraction and their application in inorganic ultratrace analysis and speciation: a review. Spectrochim Acta—Part B At Spectrosc 64:1–15. https://doi.org/10.1016/j.sab.2008.10.042
62. García-Figueroa A, Lavilla I, Bendicho C (2019) Speciation of CdTe quantum dots and Te(IV) following oxidative degradation induced by iodide and headspace single-drop microextraction combined with graphite furnace atomic absorption spectrometry. Spectrochim Acta Part B 158:105631. https://doi.org/10.1016/j.sab.2019.06.001
63. Gil S, de Loos-Vollebregt MTC, Bendicho C (2009) Optimization of a single-drop microextraction method for multielemental determination by electrothermal vaporization inductively coupled plasma mass spectrometry following in situ vapor generation. Spectrochim Acta Part B At Spectrosc 64:208–214. https://doi.org/10.1016/j.sab.2008.12.002
64. Almeida JS, Anunciação TA, Brandão GC et al (2015) Ultrasound-assisted single-drop microextraction for the determination of cadmium in vegetable oils using high-resolution continuum source electrothermal atomic absorption spectrometry. Spectrochim Acta Part B At Spectrosc 107:159–163. https://doi.org/10.1016/j.sab.2015.03.002
65. Akhtar A, Kazi TG, Afridi HI et al (2020) Simultaneous preconcentration of toxic elements in eye makeup products through single drop ionic liquid based non-dispersive microextraction method using narrow glass column: multivariate application. Microchem J 157:104963. https://doi.org/10.1016/j.microc.2020.104963
66. Wen X, Deng Q, Wang J et al (2013) A new coupling of ionic liquid based-single drop microextraction with tungsten coil electrothermal atomic absorption spectrometry. Spectrochim Acta A Mol Biomol Spectrosc 105:320–325. https://doi.org/10.1016/j.saa.2012.12.040
67. López-García I, Rivas RE, Hernández-Córdoba M (2010) Liquid-phase microextraction with solidification of the organic floating drop for the preconcentration and determination of mercury traces by electrothermal atomic absorption spectrometry. Anal Bioanal Chem 396:3097–3102. https://doi.org/10.1007/s00216-010-3500-7
68. Sakanupongkul A, Sananmuang R, Udnan Y et al (2019) Speciation of mercury in water and freshwater fish samples by a two-step solidified floating organic drop microextraction with electrothermal atomic absorption spectrometry. Food Chem 277:496–503. https://doi.org/10.1016/j.foodchem.2018.10.131
69. Guo X, He M, Chen B, Hu B (2012) Solidified floating organic drop microextraction combined with ETV-ICP-MS for the determination of trace heavy metals in environmental water samples. Talanta 94:70–76. https://doi.org/10.1016/j.talanta.2012.02.053
70. Pena F, Lavilla I, Bendicho C (2008) Immersed single-drop microextraction interfaced with sequential injection analysis for determination of Cr(VI) in natural waters by electrothermal-atomic absorption spectrometry. Spectrochim Acta Part B At Spectrosc Spectrosc 63:498–503. https://doi.org/10.1016/j.sab.2008.01.005
71. Anthemidis AN, Adam ISI (2009) Development of on-line single-drop micro-extraction sequential injection system for electrothermal atomic absorption spectrometric determination of trace metals. Anal Chim Acta 632:216–220. https://doi.org/10.1016/j.aca.2008.10.078
72. Mitani C, Kotzamanidou A, Anthemidis AN (2014) Automated headspace single-drop microextraction via a lab-in-syringe platform for mercury electrothermal atomic absorption spectrometric determination after in situ vapor generation. J Anal At Spectrom 29:1491–1498. https://doi.org/10.1039/c4ja00062e
73. Xiao Q, Hu B, He M (2008) Speciation of butyltin compounds in environmental and biological samples using headspace single drop microextraction coupled with gas chromatography-inductively coupled plasma mass spectrometry. J Chromatogr A 1211:135–141. https://doi.org/10.1016/j.chroma.2008.09.089

74. Şahin ÇA, Tokgöz I (2010) A novel solidified floating organic drop microextraction method for preconcentration and determination of copper ions by flow injection flame atomic absorption spectrometry. Anal Chim Acta 667:83–87. https://doi.org/10.1016/j.aca.2010.04.012
75. Fakhriyan G, Mousavi HZ, Sajjadi SM (2016) Speciation and determination of Cr(III) and Cr(VI) by directly suspended droplet microextraction coupled with flame atomic absorption spectrometry: an application of central composite design strategy as an experimental design tool. Anal Methods 8:5070–5078. https://doi.org/10.1039/c6ay00566g
76. Liu J-F, Chi Y-G, Jiang G-B (2005) Screening the extractability of some typical environmental pollutants by ionic liquids in liquid-phase microextraction. J Sep Sci 28:87–91. https://doi.org/10.1002/jssc.200401805
77. Wang Y, Luo X, Tang J, Hu X (2011) Determination of Se(IV) using solidified floating organic drop microextraction coupled to ultrasound-assisted back-extraction and hydride generation atomic fluorescence spectrometry. Microchim Acta 173:267–273. https://doi.org/10.1007/s00604-011-0574-7
78. Yuan CG, Wang J, Jin Y (2012) Ultrasensitive determination of mercury in human saliva by atomic fluorescence spectrometry based on solidified floating organic drop microextraction. Microchim Acta 177:153–158. https://doi.org/10.1007/s00604-012-0768-7
79. Pytlakowska K, Sitko R (2012) Directly suspended droplet microextraction combined with energy-dispersive X-ray fluorescence spectrometry to determine nano levels of phosphate in surface water. J Anal At Spectrom 27:460–465. https://doi.org/10.1039/c2ja10313c
80. Oskolok KV, Monogarova OV, Alov NV (2018) Determination of mercury(II) in drinking water by total reflection X-ray fluorescence spectrometry and liquid–liquid microextraction. Anal Lett 51:2457–2467. https://doi.org/10.1080/00032719.2017.1423078
81. Aguirre MA, Legnaioli S, Almodóvar F et al (2013) Elemental analysis by surface-enhanced laser-induced breakdown spectroscopy combined with liquid-liquid microextraction. Spectrochim Acta—Part B At Spectrosc 79–80:88–93. https://doi.org/10.1016/j.sab.2012.11.011
82. Aguirre MA, Nikolova H, Hidalgo M, Canals A (2015) Hyphenation of single-drop microextraction with laser-induced breakdown spectrometry for trace analysis in liquid samples: a viability study. Anal Methods 7:877–883. https://doi.org/10.1039/c4ay02218a
83. Pena-Pereira F, Costas-Mora I, Romero V et al (2011) Advances in miniaturized UV-vis spectrometric systems. TrAC Trends Anal Chem 30:1637–1648. https://doi.org/10.1016/j.trac.2011.04.018
84. Senra-Ferreiro S, Pena-Pereira F, Lavilla I, Bendicho C (2010) Griess micro-assay for the determination of nitrite by combining fibre optics-based cuvetteless UV-Vis micro-spectrophotometry with liquid-phase microextraction. Anal Chim Acta 668:195–200. https://doi.org/10.1016/j.aca.2010.04.038
85. Sáenz M, Alvarado J, Pena-Pereira F et al (2011) Liquid-phase microextraction with in-drop derivatization combined with microvolume fluorospectrometry for free and hydrolyzed formaldehyde determination in textile samples. Anal Chim Acta 687:50–55. https://doi.org/10.1016/j.aca.2010.12.006
86. Costas-Mora I, Romero V, Pena-Pereira F et al (2012) Quantum dots confined in an organic drop as luminescent probes for detection of selenium by microfluorospectrometry after hydridation: study of the quenching mechanism and analytical performance. Anal Chem 84. https://doi.org/10.1021/ac300221s
87. Tang S, Qi T, Yao Y et al (2020) Magnetic three-phase single-drop microextraction for rapid amplification of the signals of DNA and microRNA analysis. Anal Chem 92:12290–12296. https://doi.org/10.1021/acs.analchem.0c01936
88. Lavilla I, Pena-Pereira F, Gil S et al (2009) Microvolume turbidimetry for rapid and sensitive determination of the acid labile sulfide fraction in waters after headspace single-drop microextraction with in situ generation of volatile hydrogen sulfide. Anal Chim Acta 647:112–116. https://doi.org/10.1016/j.aca.2009.05.035
89. Zaruba S, Vishnikin AB, Škrlíková J, Andruch V (2016) Using an optical probe as the microdrop holder in headspace single drop microextraction: determination of sulfite in food samples. Anal Chem 88:10296–10300. https://doi.org/10.1021/acs.analchem.6b03129

90. Zaruba S, Vishnikin AB, Škrlíková J et al (2017) A two-in-one device for online monitoring of direct immersion single-drop microextraction: an optical probe as both microdrop holder and measuring cell. RSC Adv 7:29421–29427. https://doi.org/10.1039/c7ra02326j
91. Šrámková I, Horstkotte B, Solich P, Sklenářová H (2014) Automated in-syringe single-drop head-space micro-extraction applied to the determination of ethanol in wine samples. Anal Chim Acta 828:53–60. https://doi.org/10.1016/j.aca.2014.04.031
92. Šrámková IH, Horstkotte B, Fikarová K et al (2018) Direct-immersion single-drop microextraction and in-drop stirring microextraction for the determination of nanomolar concentrations of lead using automated Lab-In-Syringe technique. Talanta 184:162–172. https://doi.org/10.1016/j.talanta.2018.02.101
93. Shahvar A, Saraji M, Shamsaei D (2018) Headspace single drop microextraction combined with mobile phone-based on-drop sensing for the determination of formaldehyde. Sensors Actuators, B Chem 273:1474–1478. https://doi.org/10.1016/j.snb.2018.07.071
94. Fan YZ, Dong JX, Zhang Y et al (2019) A smartphone-coalesced nanoprobe for high selective ammonia sensing based on the pH-responsive biomass carbon nanodots and headspace single drop microextraction. Spectrochim Acta—Part A Mol Biomol Spectrosc 219:382–390. https://doi.org/10.1016/j.saa.2019.04.073
95. Tang S, Qi T, Xia D et al (2019) Smartphone nanocolorimetric determination of hydrogen sulfide in biosamples after silver-gold core-shell nanoprism-based headspace single-drop microextraction. Anal Chem 91:5888–5895. https://doi.org/10.1021/acs.analchem.9b00255
96. Sudhir P-R, Wu H-F, Zhou Z-C (2005) Identification of peptides using gold coupled with AP-MALDI mass spectrometry. Anal Chem 77:7380–7385. https://doi.org/10.1021/ac051162m
97. Sudhir P-R, Shrivas K, Zhou Z-C, Wu H-F (2008) Single drop microextraction using silver nanoparticles as electrostatic probes for peptide analysis in atmospheric pressure matrix-assisted laser desorption/ionization mass spectrometry and comparison with gold electrostatic probes and silver hydrophobic. Rapid Commun Mass Spectrom 22:3076–3086. https://doi.org/10.1002/rcm.3710Single
98. Ahmad F, Wu H-F (2011) Characterization of pathogenic bacteria using ionic liquid via single drop microextraction combined with MALDI-TOF MS. Analyst 136:4020–4027. https://doi.org/10.1039/c1an15350a
99. Shastri L, Abdelhamid HN, Nawaz M, Wu HF (2015) Synthesis, characterization and bifunctional applications of bidentate silver nanoparticle assisted single drop microextraction as a highly sensitive preconcentrating probe for protein analysis. RSC Adv 5:41595–41603. https://doi.org/10.1039/c5ra04032a
100. Fang L, Deng J, Yu Y et al (2016) Coupling liquid-phase microextraction with paper spray for rapid analysis of malachite green, crystal violet and their metabolites in complex samples using mass spectrometry. Anal Methods 8:6651–6656. https://doi.org/10.1039/c6ay01466f
101. Sun WH, Wei Y, Guo XL et al (2020) Nanoliter-scale droplet-droplet microfluidic microextraction coupled with MALDI-TOF mass spectrometry for metabolite analysis of cell droplets. Anal Chem 92:8759–8767. https://doi.org/10.1021/acs.analchem.0c00007
102. Timofeeva I, Medinskaia K, Nikolaeva L et al (2016) Stepwise injection potentiometric determination of caffeine in saliva using single-drop microextraction combined with solvent exchange. Talanta 150:655–660. https://doi.org/10.1016/j.talanta.2016.01.001
103. Jahromi Z, Mostafavi A, Shamspur T, Mohamadim M (2017) Magnetic ionic liquid assisted single-drop microextraction of ascorbic acid before its voltammetric determination. J Sep Sci 40:4041–4049. https://doi.org/10.1002/jssc.201700664
104. Jiang Y, Dong X, Li Y et al (2021) An environmentally-benign flow-batch system for headspace single-drop microextraction and on-drop conductometric detecting ammonium. Talanta 224:121849. https://doi.org/10.1016/j.talanta.2020.121849
105. Verma D, Verma SK, Deb MK (2009) Single-drop micro-extraction and diffuse reflectance Fourier transform infrared spectroscopic determination of chromium in biological fluids. Talanta 78:270–277. https://doi.org/10.1016/j.talanta.2008.11.020
106. Verma D, Deb MK (2012) Determination of vanadium (V) employing a new combined single drop micro-extraction and diffuse reflectance Fourier transform infrared spectroscopy technique. Int J Environ Anal Chem 92:59–75. https://doi.org/10.1080/03067310903410998

107. Sen BK, Tiwari S, Deb MK, Pervez S (2015) Nanogram level quantification of molybdenum(VI) by novel hyphenated SDME/DRS-FTIR in human biological fluid. Anal Methods 7:9474–9481. https://doi.org/10.1039/c5ay01801c
108. Chandrawanshi S, Verma SK, Deb MK (2018) Collective ion-pair single-drop microextraction attenuated total reflectance fourier transform infrared spectroscopic determination of perchlorate in bioenvironmental samples. J AOAC Int 101:1145–1155. https://doi.org/10.5740/jaoacint.17-0188
109. Tiwari S, Deb MK (2019) Modified silver nanoparticles-enhanced single drop microextraction of tartrazine in food samples coupled with diffuse reflectance Fourier transform infrared spectroscopic analysis. Anal Methods 11:3552–3562. https://doi.org/10.1039/c9ay00713j
110. Kurrey R, Deb MK, Shrivas K (2019) Surface enhanced infra-red spectroscopy with modified silver nanoparticles (AgNPs) for detection of quaternary ammonium cationic surfactants. New J Chem 43:8109–8121. https://doi.org/10.1039/c9nj01795j
111. Aguilera-Herrador E, Lucena R, Cárdenas S, Valcárcel M (2009) Ionic liquid-based single drop microextraction and room-temperature gas chromatography for on-site ion mobility spectrometric analysis. J Chromatogr A 1216:5580–5587. https://doi.org/10.1016/j.chroma.2009.05.071
112. Márquez-Sillero I, Aguilera-Herrador E, Cárdenas S, Valcárcel M (2011) Determination of 2,4,6-tricholoroanisole in water and wine samples by ionic liquid-based single-drop microextraction and ion mobility spectrometry. Anal Chim Acta 702:199–204. https://doi.org/10.1016/j.aca.2011.06.046
113. Márquez-Sillero I, Cárdenas S, Valcárcel M (2011) Direct determination of 2,4,6-tricholoroanisole in wines by single-drop ionic liquid microextraction coupled with multicapillary column separation and ion mobility spectrometry detection. J Chromatogr A 1218:7574–7580. https://doi.org/10.1016/j.chroma.2011.06.032
114. Li S (2019) Preconcentration and determination of psychotropic drugs in urine samples by ion mobility spectrometry with electrospray ionization coupling on-line single-drop liquid-liquid-liquid microextraction. J Anal Methods Chem 8561801. https://doi.org/10.1155/2019/8561801
115. Costas-Mora I, Romero V, Lavilla I, Bendicho C (2013) Solid-state chemiluminescence assay for ultrasensitive detection of antimony using on-vial immobilization of CdSe quantum dots combined with liquid–liquid–liquid microextraction. Anal Chim Acta 788:114–121. https://doi.org/10.1016/j.aca.2013.06.007
116. Santos EB, Valsecchi C, Gonçalves JLS et al (2019) Coupling single-drop microextraction with SERS: a demonstration using p-MBA on gold nanohole array substrate. Sensors 19:4394. https://doi.org/10.3390/s19204394
117. Abbasi-Ahd A, Shokoufi N, Kargosha K (2017) Headspace single-drop microextraction coupled to microchip-photothermal lens microscopy for highly sensitive determination of captopril in human serum and pharmaceuticals. Microchim Acta 184:2403–2409. https://doi.org/10.1007/s00604-017-2266-4
118. Emaus MN, Clark KD, Hinners P, Anderson JL (2018) Preconcentration of DNA using magnetic ionic liquids that are compatible with real-time PCR for rapid nucleic acid quantification. Anal Bioanal Chem 410:4135–4144. https://doi.org/10.1007/s00216-018-1092-9
119. Kiszkiel-Taudul I, Starczewska B (2019) Single drop microextraction coupled with liquid chromatography-tandem mass spectrometry (SDME-LC-MS/MS) for determination of ranitidine in water samples. Microchem J 145:936–941. https://doi.org/10.1016/j.microc.2018.12.015
120. Neri TS, Rocha DP, Muñoz RAA et al (2019) Highly sensitive procedure for determination of Cu(II) by GF AAS using single-drop microextraction. Microchem J 147:894–898. https://doi.org/10.1016/j.microc.2019.04.014
121. Chullasat K, Huang Z, Bunkoed O et al (2020) Bubble-in-drop microextraction of carbamate pesticides followed by gas chromatography-mass spectrometric analysis. Microchem J 155:104666. https://doi.org/10.1016/j.microc.2020.104666

122. Abolghasemi MM, Piryaei M, Imani RM (2020) Deep eutectic solvents as extraction phase in head-space single-drop microextraction for determination of pesticides in fruit juice and vegetable samples. Microchem J 158:105041. https://doi.org/10.1016/j.microc.2020.105041
123. Padilha J da S, Pedrozo-Peñafiel MJ, Azevedo MFMF, et al (2022) Silver-modified nitrogen-doped graphene quantum dots as a sensor for formaldehyde in milk using headspace micro-extraction on a single-drop of aqueous nanoparticles dispersion. Anal Chim Acta 1232:340479. https://doi.org/10.1016/j.aca.2022.340479
124. Purgat K, Olejarz P, Kośka I et al (2020) Determination of homocysteine thiolactone in human urine by capillary zone electrophoresis and single drop microextraction. Anal Biochem 596:113640. https://doi.org/10.1016/j.ab.2020.113640
125. Nunes LS, Korn MGA, Lemos VA (2021) A novel direct-immersion single-drop microextraction combined with digital colorimetry applied to the determination of vanadium in water. Talanta 224:121893. https://doi.org/10.1016/j.talanta.2020.121893
126. Li M, Gu C, Luo L et al (2019) Determination of trace methanesulfonates in drug matrix using derivatization and headspace single drop microextraction followed by high-performance liquid chromatography with ultraviolet detection. J Chromatogr A 1591:131–137. https://doi.org/10.1016/j.chroma.2019.01.038
127. Jeong S, Valdez JE, Miękus N et al (2021) Facile and highly efficient three-phase single drop microextraction in-line coupled with capillary electrophoresis. J Chromatogr A 1655:462520. https://doi.org/10.1016/j.chroma.2021.462520
128. Farooq MQ, Zeger VR, Anderson JL (2021) Comparing the extraction performance of cyclodextrin-containing supramolecular deep eutectic solvents versus conventional deep eutectic solvents by headspace single drop microextraction. J Chromatogr A 1658:462588. https://doi.org/10.1016/j.chroma.2021.462588
129. Qin B, Wang X, Tang L et al (2022) Comparative study of headspace and headspace single drop microextraction combined with GC for the determination of methanol in wine. J Chromatogr A 1673:463079. https://doi.org/10.1016/j.chroma.2022.463079
130. Saraji M, Javadian S (2019) Single-drop microextraction combined with gas chromatography-electron capture detection for the determination of acrylamide in food samples. Food Chem 274:55–60. https://doi.org/10.1016/j.foodchem.2018.08.108
131. Qi T, Xu M, Yao Y et al (2020) Gold nanoprism/Tollens' reagent complex as plasmonic sensor in headspace single-drop microextraction for colorimetric detection of formaldehyde in food samples using smartphone readout. Talanta 220:121388. https://doi.org/10.1016/j.talanta.2020.121388
132. Yıldırım S, Cocovi-Solberg DJ, Uslu B et al (2022) Lab-In-Syringe automation of deep eutectic solvent-based direct immersion single drop microextraction coupled online to high-performance liquid chromatography for the determination of fluoroquinolones. Talanta 246:123476. https://doi.org/10.1016/j.talanta.2022.123476
133. Abreu DCP, Botrel BMC, Bazana MJF et al (2019) Development and comparative analysis of single-drop and solid-phase microextraction techniques in the residual determination of 2-phenoxyethanol in fish. Food Chem 270:487–493. https://doi.org/10.1016/j.foodchem.2018.07.136
134. Jain A, Soni S, Verma KK (2021) Combined liquid phase microextraction and fiber-optics-based cuvetteless micro-spectrophotometry for sensitive determination of ammonia in water and food samples by the indophenol reaction. Food Chem 340:128156. https://doi.org/10.1016/j.foodchem.2020.128156
135. Ma Z, Zhao T, Cui S et al (2022) Determination of ethyl carbamate in wine by matrix modification-assisted headspace single-drop microextraction and gas chromatography–mass spectrometry technique. Food Chem 373:131573. https://doi.org/10.1016/j.foodchem.2021.131573
136. Yao Y, Kuang J, Ju J et al (2022) Solvent-free magnetic-tip microextraction into a single drop for fluorescence sensing. Sens Actuators B Chem 352:131044. https://doi.org/10.1016/j.snb.2021.131044

137. Bokhon F, Daryanavard SM, Gholamshahzadeh A, Tezerji AK (2021) Application of experimental design for determination of methanol and ethanol in transformer industrial oils using headspace single-drop microextraction. Anal Bioanal Chem Res 8:27–38. https://doi.org/10.22036/abcr.2020.233914.1510
138. Villarino N, Pena-Pereira F, Lavilla I, Bendicho C (2022) Waterproof cellulose-based substrates for in-drop plasmonic colorimetric sensing of volatiles: application to acid-labile sulfide determination in waters. ACS Sensors 7:839–848. https://doi.org/10.1021/acssensors.1c02585
139. Bocelli MD, Vargas Medina DA, García Rodríguez JP et al (2022) Determination of parabens in wastewater samples via robot-assisted dynamic single-drop microextraction and liquid chromatography–tandem mass spectrometry. Electrophoresis 43:1567–1576. https://doi.org/10.1002/elps.202100390
140. Wu L, Li Z (2021) Continuous-flow microwave-assisted extraction coupled with online single drop microextraction prior to GC-MS for determination of amide herbicides in rice samples. J Sep Sci 44:870–878. https://doi.org/10.1002/jssc.202001092
141. Mafra G, Will C, Huelsmann R et al (2021) A proof-of-concept of parallel single-drop microextraction for the rapid and sensitive biomonitoring of pesticides in urine. J Sep Sci 44:1961–1968. https://doi.org/10.1002/jssc.202001157
142. Poliwoda A, Zielińska K, Wieczorek PP (2020) Direct analysis of psilocin and muscimol in urine samples using single drop microextraction technique in-line with capillary electrophoresis. Molecules 25:1566. https://doi.org/10.3390/molecules25071566
143. Purgat K, Kośka I, Kubalczyk P (2021) The use of single drop microextraction and field amplified sample injection for CZE determination of homocysteine thiolactone in urine. Molecules 26:5687. https://doi.org/10.3390/molecules26185687
144. Bagheri N, Saraji M (2019) Combining gold nanoparticle-based headspace single-drop microextraction and a paper-based colorimetric assay for selenium determination. Anal Bioanal Chem 411:7441–7449. https://doi.org/10.1007/s00216-019-02106-4
145. Triaux Z, Petitjean H, Marchioni E et al (2020) Deep eutectic solvent–based headspace single-drop microextraction for the quantification of terpenes in spices. Anal Bioanal Chem 412:933–948. https://doi.org/10.1007/s00216-019-02317-9
146. Nunes LS, das Graças Andrade Korn M, Lemos VA (2020) Direct immersion single-drop microextraction and continuous-flow microextraction for the determination of manganese in tonic drinks and seafood samples. Food Anal Methods 13:1681–1689.https://doi.org/10.1007/s12161-020-01794-4
147. Chen Y, Liu H, Peng J et al (2020) Determination of polybrominated diphenyl ethers and metabolites by single-drop microextraction and GC–MS/MS. SN Appl Sci 2:986. https://doi.org/10.1007/s42452-020-2642-2
148. Kim J, Choi K, Chung DS (2019) Synergistic coupling of in-line single-drop microextraction and on-line large-volume sample stacking for capillary electrophoresis/mass spectrometry. Anal Bioanal Chem 411:1067–1073. https://doi.org/10.1007/s00216-018-1535-3
149. Li Y, Zhang L, Wu L et al (2018) Purification and enrichment of polycyclic aromatic hydrocarbons in environmental water samples by column clean-up coupled with continuous flow single drop microextraction. J Chromatogr A 1567:81–89. https://doi.org/10.1016/j.chroma.2018.07.013
150. Tolessa T, Tan ZQ, Yin YG, Liu JF (2018) Single-drop gold nanoparticles for headspace microextraction and colorimetric assay of mercury (II) in environmental waters. Talanta 176:77–84. https://doi.org/10.1016/j.talanta.2017.07.097
151. Li X, Li H, Ma W et al (2018) Determination of patulin in apple juice by single-drop liquid-liquid-liquid microextraction coupled with liquid chromatography-mass spectrometry. Food Chem 257:1–6. https://doi.org/10.1016/j.foodchem.2018.02.077
152. Pena-Pereira F, García-Figueroa A, Lavilla I, Bendicho C (2018) Ratiometric detection of total bromine in E-waste polymers by colloidal gold-based headspace single-drop microextraction and microvolume spectrophotometry. Sensors Actuators B Chem 261:481–488. https://doi.org/10.1016/j.snb.2018.01.107

153. Huang X, Guan M, Lu Z, Hang Y (2018) Determination of trace antimony (III) in water samples with single drop microextraction using BPHA-[C4mim][PF6] system followed by graphite furnace atomic absorption spectrometry. Int J Anal Chem 8045324. https://doi.org/10.1155/2018/8045324
154. Fernández E, Vidal L, Canals A (2018) Hydrophilic magnetic ionic liquid for magnetic headspace single-drop microextraction of chlorobenzenes prior to thermal desorption-gas chromatography-mass spectrometry. Anal Bioanal Chem 410:4679–4687. https://doi.org/10.1007/s00216-017-0755-2
155. Martín-Alonso M, Pena-Pereira F, Lavilla I, Bendicho C (2018) Gold nanorods for in-drop colorimetric determination of thiomersal after photochemical decomposition. Microchim Acta 185:221. https://doi.org/10.1007/s00604-018-2760-3

Hollow-Fibre Liquid-Phase Microextraction

Frederik André Hansen and Stig Pedersen-Bjergaard

Abstract This chapter discusses microextraction using liquid membranes immobilized in a porous support membrane, including two- and three-phase hollow fibre liquid-phase microextraction (HF-LPME), 96-well LPME (or parallel artificial liquid membrane extraction (PALME)), and electromembrane extraction (EME). These techniques are essentially two- or three-phase liquid extraction systems, but downscaled to the level where the consumption of organic solvent per sample is less than 10 μL. Such microextraction systems are interesting for several reasons. First, they are ideal for green sample preparation, and therefore they are expected to be important in the near future in the context of sustainability. In addition, due to size and technical arrangement, they are easily implemented in microchip systems. Recently, several research papers have been investigating such microchip systems in combination with smartphone detection. This research has the potential to move analytical measurements out of today's specialized laboratories. The fundamentals are discussed, to underline that microextraction with liquid membranes can be performed in partition-based systems, or in systems controlled by an external electrical field. In addition, this chapter discusses novel developments and new applications, based on examples from recent literature. The chapter is not comprehensive but is intended to give a flavour of the field.

Keywords Sample preparation · Microextraction · Liquid-phase microextraction · Hollow-fibre · Electromembrane extraction

F. A. Hansen · S. Pedersen-Bjergaard (✉)
Department of Pharmacy, University of Oslo, Blindern, P.O. Box 1068, 0316 Oslo, Norway
e-mail: stigpe@farmasi.uio.no

F. A. Hansen
e-mail: f.a.hansen@farmasi.uio.no

S. Pedersen-Bjergaard
Department of Pharmacy, Faculty of Health and Medical Sciences, University of Copenhagen, Universitetsparken 2, 2100 Copenhagen, Denmark

M. Á. Rodríguez-Delgado et al. (eds.), *Microextraction Techniques*, Integrated Analytical Systems, https://doi.org/10.1007/978-3-031-50527-0_8

Abbreviations

EME	Electromembrane extraction
HF-LPME	Hollow-fibre liquid-phase microextraction
LPME	Liquid-phase microextraction
SDME	Single drop microextraction
SPME	Solid-phase microextraction

1 Introduction

The interest for miniaturized extraction systems was initiated by the invention of solid-phase microextraction (SPME) in 1990 [1]. In SPME, a thin fibre covered with a polymeric coating serves as extraction phase. The fibre can be immersed directly in the sample for extraction or can be inserted in the headspace for extraction of volatile compounds. Polydimethylsiloxane (PDMS) is the most popular coating, but a range of commercial alternatives are available. SPME is often combined with gas chromatography (GC), after thermal desorption of the extracted material. SPME can also be combined with liquid chromatography (LC), and in such cases the extracted material is desorbed from the fibre using organic solvent. SPME was commercialized a few years after the first scientific paper and has gained substantial interest. SPME is solvent-free, automated, and enables soft extraction.

In parallel with the development of SPME, different approaches to liquid-phase microextraction were introduced. In 1996, single-drop microextraction (SDME) was introduced [2, 3]. In SDME, target analytes are extracted from the sample, and into a small droplet of organic solvent located at the needle tip of a micro-syringe. The volume is only a few micro litres, and the droplet can be injected directly into GC after extraction. Equipment for SDME is very simple and includes a micro-syringe and a magnetic stirrer. SDME is simple and inexpensive, but the droplet is unprotected and may be lost in the sample during extraction. Therefore, hollow-fibre liquid-phase microextraction (HF-LPME) was introduced in 1999, where the droplet was located inside the lumen of a hollow fibre as illustrated in Fig. 1 [4]. The purpose of the hollow fibre was to mechanically protect the droplet. The droplet (termed acceptor) was in liquid contact with a thin layer of organic solvent immobilized in the porous wall of the hollow fibre, and this was again in liquid contact with the sample. Thus, analytes were extracted from the sample, through the liquid membrane located in the wall of the hollow fibre, and into the acceptor inside the lumen of the hollow fibre.

HF-LPME has been conducted in 96-well systems [5], but unfortunately no such systems are commercially available. However, the idea of 96-well HF-LPME with commercial equipment was realized using commercial 96-well filter plates [6]. In this configuration, originally termed parallel artificial liquid membrane extraction (PALME), the samples are pipetted into separate wells in a 96-well sample plate (Fig. 2a). The membrane solvent is then pipetted onto the filters in a 96-well filter

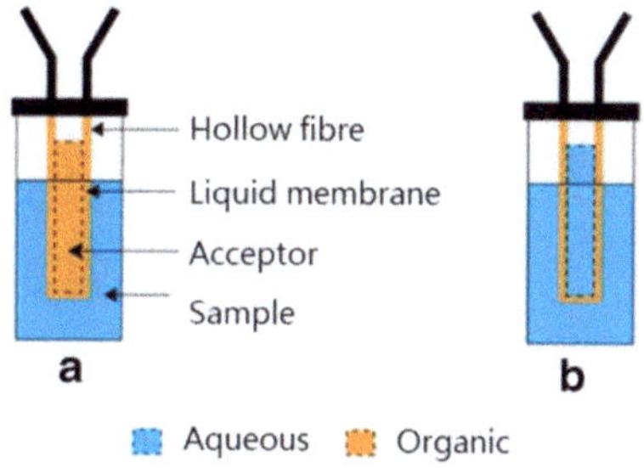

Fig. 1 Principles of two- and three-phase hollow-fibre liquid-phase microextraction (HF-LPME)

plate (Fig. 2b). The membrane solvent is immobilized by capillary forces in the pores of the filter, and by such the liquid membrane is established. Finally, the reservoir above each filter in the filter plate is filled with acceptor (Fig. 2c), the filter plate and the sample plate are clamped, and extraction is facilitated by agitation of the clamped plates (Fig. 2d). Although the 96-well plates used are not intended for 96-well LPME, they are excellent for this purpose [6].

Mass transfer in HF-LPME is based on passive diffusion and is driven by partition coefficients. For this reason, extraction is relatively slow, and extraction times may be up to 45 min to reach equilibrium. To improve kinetics, electromembrane extraction (EME) was introduced in 2006 [7]. The setup for EME is very similar to HF-LPME or 96-well LPME, but electrodes are placed in the sample and acceptor, and are connected to an external power supply (Fig. 3). In this way, an electrical field can be sustained across the liquid membrane, and positively and negatively charged compounds can be extracted across the liquid membrane.

This chapter will focus on HF-LPME, 96-well LPME, and EME. Related approaches are found in literature, including membrane bag-assisted-liquid-phase

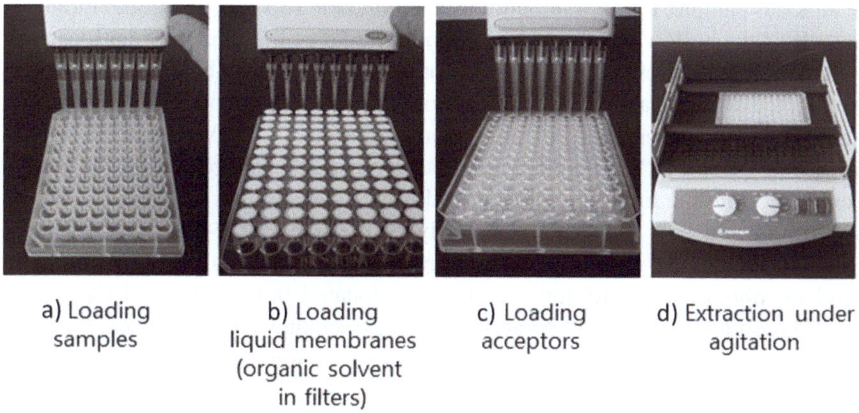

Fig. 2 96-well LPME equipment and procedure

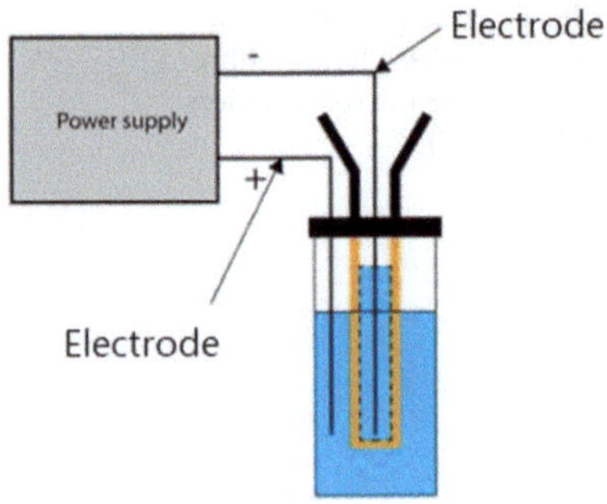

Fig. 3 Principle of electromembrane extraction (EME)

microextraction [8], solvent bar microextraction [9], solvent stir-bar microextraction [10], in-line supported liquid membrane extraction in capillary electrophoresis [11], hollow fibre-supported dispersive liquid–liquid microextraction [12], and micro-fluidic liquid–liquid microextraction [13]. Literature also include important approaches not based on SLMs, such as dispersive liquid–liquid microextraction [14], but these techniques are not discussed in this chapter.

2 Fundamentals

In HF-LPME and 96-well LPME, mass transfer across the liquid membrane is due to passive diffusion, and transfer across the phase-boundaries in and out of the organic liquid membrane is controlled by partition coefficients. HF-LPME can be performed either in two-phase or three-phase mode (Fig. 1). In two-phase mode, the liquid membrane and the acceptor is an organic solvent. Thus, the analyte is extracted from the aqueous sample, through the immobilized organic solvent (liquid membrane) in the wall of the hollow fibre, and into the bulk liquid of the same solvent located in the lumen of the hollow fibre. The acceptor is organic, and the bulk part of this may be collected after extraction and injected into gas chromatography (GC). Thus, two-phase HF-LPME is very convenient in combination with GC. Two-phase HF-LPME can also be combined with liquid chromatography (LC). In those cases, the analyte is extracted into the liquid membrane, followed by desorption into a polar organic solvent and injection in LC. For LC analysis, however, three-phase HF-LPME is more convenient. Here, the analyte is extracted from aqueous sample, through the liquid membrane, and into aqueous acceptor located in the lumen of the hollow fibre. Three-phase LPME can be performed for basic and acidic analytes. For basic analytes, the sample is made alkaline, to suppress the ionization of the analyte, and by such increase the partition into the organic solvent. The acceptor is acidic, and the analyte is protonated in contact with the acceptor, and trapped there. Bases are thus extracted from high to low pH across the liquid membrane. For acidic analytes, the pH gradient is reversed, and they are extracted from low pH to high pH.

In EME, mass transport across the liquid membrane is mainly by electrokinetic migration. For this reason, EME is faster than LPME under optimized conditions [15]. Mass transfer across the phase-boundaries in and out of the organic liquid membrane is controlled by electro-partition. Basic analytes are extracted as cations, and pH in both sample and acceptor should be neutral or acidic, to ensure that the analytes are completely protonated during extraction. The negative electrode (cathode) is placed in the acceptor and the positive electrode (anode) is placed in the sample. Acidic analytes are extracted as anions; the direction of the electrical field is reversed, and pH is neutral or alkaline to ensure the analytes are deprotonated during extraction.

The liquid membrane is a crucial component in both LPME and EME systems. The liquid membrane separates the sample and acceptor, and mass transfer is to a large degree controlled by the chemical composition and properties of the liquid membrane both in LPME and EME. The membrane solvent should be immiscible with water, and solubility in water should not exceed 0.5 mg/mL. The reason for this is to avoid leakage of membrane solvent into the acceptor during extraction. In addition, the membrane solvent should be non-volatile to avoid evaporative losses during extraction. Typical liquid membranes in two-phase LPME are dihexyl ether, dodecyl acetate, and octanol, but a large range of other solvents have also been reported to be successful [16]. Mass transfer in LPME is affected by the chemical composition and polarity of the membrane solvent, but also viscosity plays a very important role. In EME, the selection of membrane solvent is more critical than in LPME, because EME involves extraction of charged species. Thus, while a large range of solvents can work in LPME, the number of efficient membrane solvents for EME is much more limited. For basic analytes, 2-nitrophenyl octyl ether (NPOE) has established as a first choice, and NPOE is efficient for bases in the log P range 2.0 to 6.0. For more polar bases, 2-undecanone can be used in the log P range 1.0 to 2.0, while bases with log $P < 1.0$ normally are extracted with NPOE mixed with di(2-ethylhexyl) phosphate (DEHP) [17]. The latter compound is added to the liquid membrane as ionic carrier. For acidic analytes, EME is normally performed with higher alcohols such as 1-octanol, 1-nonanol, and 1-decanol, and polar acids are in similar way extracted with higher alcohols mixed with ionic carriers [17].

The pH-value in the sample is an important experimental parameter. In LPME, target analytes should be in neutral form in order to transfer across the liquid membrane. Therefore, acidic analytes require low pH in the sample, while basic analytes requires high pH. For EME, the analyte should be in ionic form; therefore, samples are acidified or neutral for extraction of bases, and is neutral or alkaline for extraction of acids. Adjustment of pH in samples solutions has been accomplished using a broad range of buffer systems, or with strong acids or bases.

Acceptor pH is equally important, and in both LPME and EME, the analyte should be ionized within the acceptor. This prevents back-diffusion into the liquid membrane, and is an important action in order to maximize extraction recovery. Aqueous solutions of formic acid, acetic acid, or ammonia are recommended when the acceptor is to be analysed by LC–MS. In cases where the analytical instrument is less sensitive to acidic or alkaline solutions, phosphate buffers and dilute solutions of hydrochloric acid or sodium hydroxide can be used.

In both LPME and EME, mass transfer in the sample solution is mainly by passive diffusion. Therefore, agitation is important. Extraction efficiency normally increases with increasing agitation rate up to 750 to 1000 rpm. Agitation rates above this level tend to be less efficient. Agitation is applied during the entire extraction process. The extraction time also plays a key role in both techniques, and recoveries increase with time until the systems are in equilibrium. The typical equilibrium time in EME is between 5 and 30 min, depending on the polarity and charge of the analyte. Extraction times are also affected by operational parameters including the volumes of sample, liquid membrane, and acceptor respectively, and by the extraction potential. Equilibrium times in LPME are longer, and typically in the range 30 to 60 min. The optimal extraction time (equilibrium time) is normally established experimentally during method development.

In EME, the electrical field serves as driving force for extraction. Therefore, the extraction efficiency increases with increasing extraction potential up to a certain limit (typically 50–100 V), from where the extraction potential is no longer the limiting factor. The optimal voltage is found by optimization experiments. EME is normally performed at optimal voltage to maximize extraction recovery. EME at lower voltages requires longer extraction time, but may increase the selectivity.

Under optimized conditions, recoveries in LPME an EME are often above 85%. Samples are normally exhausted for analyte, but small amounts of analyte may be trapped in the liquid membrane upon equilibrium. Therefore, extractions with recovery at 85% or higher are considered exhaustive. For polar analytes ($\log P < 0$), recoveries are often lower than 85%, due to poor partition into the liquid membrane. When compared to SPME, extraction recoveries are normally higher with LPME and EME. Enrichment can be obtained if the volume of sample exceeds the volume of the acceptor. While most EME systems are operated with low enrichment (< five times), 27.000 times enrichment has been reported with LPME, where pharmaceuticals were extracted from 2 litres of sea water [18]. Due to the lipophilic nature of the liquid membrane, and due to the electrical field, selectivity is high in EME. Selectivity can easily be tuned by the direction and magnitude of the electrical field, by the chemical composition of the liquid membrane, and by pH in the sample and acceptor. Also LPME provides high selectivity, due to the discriminative nature of liquid membrane and the pH gradient. The consumption of organic solvent is less than 10 μL per sample, and LPME and EME are therefore future candidates for green and sustainable sample preparation.

3 Novel Developments

HF-LPME has since its introduction been demonstrated to have several advantages over traditional liquid–liquid extraction (LLE), namely the miniaturization of the extraction process which has enabled a substantially reduced consumption of organic solvents, as well as provide clean extracts and potentially great enrichment of analytes. HF-LPME however also has several disadvantages that have limited its adoption outside of academia; for example slower extraction kinetics compared to LLE. HF-LPME (Fig. 1) also requires a lot of manual handling in preparation of extraction, and therefore has limited applicability for automated and/or high-throughput sample preparation configurations. These limitations have however been addressed in various ways in recent years. HF-LPME is fundamentally a non-exhaustive technique, meaning that extraction recoveries rarely reach 100%. Strategies to improve the extraction efficiency and range of compounds that can be extracted with high efficiency have therefore also been explored. This includes the investigation of new membrane materials, and solvents to be used as SLM. A major trend in recent years has likewise been the development of environmentally friendly, or so-called "green", solvents and materials. These novel developments are discussed in more detail in the following sections. For each subject, a few representative examples are given, with emphasis on the most relevant aspect for the specific subject. For more detailed examples of HF-LPME applications, the reader is referred to Sect. 4.

3.1 External Force-Assisted HF-LPME and EME

As previously discussed, one limitation of HF-LPME is the slow extraction kinetics. Over time, several approaches involving the application of external force to the extraction system have therefore been developed, one example being the application of an electric field across the SLM to stimulate mass transport (termed electromembrane extraction (EME). Another example is increased temperature. In HF-LPME and EME, the formation of boundary diffusion layers also constitute a mechanism for slow kinetics. Specifically, as analytes are extracted into the SLM a thin layer of sample solution surrounding the SLM becomes depleted, and must therefore be replenished by new analyte molecules to make the extraction progress. This process is diffusion-driven and therefore slow. Similarly, a boundary layer with high analyte concentration is formed on the acceptor-side of the SLM, creating a gradient that slows the transport out of the SLM. Good convection of the solutions is therefore important to ensure that these boundary layers are as thin as possible. In HF-based extractions, stirring the sample solution with a magnetic stirrer is a very common approach; however, other novel approaches have also been introduced.

Vortexing the sample instead of stirring has been suggested as a method for improving kinetics. Wang et al. for example compared vortex-assisted (VA)-HF-LPME to stirring-based HF-LPME, and found the VA-HF-LPME system to reach extraction equilibrium (88–94% recovery) after only two minutes [19]. The stirring-based system, on the other hand, had only achieved 6–10% recovery in the same time, but reached 86–95% after 40 min. If reviewing the academic literature, one however quickly finds that vortexing only occasionally is used for agitation in HF-LPME. There may be several good reasons for this, namely that it typically requires the operator to manually hold the sample as it is being vortexed, whereas a stirring-based system can be left while the extraction progresses. The strong shear forces of sample against SLM may also put greater requirements on the SLM stability; else, parts of the SLM solvent may be dissolved or emulsified in the sample.

Ultrasound has also been suggested to improve extraction kinetics, and was likewise compared to stirring by Wang et al. [19]. Compared to stirring, ultrasound-assisted HF-LPME had, like VA-HF-LPME, much faster kinetics with 76–82% recovery after two minutes. However, there was also a clear loss of SLM integrity that made the sample solution clouded. Compared to stirring and vortexing, ultrasound however has the advantage of also providing convection to the acceptor solution as well, which may help decrease the thickness of the boundary layer on the acceptor side of the SLM. Ultrasound has also recently been applied in EME, with the dual aim of decreasing the thickness of the diffusion boundary layer and the electrical double layer [20, 21]. The latter arises from ions that cannot enter the SLM due to poor partitioning, such as salts or buffer ions, and therefore accumulate in the boundary layers in either side of the SLM. Due to the electric field, these ions will have the same charge as the analyte(s), and therefore exert a charge-repulsive effect on approaching analyte ions, which limits their partitioning into the SLM. In the report be Seyfinejad et al. [20], the impact of ultrasound power was investigated. It was found that increasing recoveries were obtained until 20 W. Above 20 W, the SLM (2-nitrophenyl octyl ether) was dissolved into the sample and acceptor solutions. However, at 20 W both the extraction kinetics and steady-state recovery were higher than a conventional stirring-based method, which was attributed to the reduced thickness of diffusion and electric double layers. This was confirmed experimentally by Shang et al. [21].

Microwave-assisted extraction (MAE) is another emerging method that has particularly been successful for solid–liquid extraction, for example of plant material. However, to the best of the authors' knowledge, no reports on microwave-assisted HF-LPME or EME have been published.

3.2 *Novel Membrane Materials*

In the vast majority of HF-LPME and EME reports published so far, polypropylene (PP) has been the preferred material for the support membrane (i.e. the hollow fiber) due to its excellent properties such as chemical inertness and being very inexpensive. However, alternative membranes and materials are currently also in development. Some of these are discussed in this section.

3.2.1 Polymer Inclusion Membranes (PIMs)

PIMs are a novel class of functionalized membranes that, among other applications, are emerging for microextraction purposes. Typically, PIMs are composed of a base polymer, in example cellulose triacetate (CTA) or polyvinyl chloride (PVC), a plasticizer, and a carrier. The base polymer serves as the backbone of the material, providing mechanical strength, while plasticizers make the membrane flexible and allow it to be formed into different shapes. Plasticizers act as the solvent, and may serve to enhance the transport analyte molecules, in tandem with carriers that enhance the analyte transport by complexation. In some cases, the carrier may also function as a plasticizer. The selection of all three components may affect the extraction properties of the membrane, though the carrier generally is considered the most important. Carriers may be acidic, basic, neutral, have chelating properties, and other characteristics, which enables quite different extraction selectivity. Practically, a PIM is prepared by dissolving the base polymer in an organic solvent and adding plasticizer and carrier, after which the mixture is cast into the desired shape. Flat sheet PIMs are most common, though hollow fiber PIMs also are made.

An example application of a PIM in EME was reported by Román-Hidalgo et al. in 2018 [22]. The PIM was synthesized by dissolving CTA in dichloromethane and adding Aliquat® 336, a technical mixture of alkyl-substituted quaternary ammonium chlorides, as carrier. The mixture was poured into a glass Petri dish and left for the dichloromethane to evaporate, after which the PIM could be peeled out and applied as a 25 μm thick, flat membrane. The final PIM composition was 29% CTA and 71% Aliquat® 336. Prior to extraction, the PIM was added 1-octanol as extraction solvent and applied in a homemade extraction device for EME of polar acidic substances. In this case, the positive charge on the ammonium group of Aliquat® 336 could form an ion-pair with the negatively charged acidic analytes, to stimulate the transfer across the PIM.

3.2.2 Alternative Membrane Materials

Various other materials have been applied as supporting membrane. Hydrophobic membranes, similar to PP, have included polytetrafluoroethylene (PTFE) and polyvinylidene fluoride (PVDF). The latter has been suggested to improve extraction kinetics [16, 23], but is still not widely used. Other, more polar, materials have also been used, for example nylon and electrospun acrylic-based nanofiber sheets. Biopolymers such as chitosan and agarose have also emerged as environmentally friendly materials.

For a more detailed overview of supporting membrane materials in HF-LPME and EME the reader is referred to Ocaña-González et al. [16].

3.3 Novel Green Solvents

In HF-LPME, dihexyl ether, 1-octanol, toluene, and dodecane are among the most commonly used extraction solvents. While these solvents have been used with great success in many applications, there are several drawbacks to these and other similar solvents. Firstly, some solvents have low boiling points, high vapor pressure, and undesirable toxicity profiles. Toluene, for example, may evaporate during preparation of extraction and expose the operator to toxic fumes. Secondly, the traditional solvents typically only have one or two different functional groups, which limits the diversity of chemical interactions that can happen between solvent and analyte(s). This consequently limits the span of analytes that can be extracted with one solvent. While this may be an advantage if selective extraction is important, it makes method development for new analytes more cumbersome since many solvents may have to be tested. The difficulty of extraction generally increases with increasing analyte polarity, since these exhibit low partitioning coefficients, resulting from a large offset in hydrophobicity between analyte and hydrophobic solvent. One way of overcoming this hydrophobic discrimination is to increase the strength and diversity of chemical interactions offered by the solvent. However, increasing strength of interactions generally also increases the melting point of a solvent, which may form a solid substance at room temperature. To mitigate this problem, different novel solvent systems have been introduced. The two most prominent are ionic liquids (ILs) and deep eutectic solvents (DESs), both of which often also are claimed to be green. However, it should be noted that one never should generalize the greenness of an entire solvent system, as it is entirely dependent on the composition of a specific solvent. Another novel solvent system often considered green are the nanostructured supramolecular solvents (SUPRASs).

While the formation of ILs, DESs, and SUPRASs relies on different principles, they share the potential to be used as task-specific "designer solvents", with properties that can be tuned to be suited for specific analytes, extraction methods, or sample matrices. This prospect and some advantages/disadvantages of each solvent system is discussed in sections below.

3.3.1 Ionic Liquids (ILs)

An IL is a salt made from a bulky organic cation and a smaller organic or inorganic anion. Some examples are given in Fig. 4. In conventional salt (e.g. NaCl), strong electrostatic interactions between the cation and anion result in a very high melting point. However, because of the asymmetric size of ions in ILs the electrostatic interactions are decreased substantially, to a point where the salt may form a liquid below 100 °C or even room temperature. Because both the cation and anion can be varied the number of potential ILs are billions or more. By varying the ions used, ILs may offer electrostatic, hydrogen bond, π-π and other π-type, dipole, dispersion, and hydrophobic interactions. Physico-chemical properties can likewise be tuned; viscosity, conductivity, density, and water-miscibility are examples that are particularly important in HF-LPME and EME. ILs are generally non-volatile, which is one of the reasons why ILs are proclaimed to be green. However, it should be noted that some ILs also have been shown to be toxic and poorly biodegradable [24]. Considerations of greenness is therefore much dependent on the toxicity of the organic solvent the IL is meant to replace.

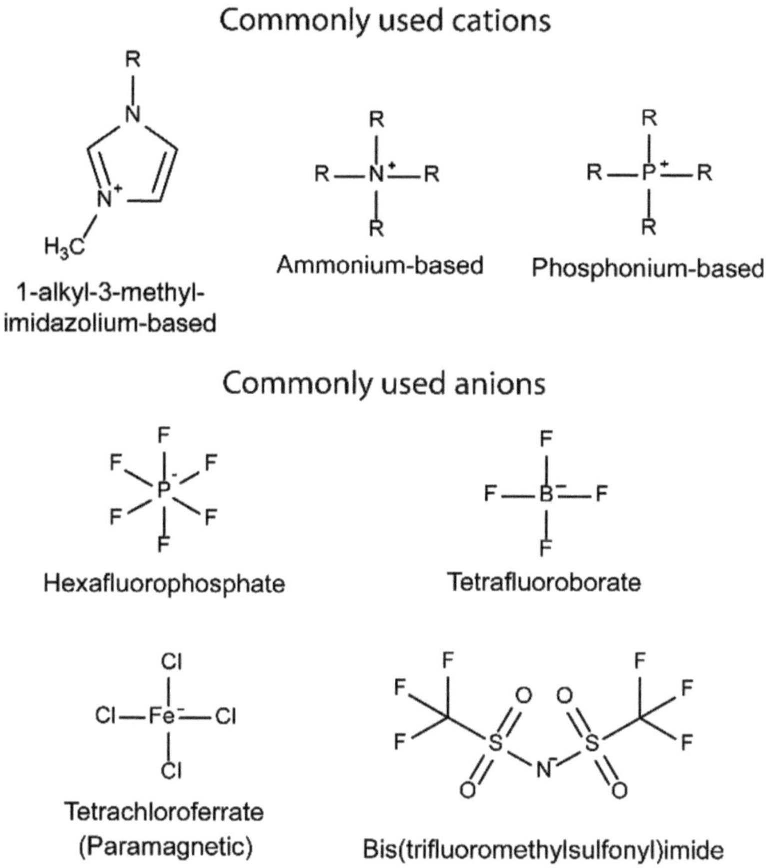

Fig. 4 Examples of frequently used cations and anions in ILs. Reprinted from Hansen and Pedersen-Bjergaard [25] with permission from American Chemical Society

ILs have been applied in different configurations, both two-phase and three-phase HF-LPME, although the widespread use has been somewhat limited. In one representative example, Wang et al. [26] developed a three-phase HF-LPME system for determination of phthalate esters in tea beverages, using 1-nonanol as SLM and the IL 1-butyl-3-methyl-imidazolium hexafluorophosphate ([BMIm]PF_6) as acceptor phase. The phthalates were subsequently quantified by HPLC–DAD. After 4 min of extraction, recoveries were near 100% and 200-fold enrichment of the analytes was achieved.

A major disadvantage of ILs in HF-LPME is their relatively high viscosity, which may hinder the penetration into the pores of the hollow fiber and slow the mass transfer kinetics. This may be one of the reasons why the use of ILs as extraction solvent has been somewhat limited in HF-LPME. ILs have also been applied as SLM in a few instances of three-phase EME [27, 28], but is generally disfavored due to high conductivity leading to current-induced instability. ILs may also pose a problem in chromatography, where they may give high background signals or cause ion suppression in electrospray ionization mass spectrometry-based analysis [29].

3.3.2 Deep Eutectic Solvents (DESs)

DESs are in many ways similar to ILs in terms of properties and potential as "designer solvents". While ILs form a liquid due to weakened electrostatic interactions, the principle of DESs is based on hydrogen bonding. A eutectic mixture is composed of two or more solid components, one hydrogen bond donor (HBD) and one acceptor (HBA), that when mixed in specific molar ratios form intermolecular hydrogen bonds that lower the melting point below that of the individual components. This is also seen from the phase diagram in Fig. 5a. If the melting point depression is large and the mixture forms a liquid at room temperature, it may be termed a deep eutectic solvent (DES) [30]. In example, the first DES discovered, in 2003, was composed of choline chloride and urea in 1:2 molar ratio, and exhibited a melting point 290 °C lower than that of choline chloride [31]. Until 2015, only hydrophilic (water-miscible) DESs were reported. The first hydrophobic and water-immiscible DESs were based on alkylated quaternary ammonium salts and fatty acids, for example tetrabutylammonium chloride (HBA) and decanoic acid (HBD) [32]. Other examples include phosphonium-based cations (Fig. 5b). Recently, hydrophobic DESs based on naturally occurring monoterpenes have also been developed [33]. Some examples of components include camphor and coumarin as HBA components and thymol, menthol, and different fatty acids as HBD components [34].

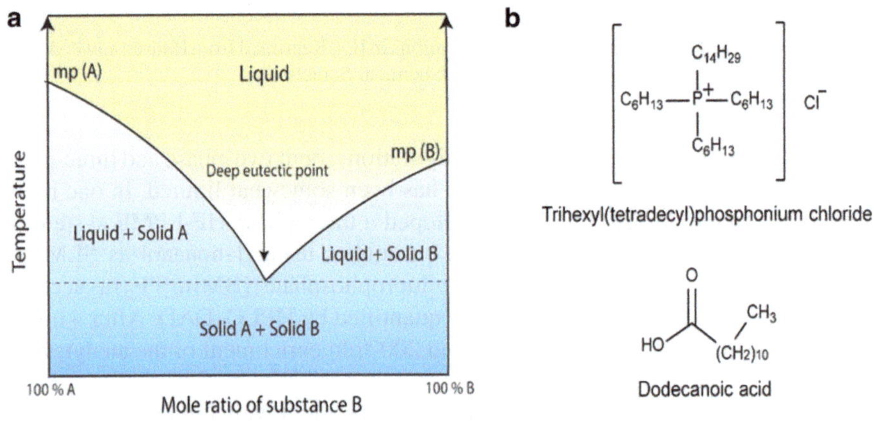

Fig. 5 **a** Illustration of a phase diagram for a deep eutectic two-component mixture. **b** Example HBA and HBD components. Adapted from Hansen and Pedersen-Bjergaard [25] with permission from American Chemical Society

DESs hold several advantages over ILs. They are easier to synthesize, less expensive and often readily available, have less toxicity and better biodegradability, and are generally less viscous then ILs [35]. The latter is an important advantage for HF-LPME. The naturally occurring DES components are particularly interesting, especially from a greenness perspective, since they are fully biodegradable and readily available at a low cost. The non-ionic structure of the above mentioned components further yields even lower viscosity, because of the weaker type of interactions. DESs also share the prospect of being "designer solvents", since many solvents of different properties and chemical interactions can be produced.

DESs have recently become popular solvents for HF-LPME and EME. Hydrophilic DESs are however limited to be used as acceptor phase for aqueous samples, because they yield unstable SLMs in contact with aqueous samples. Seidi et al. for example used a DES of choline chloride (HBA) and ethylene glycol (HBD) in 1:4 molar ratio as acceptor solution, with an SLM of 1-octanol with cetyltrimethylammonium bromide (CTAB) (carrier-mediated HF-LPME), to extract raloxifine and ethinylestradiol from wastewater prior to HPLC–UV analysis [36]. Zhang et al. [37] used a hydrophobic DES of tetrabutylammonium chloride (HBA) and hexanoic acid (HBD) in 1:3 molar ratio for HF-LPME of some cinnamic acid derivatives from traditional Chinese medicine and human plasma. Because this DES was hydrophobic, it could be used as the SLM solvent, while an 85% solution of the DES in methanol was used as acceptor solution. The DES was reported to be fairly viscous, which hindered the mass transfer of analytes. However, by increasing the temperature during extraction to 55 °C the viscosity was reduced and the mass transfer was improved.

Different hydrophobic DESs have also been applied in EME. Because an electric field is applied across the SLM to stimulate the extraction in EME, the SLM solvent can only be slightly conductive to avoid excessive current and electrolysis. Thus, only non-ionic DESs have been reported. Solvents with coumarin or 6-methylcoumarin as HBA and thymol as HBD have in particular been successful for extraction of both non-polar drugs [38], polar drugs and endogenous metabolites [39, 40], and peptides [41] in human plasma samples. For these examples, the SLM was based on flat membranes either in 96-well format or in prototype equipment based on conductive vials. The EME systems were three-phase with aqueous sample and acceptor solutions, and 4–10 μL DES in the SLM. The wide range of analyte properties that have been successfully extracted with this DES is likely due to its structure being highly aromatic, thereby providing different π-type interactions in addition to hydrogen bonding.

3.3.3 Nanostructured Supramolecular Solvents (SUPRASs)

SUPRASs represent a third category of green solvents under development. SUPRASs are composed of amphiphilic surfactants (ionic or non-ionic) that in solution form nanostructured micelles or vesicles when above the critical micelle concentration (CMC). The formation of a SUPRAS happens by a change in environmental conditions that induces further aggregation and self-assembly of the nanostructures, which leads into phase-separation of the supramolecular aggregates from the bulk solution (Fig. 6). The process is termed coacervation, and the new supramolecular phase is termed a SUPRAS. The environmental change is made by addition of a coacervation reagent, which may be acids/bases (pH), organic solvent, or salts (ionic strength). Alternatively, temperature may be changed to induce coacervation. After assembly of aggregates, the SUPRAS structure is held together by non-covalent bonding. For application in HF-LPME, the SUPRAS is typically formed in a separate container and collected by pipette, before application to the hollow fiber. However, for other extraction techniques such as dispersive liquid–liquid microextraction (DLLME), the coacervation process may be performed in situ (in the sample solution itself). Here, the matrix may itself act as the coacervation agent, for example the high salt levels found in urine samples.

The microscopic structure of SUPRAS provide them with very interesting properties, from an extraction point-of-view. The assembled SUPRAS has microenvironments with widely different properties, for example hydrophilic areas associated with the head of surfactants and water incorporated in the structure. Hydrophobic areas are on the other hand associated with the tail of the surfactants. SUPRASs thus have great potential for multi-residue extraction of analytes across a wide range of hydrophilicity. Similarly to ILs and DESs, the chemical interactions provided by a SUPRAS can be designed based on the type of surfactants used to make the solvent. Examples of chemical functionality for the polar head of surfactants include carboxylic acids, sulfonates, alcohols, ammonium and pyridinium ions. Most SUPRAS can thus be considered environmentally friendly, also considering

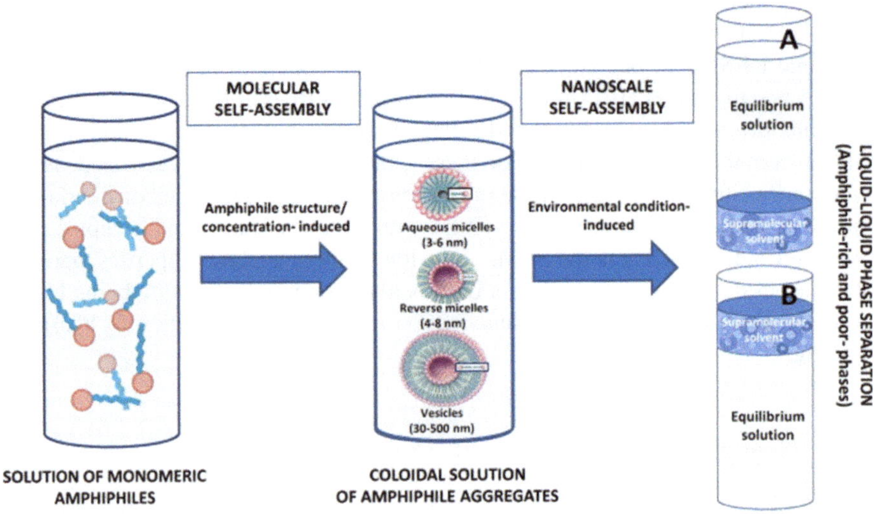

Fig. 6 Formation process of SUPRASs. The final SUPRAS may have higher or lower density than the bulk solvent, which means it may sink or float. Reprinted from Ballesteros-Gómez et al. [42] with permission from Elsevier

that the formation is spontaneous and requires no energy input, albeit that organic solvents used to induce the coacervation are disfavored in this aspect. Another desirable feature of SUPRASs is a large surface area, because the coacervate remains as droplets. The mass transfer of extraction is therefore fast. Parts of the surface area is located within the internal structures of the droplets, and SUPRASs may therefore also possess restricted access properties (RAM). Macromolecules such as proteins may therefore be discriminated from extraction. The size of cavities can be tuned by different coacervation parameters, for example the amount of organic solvent used.

One limitation of SUPRASs is related to compatibility with HPLC and GC analysis. SUPRASs are very rich in non-volatile surfactants, and a back-extraction step into a volatile solvent may therefore be necessary prior to GC analysis. Alternatively, headspace GC may be utilized if the surfactants are thermally stable. SUPRAS are directly compatible with HPLC, as the nanostructures generally disassemble in the organic/aqueous mobile phase, and are retained and eluted as any compound normally would. However, the micelle or vesicle structure may be retained if the mobile phase is primarily aqueous, and may thus act as a pseudo-phase that disturbs the chromatography. Hyphenation of HPLC to MS is however more problematic, since the surfactants may foul the ion source and cause ion suppression. Chromatographic separation from the analytes is thus required, so the surfactants may be directed to the waste container.

There are several examples of SUPRASs applied in HF-LPME. However, because the SUPRAS is in an equilibrium with surrounding solutions, it may dissolve into the sample if the internal binding forces are too weak and the sample-to-SLM volume ratio is very large (as is typically the case in HF-LPME). HF-LPME has thus not been the primary microextraction technique SUPRASs have been applied for. However, a few applications have been developed in this way. One example of the successful application was presented by Rezaei et al. in 2013 [43]. In this work, a SUPRAS based on tetrabutylammonium and decanoic acid was used in a two-phase HF-LPME system for extraction of benzodiazepines from samples of fruit juice, human plasma, and urine. Quantitation was accomplished by HPLC–DAD analysis. The driving force of the extraction was proposed to be a mix of hydrogen bonding, cation-π, and hydrophobic interactions. Another example was reported by Li et al. in 2020 [44]. In this, the authors employed a magnetic solvent bar with a SUPRAS immobilized into the HF-wall to extract NSAID drugs from human serum, followed by LC–MS/MS determination. The solvent bar was prepared by fitting a stainless steel needle into the lumen of the fiber, after which the ends were closed with tweezers, and the fiber was immersed in the SUPRAS for 1 min to impregnate the pores. The SUPRAS was made from hexafluoroisopropanol and decanol. Extraction was performed by magnetically stirring the solvent bar in the sample for 33 min, after which the now analyte-enriched SUPRAS was eluted from the HF-walls using ethyl acetate and ultrasound. The eluate was subsequently evaporated to dryness, reconstituted, and injected into the LC–MS system.

Examples of SUPRASs applied for EME have so far not been reported. For more extensive overview of the SUPRAS concept and its applications, the reader is encouraged to see a recent review by Rubio [45].

3.4 *Novel Sorbents*

While substantial effort has gone into finding efficient and green solvents for HF-LPME, modification of HF-LPME systems by solid sorbents has also been investigated. The main idea of modifying the extraction system with solid sorbents is to improve extraction kinetics and/or selectivity by operating in mixed-mode. Practically, the addition of sorbent material can be done by dispersing into the acceptor solution, in which case an elution step must be performed after extraction. Alternatively, the sorbent may be added to the SLM itself, either by dispersing the sorbent in the SLM solvent or by decorating the surface and/or pores of the fiber with sorbent prior to immobilization. The latter may be done by submersing the hollow fiber into a solution of sorbent dispersed in a volatile solvent, and treating the solution with ultrasound to stimulate the movement of sorbent particles into the pores. Finally, the volatile solvent can be evaporated to leave a dry and decorated fiber. By keeping the sorbents inside the hollow fiber there is thus not any need for post-extraction elution steps.

Many different sorbent chemistries have over time been applied in HF-LPME. In the following sections, we discuss some of the most predominant materials.

3.4.1 Molecularly Imprinted Polymers (MIPs)

Much like ILs, DESs, and SUPRASs may be designed and applied with the aims of achieving specific analyte-solvent interactions, solid sorbents can be designed to offer desired chemical interactions, for example hydrogen bonding, π-π stacking, dispersion, dipole, etc. However, while these interactions may provide some selectivity they are fundamentally non-selective towards specific analytes, and the clean-up from matrix components may thus be unsatisfactory. Molecularly imprinted polymers (MIPs), on the other hand, have potential to offer much greater extraction selectivity, even between analytes with closely related structures/properties. MIPs are synthesized by polymerizing functional monomers (for analyte interaction) and cross-linking monomers (for cross-linking polymer) in the presence of a template molecule. The template is incorporated into the polymer by non-covalent interactions, and can subsequently be removed by appropriate washing steps, leaving cavities of the size, shape, and functionality of the analyte template. During extraction, the MIP functions as a restricted access material (RAM) to enable very selective extraction of target analytes. The template molecule may be the target analyte itself, but a structural analog that can be distinguished from the analyte in the analysis is often desirable to avoid contamination from leaching template molecules. Isotopes of the analyte may for example be distinguished by mass spectrometry. In addition to providing selective extraction, MIPs also have the benefit of being relatively inexpensive to prepare.

MIPs have been applied both for HF-LPME and EME, which has enabled greater extraction selectivity and efficiency. However, MIPs are more commonly applied in dispersive solid-phase extraction, where they are capable of extracting directly in the sample matrix. In this case, extraction recoveries are subject to decrease in the presence of a complex matrix due to blockage of the MIP binding sites. This may however be avoided by protecting the MIPs with an SLM that prevents for example proteins and larger suspended particles from getting contact with the MIP. Such a system can further offer very good clean-up because of dual-mechanism selectivity. An early example of the application of MIPs in HF-LPME was reported by Nemulenzi et al. [46], who extracted the hormone 17β-estradiol from wastewater (37 mL) using a two-phase flat membrane-based system with hexane:ethyl acetate in 3:2 ratio as SLM and acceptor solution [46]. 30 mg MIPs were suspended in the acceptor solution (2.5 mL). During 60 min of extraction, the analyte was extracted into the acceptor solution and partially bonded to the MIP particles. However, by subsequently transferring the entire acceptor solution to a vial and treating it with ultrasound for 15 min, the binding efficiency could be increased. Elution of analyte from the MIP was performed with acetonitrile, which was partially evaporated prior to injection on HPLC (to increase enrichment factor). The improved selectivity of the SLM-MIP-based method compared to simple SLM extraction was evident by cleaner chromatograms.

MIPs have also been dispersed in the SLM solvent itself in three-phase systems. In example, Yaripour et al. suspended MIPs in 1-octanol prior to immobilization in a hollow fiber, to perform EME of phenytoin and phenobarbital from biological fluids [47]. However, in the example, the extraction recovery was worse with MIPs added to the SLM, compared to the pure solvent. This may have been due to too strong bonding between the analytes and MIPs, which may have caused trapping inside the SLM.

3.4.2 Carbon-Based Nanoparticles

Another class of novel sorbent phases are carbon-based nanoparticles, such as carbon nanotubes (CNTs), graphite/graphene and related derivatives. These materials are attractive sorbents due to properties such as large surface-to-volume ratios and ability to provide different chemical interactions, for example hydrophobic interactions and π-π stacking. The materials can however also be functionalized to provide polar interactions, for example dipole–dipole and hydrogen bonding. For instance, graphene may by oxidized to graphene oxide (GO) to include epoxy, COOH, and OH groups on the surface. GO is sometimes chemically reduced to yield reduced graphene oxide (RGO), which contains fewer polar groups on its surface.

In 2020, Worawit et al. reported an application of HF-LPME for extraction of trihalomethanes (for example chloroform) in different water samples. Here, the effect of different carbon-based materials added to individual polypropylene hollow fibers was investigated [48]. This included graphite, graphene, and multi-walled carbon nanotubes (MWCNTs). The hollow fibers were prepared in advance by loading the sorbents into the pores. Preparation for extraction therefore only comprised the immobilization of 1-octanol as SLM solvent. All three materials provided equal or better extraction recoveries than a pure 1-octanol SLM, with the greatest improvement seen for graphite. This confirmed the potential benefit of operating such a mixed-mode extraction.

3.4.3 Metallic Nanoparticles

Metallic nanoparticles have also been used to decorate hollow fibers for HF-LPME and EME. Their benefits resemble that of carbon-based nanoparticles, including the potential for functionalization, and they are generally applied in a similar way. Various metals and metal oxides have so far been used to make the particles, including gold (Au), silver (Ag), copper (Cu), iron (Fe), and titanium dioxide (TiO_2). The nanoparticles often enhance the extraction non-specifically (e.g. by enhancing surface area), but may also be selected to have more specific interactions with the analyte(s). In example, a hollow fiber decorated with copper nanoparticles (CuNPs) was applied for EME of thiouracil from urine [49]. CuNPs were in this case selected because they have high affinity for sulfhydryl compounds, such as thiols. Compared to AuNPs or AgNPs, CuNPs are further less expensive.

3.5 Automation and High Throughput

Automation and high throughput systems both represent one principle within the "Ten principles of green sample preparation" [50]. Automated systems are desirable because they free up personnel that otherwise would have performed the procedures manually, and any exposure of the operator to potentially toxic chemical is substantially reduced or eliminated. High throughput is desirable from a green chemistry point-of-view because the total analysis time, and thereby the operator's exposure, as well as the energy consumption and cost typically is reduced for each sample. From a more practical point-of-view, high throughput also means that more samples can be processed in each working day. High throughput may be achieved by either very short analysis time of successive samples, or by analyzing multiple samples in parallel (for example in 96-well format).

Considering classical HF-LPME, a typical procedure will, for each sample, involve manual preparation and cleaning of the hollow fiber, immobilization of the SLM, pipetting sample and acceptor solutions, immersing HF in sample and waiting during extraction, followed by removal of the analyte-enriched acceptor solution. It is clear that this is neither easily automated nor applicable for high throughput. However, research to transfer the principle of HF-LPME into suitable technical formats is currently ongoing, and in this section, we shall discuss some examples.

3.5.1 Approaches for Automation

In general, a system can be said to exhibit some degree of automation if one or more discrete steps in a procedure are performed without requiring any human intervention. In literature, terms such as "fully" or "semi" automated are frequently used. In accordance with established correct IUPAC terminology, a fully automated system involves the complete processing of consecutive samples without any human intervention [51]. Semi-automated systems, on the other hand, can perform one or more discrete steps automatically, but require some form of human intervention in the process. This may for example be manual transfer or replenishment of solutions between samples, or preparation of a new HF-SLM. The typical analytical workflow is illustrated in Fig. 7.

In this workflow, a method's degree of automation is normally considered from the raw, untreated sample (after sampling), to the detection (e.g. HPLC) has been performed. Data processing and interpretation is also sometimes included in the automated workflow. Considering extractions, this step can be performed *off-line* (i.e. not automated), *at-line, on-line, or in-line. At-line* refers to procedures performed in close proximity to the analytical instrument, with an automated transfer of the prepared sample, for example by robotics. In *on-line* systems, the sample preparation step is typically connected to the analytical instrument by means of tubing and a modulator (for example a switch-valve with a loop). Finally, *in-line* systems couple a continuous flow-system with an extraction device to a detector, without any flow-interruptions. Fully automated *in-line* HF-LPME is however rarely used when dealing with multiple samples, since it is mainly suitable for continuous process monitoring.

There are several practical challenges for automating HF-LPME and EME. Firstly, to avoid memory effects (i.e. carry-over and changing performance over time) it is normal to use a new HF-SLM for each extraction. Immobilization of the SLM solvent and placement of the fiber into the sample is however difficult to automate. Most literature reports have therefore comprised semi-automated systems, with the HF preparation steps performed manually. Secondly, handling of acceptor volumes of 5–20 μL inside the HF lumen is likewise difficult for robotic instrumentation. One example of this was reported by Cabal et al. in 2019 [52]. In the example, the authors constructed a semi-automated HF-LPME-HPLC-FLD (fluorescence detection) system based on a programmable syringe for loading and removal of acceptor solution from the lumen of the hollow fiber, and further transfer to the HPLC analysis. To prepare for operation, 8 mL urine sample was manually filled into a 10 mL

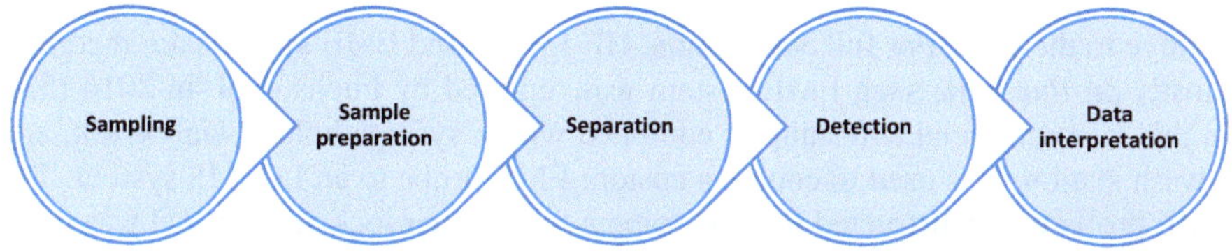

Fig. 7 The typical five steps in an analytical workflow

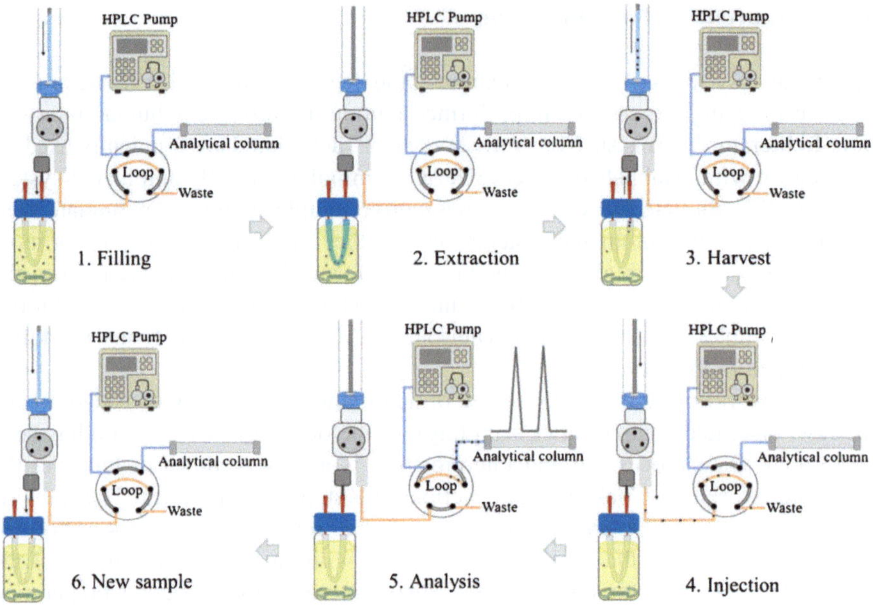

Fig. 8 Schematic illustration of the semi-automated extraction and HPLC analysis. Reprinted from Cabal et al. [52] with permission from Springer Nature

vial with a stirrer, and a 10 cm long hollow fiber was introduced. One end was fitted manually to the syringe, via a three-way solenoid valve, which was also connected to a 6-port switch valve equipped with a 2 μL loop and coupled to the analytical column of the HPLC. The extraction steps are illustrated in Fig. 8.

As the first step, 1-octanol (acceptor solution) was drawn into the syringe and used to fill the pores and lumen of the hollow fiber (two-phase HF-LPME). Extraction of some amphetamines from the urine was then performed for 14 min (not to steady-state), after which the analyte-enriched acceptor solution automatically was withdrawn from the HF-lumen, and flushed to the loop by the syringe. The 6-port valve was then switch to flush the loop with mobile phase towards the analytical column, and data acquisition was triggered. While the chromatographic separation was performed, the syringe underwent some washing steps and a new urine sample and hollow fiber was prepared for the next extraction. The present report is an example of a semi-automated extraction and chromatographic system, requiring for each sample manual preparation of the hollow fiber and manual fitting of the syringe to the fiber. For full automation, HF-LPME and EME systems are therefore mostly *on-line*. One such EME system was reported by Fuchs et al. in 2016 [53]. In this, a commercial autosampler equipped with a syringe holder, vial-wrack, and a wash station, was used to couple a custom EME probe to an LC–MS system. The EME probe was constructed by modifying a female luer lock adapter that was fitted to a 1 mL glass syringe (Fig. 9a). A hollow fiber with fluidic connections was then fixed inside the lumen of the luer lock adapter; specifically, a fused silica capillary

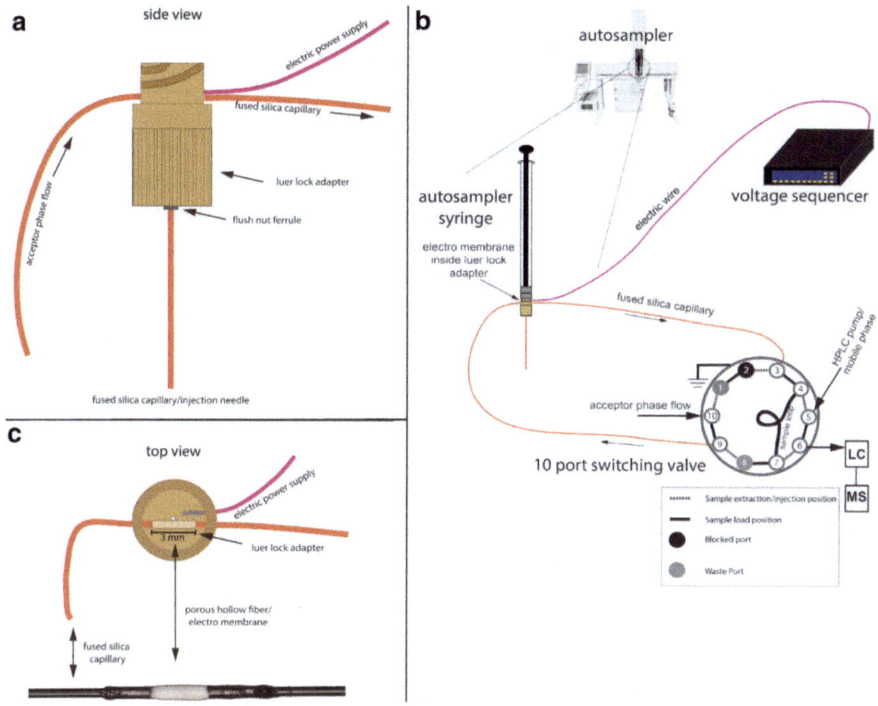

Fig. 9 **a** Side-view illustration of EME probe. **b** Top view illustration with micrograph of the hollow fiber fixed to fused silica capillaries. **c** Schematic representation of the *on-line* hyphenation of EME to LC–MS. Adapted from Fuchs et al. [53] with permission from American Chemical Society

was inserted into each end of a short hollow fiber segment that subsequently was fixed to the capillaries by heat-shrink tubing and heating (which also collapsed the HF pores), leaving a 3 mm segment with intact pores (Fig. 9b).

The capillaries were connected to a 10-port switch valve that was set up with a 20 μL loop, flow of acceptor solution, and connection to an LC–MS instrument (Fig. 9c). A metal wire connected to a programmable voltage-sequencer was also inserted into the luer lock to provide an electric field across the SLM, with the 10-port valve serving as counter electrode (grounded). Before each working day, the SLM was prepared by immobilizing 2-nitrophenyl octyl ether (NPOE) in the pores of the 3 mm free HF segment, and the lumen was flushed with acidified aqueous acceptor solution (thus three-phase system). The extraction sequence for each sample was as follows: the syringe needle, with the modified adapter, was immersed into the sample solution and 100 μL was drawn up and down twice, thus passing the SLM inside the adapter lumen. A potential of 200 V was simultaneously applied to stimulate the extraction of protonated basic analytes through the SLM. The acceptor solution was kept stagnant during this step. Thereafter, the 10-port valve was switched to flush the analyte-enriched acceptor solution into the loop. Once loaded onto the loop, the valve was switched again to flush mobile phase towards the column and mass spectrometer.

Meanwhile, the syringe needle moved to a wash port and washing was performed by repeated drawing and injecting of an aqueous acidic wash solution, while a potential of −200 V was applied to back-extract analytes trapped inside the SLM. Once the LC–MS separation was complete, the system would automatically progress to the next sample. The autosampler, 10-port valve, and voltage-sequencer were all controlled from the LC–MS by an inbuilt relay function. The total analysis time, including EME and LC–MS was only 5.5 min, and could therefore be considered as fairly high throughput. Considering that the same SLM was used for an entire sample sequence, potential memory effects were evaluated. Regarding stability, the system was reported to give stable signals for an entire day of work (>50 extractions), which may be attributed to a very low water solubility of NPOE. Carry-over was determined to be 0.9% when the method was applied for determination of methadone and its metabolite EDDP during an in vitro metabolism study. An interesting feature of the system was that only 4% recovery (termed soft extraction) was obtained during the extraction. The system can thus be useful when there may be a wish for analyzing the sample multiple times, when it is undesirable to disturb the chemical equilibrium of a biological system. For example during drug metabolism studies where soft extraction may be used to probe the drug and metabolite concentration over time. Soft extraction may however be disadvantageous if highly sensitive analysis is required.

3.5.2 Approaches for High Throughput

While high throughput can be achieved by analyzing samples in sequence with very short analysis time per sample, as exemplified in the above section, a more common approach is to analyze samples in parallel, due to typical duration of HF-LPME or EME extractions. The latter is commonly performed in 96-well format. Some examples of hollow fiber-based 96-well formats have been reported [54, 55], but these have required highly skilled operators performing considerable manual handling, and are therefore not very user-friendly. Alternatively, 96-well formats of HF-LPME and EME based on flat membranes have been developed [56–58]. These were discussed in Sect. 1. An example application of 96-well LPME is also given in Sect. 4.

Another strategy towards high throughput analysis based on HF-LPME was reported by Miková et al. in 2020, for *at-line* elution and extraction of analytes from dried blood spots (DBS) ahead of capillary electrophoresis—diode array detection (CE-DAD) quantitation [59]. DBSs are currently emerging as a new method for blood sampling that is inexpensive, as well as more convenient and less invasive to the donor than conventional blood sampling. Typically, a few drops of blood from a finger are place on a DBS card (paper) and dried, and may thereafter simply be mailed to the analytical laboratory. For analysis of DBSs, a typical workflow will comprise punching an area of the DBS and performing an elution of the dried blood by shaking with an elution solvent. This is followed by sample clean-up procedures and quantitative determination by for example HPLC. To improve the throughput of the procedure, the authors developed a simple and inexpensive device with a 9 mm long hollow fiber closed on one end and fixed to a pipette tip in the other end. While the fabrication was performed manually, devices were made in larger batches and stored until use. Prior to extraction, the SLM solvent was immobilized in the HF pores for 10 s, and 5 μL acceptor solution was pipetted into the lumen. The workflow is illustrated in Fig. 10.

The punched DBS was added to a CE-vial with 550 μL aqueous elution/donor solution, and the HF-device was placed into the CE-vial. The vial was then shaken for 10 min for simultaneous DBS elution and HF-LPME of some model basic drugs. Shaking was performed for up to 50 parallel sample vials that subsequently were transferred manually to the CE-instrument's autosampler and analyzed. The extraction device was designed to allow immersion of the inlet capillary into the lumen of the HF, which enabled direct injection of the acceptor solution (Fig. 10f–g). The total CE-analysis time was 8 min, including regeneration steps between samples. For parallel DBS analysis, the total analysis time was thus approximately 10 min per sample. Despite the efficient combination of multiple sample preparation steps,

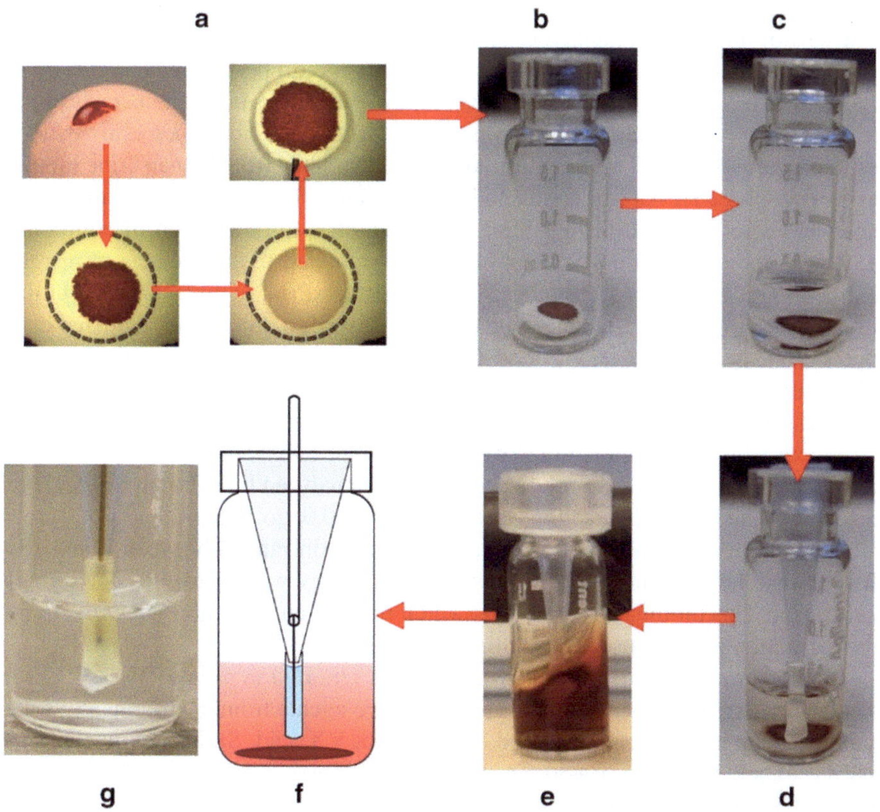

Fig. 10 The workflow of *at-line* DBS analysis by HF-LPME-CE-DAD. Reprinted from Miková et al. [59] with permission from American Chemical Society

the use of non-commercial equipment does represent a limitation. However, there is potential to make this system fully automated.

4 Main Application

A large number of applications have been reported with two- and three-phase LPME, and with EME. LPME is well suited for extraction of analytes of low polarity from aqueous samples; neutral analytes are extracted with two-phase LPME, while acids and bases are extracted either with two- or three-phase LPME. With EME, both non-polar and polar acids and bases can be extracted, extractions are faster. Extraction of pharmaceuticals and drugs of abuse from biological fluids, organic micro-pollutants from environmental waters, and contaminants from food and beverages are typical applications for LPME and EME. It is outside the scope of this chapter to review all these, and readers with interest in a particular application should find more information in review articles. In the following, a few very recent applications are discussed. The idea of this is to give a flavour of current development and directions for LPME and EME, and to exemplify experimental conditions and performance.

96-Well technology is of high interest because it enables a large number of samples to be processed simultaneously. In one recent article, the hormones estrone, 17 β-estradiol, estriol, and 17 α-ethinylestradiol were extracted from human urine samples using two-phase HF-LPME conducted in a laboratory-built 96-well system [5]. The analytes were extracted from 1.5 mL urine samples, after addition of addition of 15% w/v sodium chloride. Sodium chloride was added to the samples to increase extraction recoveries, based on the salting out effect. Prior to extraction, the hollow fibres were dipped into 1-octanol, and the organic solvent quickly immobilized in the pores of the hollow fibre and formed a liquid membrane. In this case, no solvent was filled into the lumen of the hollow fibre. The hollow fibres were then inserted in the samples for 45 min, and the target analytes were extracted into the liquid membrane. After extraction, the hollow fibres were dipped into methanol for 10 min to desorb the analytes. Finally, the desorbed analytes were analysed by LC with fluorescence detection. The limits of detection were in the range 0.03 to 15 μg/L, intra-day precision was between 1 to 13% RSD, inter-day precision ranged between 7 and 18% RSD. Since hollow fibres are low-cost materials, they were used only for a single extraction. This eliminates carry over, which may be an issue in microextraction if the extraction phase is used for multiple extractions.

In another recent paper, 96-well LPME was reported in a three-phase PALME system for extraction of organophosphorus nerve agents from environmental water samples. The nerve agents included Soman acid, VX acid, Sarin acid, cyclohexyl-sarin acid, and Russian VX acid. These are acidic compounds with log P in the range from –0.5 to 0.8. Due to their polarity, partition into organic solvents is limited. For this reason, samples (350 μL) were pH adjusted to 1.0, and 30% sodium chloride was added to the samples. The pH adjustment served to keep the analytes 100% in their neutral form, while salting out increased their partition into the liquid membrane. The latter comprised 4.0 μL of 1-octanol. The acceptor was 50 μL of sodium hydroxide solution with pH 14, and strong alkaline conditions served to minimize back-diffusion into the liquid membrane. Extractions were conducted for 60 min under agitation at 1000 rpm. After EME, the acceptors were analysed by LC–MS/MS. LOQs were in the range 0.009–1.141 ng/mL, linearity was obtained for all analytes with $r^2 > 0.99$, and precision was within 11% RSD. In spite of the fact that all the analytes were polar, the majority of the nerve agents were extracted exhaustively. Due to the high throughput capability of 96-well LPME, 192 samples were processed in 120 min (37.5 s per sample). Unlike previous application, 96-well LPME was in this case conducted with commercially available plates.

In addition to bioanalytical and environmental applications as discussed above, HF-LPME has been used also for food and beverages applications. In one example, HF-LPME was used to study migration of fluorescent whitening agents from plastic food contact materials [60]. This study involved detection of 4,4-bis(5-methylbenzoxazol-2-yl)stilbene, 1,4-bis(2-benzoxazolyl) naphthalene, 2,5-bis(5′-tert-butyl-2-benzoxazolyl) thiophene, 4,4′-bis(2-benzoxazolyl)stilbene, and 1,2-bis(5-methyl-2-benzoxazole)-ethylene. Extractions were conducted from different food simulants, including a 4% acetic acid aqueous solution, 20% aqueous ethanol solution, and isooctane. Hollow fibres of polypropylene were modified with sepiolite nanoparticles to enhance mass transfer into the liquid membrane. The liquid membrane was 15 μL of trichloromethane mixed with n-hexane in the ratio 1:1. Extraction was performed for 30 min under strong agitation. After extraction, the hollow fibres were transferred to centrifuge tubes containing acetonitrile, and the analytes were desorbed under ultrasonic treatment. Finally, the analytes were determined by UPLC-MS/MS. Due to the low volume of liquid membrane, high enrichment (71–205) was obtained. Correspondingly, limits of detection were in the range 0.3 to 0.9 ng/kg, and calibration curves were linear ($r^2 > 0.99$). The intra-day and inter-day recoveries ranged between 83 and 112%, and precision was within 12%.

Unlike LPME, dedicated commercial equipment exists for EME [61]. The commercial equipment is based on the use of vials in conducting polymer as illustrated in Fig. 1. In this way the vials are used as electrodes for coupling the electrical field. In a recent paper, the commercial device was tested in a routine laboratory for clinical pharmacology, and performance was compared with a reference method for the determination of psychoactive drugs [62]. The sample preparation was based on EME, while the final analysis was done by UHPLC-MS/MS. The psychoactive drugs (analytes) included alimemazine, amitriptyline, atomoxetine, clomipramine, doxepin, duloxetine, fluvoxamine, levomepromazine, nortriptyline, and trimipramine, and the metabolites desmethyl clomipramine and dimethyl doxepin. The analytes are all bases, and are relatively hydrophobic. Extractions were conducted from serum samples, which were pipetted into sample vials. Prior to extraction, the samples (100 μL) were diluted 1:3 with 0.1% formic acid. Acidification of the samples ensured that the analytes were fully protonated in the sample. This is important in order to ensure their migration in the electrical field. The liquid membrane was NPOE, as this is the recommended liquid membrane for mono- and dibasic analytes in the log *P* range from 2.0 to 6.0. The volume of NPOE was 9 μL, and this was pipetted directly onto the solid support membrane. The acceptor was 300 μL of 0.1 M formic acid, and was filled into acceptor vials. Due to the acidic conditions, the analytes remained protonated in the acceptor, and therefore they were prevented from back-diffusion into the liquid membrane. Extractions were performed at 50 V for 15 min under agitation at 875 rpm. The positive electrode was coupled to the sample vials, while the negative electrode was coupled to the acceptor vials. With the commercial device, up to ten samples were extracted in parallel. The extraction potential of 50 V was the optimal voltage as determined experimentally, and further increase above 50 V did not improve the mass transfer. The extraction time was also established based on optimization experiments, and at 15 min extractions were more or less complete. After EME, the acceptors were analysed directly with UHPLC-MS/MS, without any evaporation and reconstitution, which is common practice after solid-phase extraction.

EME was highly efficient is this system, and recoveries for the analytes ranged between 75 and 117%. Extractions were thus either exhaustive or near-exhaustive. The EME method showed excellent precision and accuracy. Bias was within ±6%, CVs for intra- and inter-day ranged from 0.9–6% and 2–6% respectively. The EME method was applied on 30 different patient samples, and quantitative data were comparable with those obtained for the same samples using a routine method. While the routine method used close to 1 mL organic solvent per sample, only 9 μL was used with EME. The routine method used phospholipid removal plates after protein precipitation to eliminate ion-suppression during UHPLC-MS/MS, and such plates are expensive. With EME, the phospholipids did not migrate across the liquid membrane, and the acceptors were therefore free of phospholipids. Phospholipid removal was thus an integrated part of the EME process.

EME has been down-scaled and implemented in microchip systems, and combined with smartphone detection several times in recent years. In one example dyes were extracted from water by EME [63]. After EME, the extracted dyes were passed into a very small bed of ion exchanger sorbent, where they were retained and pre-concentrated based on micro solid phase extraction (μ-SPE). The colours were measured using a RGB colorimetric application on a smartphone (Fig. 11). Both EME and μ-SPE were integrated in the microchip. Erythrosine and crystal violet were used as acidic and basic model analytes, respectively. Erythrosine was extracted using 1-octanol as liquid membrane, while crystal violet was extracted using NPOE mixed with di(2-ethylhexyl) phosphate (DEHP) as liquid membrane. DEHP served as ion-pair reagent in the liquid membrane, and facilitated the mass transfer of crystal violet. Replicate extractions and smartphone measurements were within 7.8% RSD, and the dyes were measureable down to the 10–15 μg/L level. Although this work is preliminary in nature, it illustrates a potential of EME in a totally new direction, and further development in this direction is expected.

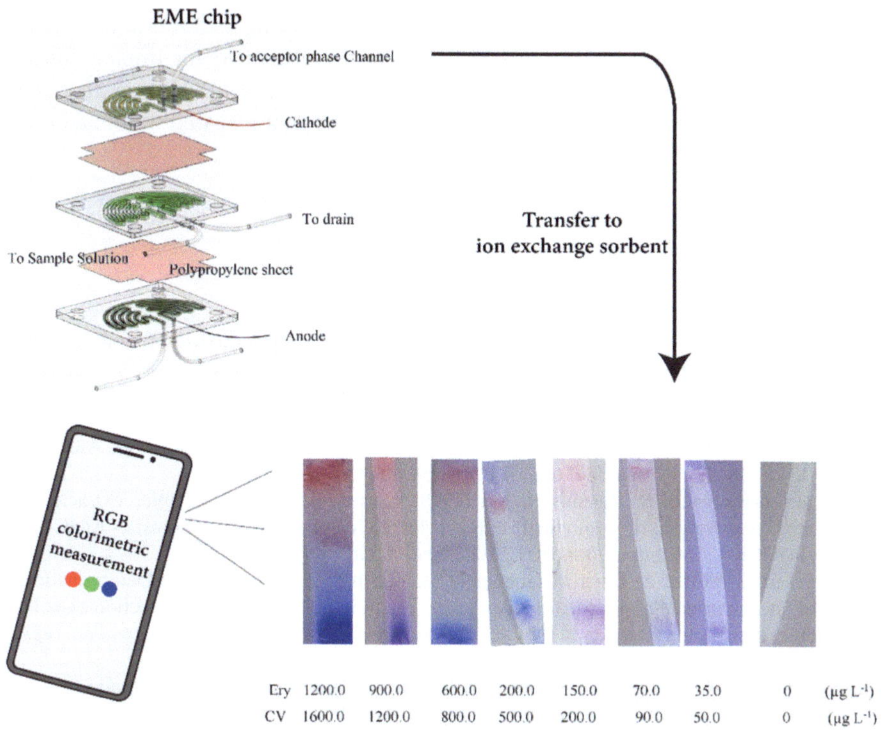

Fig. 11 Illustration of EME chip combined with μ-SPE and smartphone-based RGB measurements of different concentrations of Erythrosine and crystal violet. Modified from Zarghampour et al. [63] with permission from Elsevier

5 Conclusions and Future Trends

Up to date, LPME has been performed with laboratory-built equipment, or in commercial 96-well plates intended for filtration. The latter are neither produced nor intended for LPME. However, recently, commercial equipment was released for EME, and this can be used both for EME and LPME. This equipment is currently limited to extraction of 10 samples in parallel, but commercial 96-well equipment for EME is in progress. Thus, in very short time, analytical chemists can buy instrumentation and consumables for EME and LPME of industrial standard. This will be an important step forward for both techniques.

Research is in progress to define a set of generic methods for EME and LPME. Based on simple molecular descriptors, including charge and polarity, optimal extraction conditions can be derived for EME and LPME of a given analyte. This, in combination with commercial equipment, will make EME and LPME much more available for analytical scientists. Given the extremely low consumption of chemicals and organic solvents, green chemistry and sustainability will be one important justification for implementation of EME and LPME. Another justification will be implementation in microchip systems. Unlike the traditional extraction and sample preparation techniques, EME and LPME are very well suited for implementation in microchip systems. For both reasons, it is the opinion of the authors that EME and LPME will be used extensively in the future, and that research in this area will increase in the years to come.

References

1. Arthur CL, Pawliszyn J (1990) Solid phase microextraction with thermal desorption using fused silica optical fibers. Anal Chem 62(19):2145–2148. https://doi.org/10.1021/ac00218a019
2. Liu H, Dasgupta PK (1996) Analytical chemistry in a drop. Solvent extraction in a microdrop. Anal Chem 68(11):1817–1821. https://doi.org/10.1021/ac960145h
3. Jeannot MA, Cantwell FF (1996) Solvent microextraction into a single drop. Anal Chem 68(13):2236–2240. https://doi.org/10.1021/ac960042z
4. Pedersen-Bjergaard S, Rasmussen KE (1999) Liquid−Liquid−Liquid microextraction for sample preparation of biological fluids prior to capillary electrophoresis. Anal Chem 71(14):2650–2656. https://doi.org/10.1021/ac990055n
5. Lopes D, Morés L, da Silva M, Schneider M, Merib J, Carasek E (2022) Determination of hormones in urine by hollow fiber microporous membrane liquid–liquid extraction associated with 96-well plate system and HPLC-FLD detection. J Chromatogr B 1207. https://doi.org/10.1016/j.jchromb.2022.123406
6. Gjelstad A, Rasmussen KE, Parmer MP, Pedersen-Bjergaard S (2013) Parallel artificial liquid membrane extraction: micro-scale liquid-liquid-liquid extraction in the 96-well format. Bioanalysis 5(11):1377–1385. https://doi.org/10.4155/bio.13.59
7. Pedersen-Bjergaard S, Rasmussen KE (2006) Electrokinetic migration across artificial liquid membranes: new concept for rapid sample preparation of biological fluids. J Chromatogr A 1109(2):183–190. https://doi.org/10.1016/j.chroma.2006.01.025
8. Goh SXL, Goh HA, Lee HK (2018) Automation of ionic liquid enhanced membrane bag-assisted-liquid-phase microextraction with liquid chromatography-tandem mass spectrometry for determination of glucocorticoids in water. Anal Chim Acta 1035:77–86. https://doi.org/10.1016/j.aca.2018.07.031
9. Jiang X, Lee HK (2004) Solvent bar microextraction. Anal Chem 76(18):5591–5596. https://doi.org/10.1021/ac040069f
10. Fashi A, Salarian AA, Zamani A (2018) Solvent-stir bar microextraction system using pure tris-(2-ethylhexyl) phosphate as supported liquid membrane: a new and efficient design for the extraction of malondialdehyde from biological fluids. Talanta 182:299–305. https://doi.org/10.1016/j.talanta.2018.02.002
11. Pantůčková P, Kubáň P (2017) In-line coupling of supported liquid membrane extraction to capillary electrophoresis for simultaneous analysis of basic and acidic drugs in urine. J Chromatogr A 1519:137–144. https://doi.org/10.1016/j.chroma.2017.08.084
12. Lopes D, Dias AN, Simão V, Carasek E (2017) Determination of emerging contaminants in aqueous matrices with hollow fiber-supported dispersive liquid-liquid microextraction (HF-DLLME) and separation/detection by liquid chromatography: diode array detection. Microchem J 130:371–376. https://doi.org/10.1016/j.microc.2016.10.011
13. Asl YA, Yamini Y, Seidi S (2016) Development of a microfluidic-chip system for liquid–phase microextraction based on two immiscible organic solvents for the extraction and preconcentration of some hormonal drugs. Talanta 160:592–599. https://doi.org/10.1016/j.talanta.2016.07.063
14. Berijani S, Assadi Y, Anbia M, Milani Hosseini MR, Aghaee E (2006) Dispersive liquid-liquid microextraction combined with gas chromatography-flame photometric detection. Very simple,

rapid and sensitive method for the determination of organophosphorus pesticides in water. J Chromatogr A 1123(1):1–9. https://doi.org/10.1016/j.chroma.2006.05.010
15. Gjelstad A, Andersen TM, Rasmussen KE, Pedersen-Bjergaard S (2007) Microextraction across supported liquid membranes forced by pH gradients and electrical fields. J Chromatogr A 1157(1):38–45. https://doi.org/10.1016/j.chroma.2007.05.007
16. Ocaña-González JA, Aranda-Merino N, Pérez-Bernal JL, Ramos-Payán M (2023) Solid supports and supported liquid membranes for different liquid phase microextraction and electromembrane extraction configurations: a review. J Chromatogr A 1691:463825. https://doi.org/10.1016/j.chroma.2023.463825
17. Huang CX, Gjelstad A, Pedersen-Bjergaard S (2016) Organic solvents in electromembrane extraction: recent insights. Rev Anal Chem 35(4):169–183. https://doi.org/10.1515/revac-2016-0008
18. Ho TS, Vasskog T, Anderssen T, Jensen E, Rasmussen KE, Pedersen-Bjergaard S (2007) 25,000-fold pre-concentration in a single step with liquid-phase microextraction. Anal Chim Acta 592(1):1–8. https://doi.org/10.1016/j.aca.2007.04.014
19. Wang P, Xiao Y, Liu W, Wang J, Yang Y (2015) Vortex-assisted hollow fibre liquid-phase microextraction technique combined with high performance liquid chromatography-diode array detection for the determination of oestrogens in milk samples. Food Chem 172:385–390. https://doi.org/10.1016/j.foodchem.2014.09.092
20. Seyfinejad B, Ozkan SA, Jouyban A (2021) Ultrasound-assisted electromembrane extraction of clonazepam from plasma and determination using capillary electrophoresis. J Chromatogr B 1181:122928. https://doi.org/10.1016/j.jchromb.2021.122928
21. Shang Q, Mei H, Feng X, Huang C, Pedersen-Bjergaard S, Shen X (2021) Ultrasound-assisted electromembrane extraction with supported semi-liquid membrane. Anal Chim Acta 1184:339038. https://doi.org/10.1016/j.aca.2021.339038
22. Román-Hidalgo C, Martín-Valero MJ, Fernández-Torres R, Bello-López MÁ (2018) Use of polymer inclusion membranes (PIMs) as support for electromembrane extraction of non-steroidal anti-inflammatory drugs and highly polar acidic drugs. Talanta 179:601–607. https://doi.org/10.1016/j.talanta.2017.11.066
23. Alsharif AMA, Tan G-H, Choo Y-M, Lawal A (2017) Efficiency of hollow fiber liquid-phase microextraction chromatography methods in the separation of organic compounds: a review. J Chromatogr Sci 55(3):378–391. https://doi.org/10.1093/chromsci/bmw188
24. Romero A, Santos A, Tojo J, Rodríguez A (2008) Toxicity and biodegradability of imidazolium ionic liquids. J Hazard Mater 151(1):268–273. https://doi.org/10.1016/j.jhazmat.2007.10.079
25. Hansen FA, Pedersen-Bjergaard S (2020) Emerging extraction strategies in analytical chemistry. Anal Chem 92(1):2–15. https://doi.org/10.1021/acs.analchem.9b04677
26. Wang J, Huang S, Wang P, Yang Y (2016) Method development for the analysis of phthalate esters in tea beverages by ionic liquid hollow fibre liquid-phase microextraction and liquid chromatographic detection. Food Control 67:278–284. https://doi.org/10.1016/j.foodcont.2016.03.015
27. Sun JN, Chen J, Shi YP (2014) Ionic liquid-based electromembrane extraction and its comparison with traditional organic solvent based electromembrane extraction for the determination of strychnine and brucine in human urine. J Chromatogr A 1352:1–7. https://doi.org/10.1016/j.chroma.2014.05.037
28. Sun JN, Shi YP, Chen J (2015) Development of ionic liquid based electromembrane extraction and its application to the enrichment of acidic compounds in pig kidney tissues. RSC Adv 5(47):37682–37690. https://doi.org/10.1039/c5ra01029b
29. De Boeck M, Dubrulle L, Dehaen W, Tytgat J, Cuypers E (2018) Fast and easy extraction of antidepressants from whole blood using ionic liquids as extraction solvent. Talanta 180:292–299. https://doi.org/10.1016/j.talanta.2017.12.044
30. Andruch V, Makoś-Chełstowska P, Płotka-Wasylka J (2022) Remarks on use of the term "deep eutectic solvent" in analytical chemistry. Microchem J 179:107498. https://doi.org/10.1016/j.microc.2022.107498

31. Abbott AP, Capper G, Davies DL, Rasheed RK, Tambyrajah V (2003) Novel solvent properties of choline chloride/urea mixtures. Chem Commun 1:70–71. https://doi.org/10.1039/B210714G
32. van Osch DJGP, Zubeir LF, van den Bruinhorst A, Rocha MAA, Kroon MC (2015) Hydrophobic deep eutectic solvents as water-immiscible extractants. Green Chem 17(9):4518–4521. https://doi.org/10.1039/C5GC01451D
33. van Osch DJGP, Dietz CHJT, van Spronsen J, Kroon MC, Gallucci F, van Sint AM, Tuinier R (2019) A search for natural hydrophobic deep eutectic solvents based on natural components. ACS Sustain Chem Eng 7(3):2933–2942. https://doi.org/10.1021/acssuschemeng.8b03520
34. Makoś P, Przyjazny A, Boczkaj G (2018) Hydrophobic deep eutectic solvents as "green" extraction media for polycyclic aromatic hydrocarbons in aqueous samples. J Chromatogr A 1570:28–37. https://doi.org/10.1016/j.chroma.2018.07.070
35. Makoś P, Słupek E, Gębicki J (2020) Hydrophobic deep eutectic solvents in microextraction techniques: a review. Microchem J 152:104384. https://doi.org/10.1016/j.microc.2019.104384
36. Seidi S, Alavi L, Jabbari A, Shanehsaz M (2019) Three-phase carrier-mediated hollow fiber microextraction based on deep eutectic solvent followed by HPLC–UV for determination of raloxifene and ethinylestradiol in pharmaceutical wastewater treatment plants. J Iran Chem Soc 16(5):1007–1018. https://doi.org/10.1007/s13738-018-01572-4
37. Zhang SM, Zhang XX, Chen X, Hu S, Bai XH (2020) Deep eutectic solvent-based hollow fiber liquid-phase microextraction for quantification of Q-markers of cinnamic acid derivatives in traditional Chinese medicines and research of their plasma protein binding rates. Microchem J 155:104696. https://doi.org/10.1016/j.microc.2020.104696
38. Hansen FA, Santigosa-Murillo E, Ramos-Payán M, Muñoz M, Leere Øiestad E, Pedersen-Bjergaard S (2021) Electromembrane extraction using deep eutectic solvents as the liquid membrane. Anal Chim Acta 1143:109–116. https://doi.org/10.1016/j.aca.2020.11.044
39. Hansen FA, Tirandaz S, Pedersen-Bjergaard S (2021) Selectivity and efficiency of electromembrane extraction of polar bases with different liquid membranes—link to analyte properties. J Sep Sci 44(13):2631–2641. https://doi.org/10.1002/jssc.202100167
40. Hay AO, Trones R, Herfindal L, Skrede S, Hansen FA (2022) Determination of methotrexate and its metabolites in human plasma by electromembrane extraction in conductive vials followed by LC-MS/MS. Adv Sample Prep 2:100011. https://doi.org/10.1016/j.sampre.2022.100011
41. Rye TK, Martinovic G, Eie LV, Hansen FA, Halvorsen TG, Pedersen-Bjergaard S (2021) Electromembrane extraction of peptides using deep eutectic solvents as liquid membrane. Anal Chim Acta 1175:338717. https://doi.org/10.1016/j.aca.2021.338717
42. Ballesteros-Gómez A, Sicilia MD, Rubio S (2010) Supramolecular solvents in the extraction of organic compounds: a review. Anal Chim Acta 677(2):108–130. https://doi.org/10.1016/j.aca.2010.07.027
43. Rezaei F, Yamini Y, Moradi M, Daraei B (2013) Supramolecular solvent-based hollow fiber liquid phase microextraction of benzodiazepines. Anal Chim Acta 804:135–142. https://doi.org/10.1016/j.aca.2013.10.026
44. Li X, Huang A, Liao X, Chen J, Xiao Y (2020) Restricted access supramolecular solvent based magnetic solvent bar liquid-phase microextraction for determination of non-steroidal anti-inflammatory drugs in human serum coupled with high performance liquid chromatography-tandem mass spectrometry. J Chromatogr A 1634:461700. https://doi.org/10.1016/j.chroma.2020.461700
45. Rubio S (2020) Twenty years of supramolecular solvents in sample preparation for chromatography: achievements and challenges ahead. Anal Bioanal Chem 412(24):6037–6058. https://doi.org/10.1007/s00216-020-02559-y
46. Nemulenzi O, Mhaka B, Cukrowska E, Ramström O, Tutu H, Chimuka L (2009) Potential of combining of liquid membranes and molecularly imprinted polymers in extraction of 17β-estradiol from aqueous samples. J Sep Sci 32(11):1941–1948. https://doi.org/10.1002/jssc.200800659
47. Yaripour S, Nojavan S, Khoshayand MR, Mohammadi A (2020) Electromembrane extraction of phenytoin from biological fluids: a survey on the effects of molecularly imprinted polymer and carbon nanotubes on extraction efficiency. Microchem J 156:104800. https://doi.org/10.1016/j.microc.2020.104800

48. Worawit C, Alahmad W, Miró M, Varanusupakul P (2020) Combining graphite with hollow-fiber liquid-phase microextraction for improving the extraction efficiency of relatively polar organic compounds. Talanta 215:120902. https://doi.org/10.1016/j.talanta.2020.120902
49. Tahmasebi Z, Davarani SSH, Asgharinezhad AA (2018) Highly efficient electrochemical determination of propylthiouracil in urine samples after selective electromembrane extraction by copper nanoparticles-decorated hollow fibers. Biosens Bioelectron 114:66–71. https://doi.org/10.1016/j.bios.2018.05.014
50. López-Lorente ÁI, Pena-Pereira F, Pedersen-Bjergaard S, Zuin VG, Ozkan SA, Psillakis E (2022) The ten principles of green sample preparation. Trends Anal Chem 148:116530. https://doi.org/10.1016/j.trac.2022.116530
51. Poole C, Mester Z, Miró M, Pedersen-Bjergaard S, Pawliszyn J (2016) Glossary of terms used in extraction (IUPAC Recommendations 2016). Pure Appl Chem 88(5):517–558. https://doi.org/10.1515/pac-2015-0903
52. Cabal LFR, Medina DAV, Costa JL, Lanças FM, Santos-Neto ÁJ (2019) Determination of ring-substituted amphetamines through automated online hollow fiber liquid-phase microextraction-liquid chromatography. Anal Bioanal Chem 411(29):7889–7897. https://doi.org/10.1007/s00216-019-02196-0
53. Fuchs D, Pedersen-Bjergaard S, Jensen H, Rand KD, Honore Hansen S, Petersen NJ (2016) Fully automated electro membrane extraction autosampler for LC-MS systems allowing soft extractions for high-throughput applications. Anal Chem 88(13):6797–6804. https://doi.org/10.1021/acs.analchem.6b01243
54. Lopes D, Dias AN, Merib J, Carasek E (2018) Hollow-fiber renewal liquid membrane extraction coupled with 96-well plate system as innovative high-throughput configuration for the determination of endocrine disrupting compounds by high-performance liquid chromatography-fluorescence and diode array detection. Anal Chim Acta 1040:33–40. https://doi.org/10.1016/j.aca.2018.07.032
55. Luo G, Li Y, Bao JJ (2016) Development and application of a high-throughput sample cleanup process based on 96-well plate for simultaneous determination of 16 steroids in biological matrices using liquid chromatography–triple quadrupole mass spectrometry. Anal Bioanal Chem 408(4):1137–1149. https://doi.org/10.1007/s00216-015-9213-1
56. Gjelstad A, Rasmussen KE, Parmer MP, Pedersen-Bjergaard S (2013) Parallel artificial liquid membrane extraction: micro-scale liquid–liquid–liquid extraction in the 96-well format. Bioanalysis 5(11):1377–1385. https://doi.org/10.4155/bio.13.59
57. Drouin N, Mandscheff JF, Rudaz S, Schappler J (2017) Development of a new extraction device based on parallel-electromembrane extraction. Anal Chem 89(12):6346–6350. https://doi.org/10.1021/acs.analchem.7b01284
58. Restan MS, Jensen H, Shen X, Huang C, Martinsen OG, Kuban P, Gjelstad A, Pedersen-Bjergaard S (2017) Comprehensive study of buffer systems and local pH effects in electromembrane extraction. Anal Chim Acta 984:116–123. https://doi.org/10.1016/j.aca.2017.06.049
59. Miková B, Dvořák M, Ryšavá L, Kubáň P (2020) Hollow fiber liquid-phase microextraction at-line coupled to capillary electrophoresis for direct analysis of human body fluids. Anal Chem 92(10):7171–7178. https://doi.org/10.1021/acs.analchem.0c00697
60. Zhou WL, Ding L, Cheng YH, Xu Z, Chen ML, Fu XS (2022) Application of an improved hollow fiber liquid phase microextraction technique coupled to LC-MS/MS to studying migration of fluorescent whitening agents from plastic food contact materials. Food Addit Contam Part A 39(7):1337–1347. https://doi.org/10.1080/19440049.2022.2066192

61. Drobnjak M, Hansen F, Øiestad EL, Løvli T, Trones R, Martinsen ØG, Pedersen-Bjergaard S (2020) Electromembrane extraction with vials of conducting polymer. LCGC North Am 38(8):435–439
62. Skaalvik TG, Øiestad EL, Trones R, Pedersen-Bjergaard S, Hegstad S (2021) Determination of psychoactive drugs in serum using conductive vial electromembrane extraction combined with UHPLC-MS/MS. J Chromatogr B 1183:122926. https://doi.org/10.1016/j.jchromb.2021.122926
63. Zarghampour F, Yamini Y, Alipanahpour Dil E, Shokrollahi A, Javadian G (2023) A new microfluidic-chip device followed by sensitive image analysis of smart phone for simultaneous determination of dyes with different acidic–basic properties. Talanta 254:124168. https://doi.org/10.1016/j.talanta.2022.124168

Dispersive Liquid–Liquid Microextraction

Alaa Bedair and Fotouh R. Mansour

Abstract Dispersive liquid–liquid microextraction (DLLME) was initially described as an effective sample preparation technology in 2006. However, researchers are still interested in making it more efficient, and ecologically friendly. The dispersion of extraction solvent in aqueous samples is the critical stage in DLLME, which is commonly accomplished using dispersive solvents. Because hazardous dispersive solvents offer a significant environmental danger, attempts have been undertaken to produce greener dispersion procedures while maintaining high extraction efficiency. When it comes to ordinary DLLME, the number of fascinating approaches for changing disperser solvents has expanded. As a result, the goal of this chapter is to provide a overview of current developments in DLLME dispersion modes. Different strategies are covered, including the employment of environmentally-benign dispersers as well as other dispersion methodologies. The most noteworthy approaches that have been implemented to date are highlighted. The problems and prospects for the future of these techniques are discussed. The chapter offer new study avenues, reinforce existing hypotheses, and discover trends among existing DLLME research papers.

Keywords Dispersive liquid–liquid microextraction · Sample preparation · Green analytical chemistry · Air-assisted · Supramolecular

---

A. Bedair
Department of Analytical Chemistry, Faculty of Pharmacy, University of Sadat City, El Sadat City 32897, Monufia, Egypt
e-mail: alaa.mostafa@fop.usc.edu.eg

F. R. Mansour (✉)
Department of Pharmaceutical Analytical Chemistry, Faculty of Pharmacy, Tanta University, Tanta 31111, Egypt
e-mail: Fotouhrashed@pharm.tanta.edu.eg

M. Á. Rodríguez-Delgado et al. (eds.), *Microextraction Techniques*, Integrated Analytical Systems, https://doi.org/10.1007/978-3-031-50527-0_9

Abbreviations

ACN	Acetonitrile
DES	Deep eutectic solvent
DLLME	Dispersive liquid–liquid microextraction
HBA	Hydrogen bond acceptor
HBD	Hydrogen bond donor
HLLME	Homogeneous liquid–liquid microextraction
LLE	Liquid–liquid extraction
LPME	Liquid phase microextraction
LLME	Liquid–liquid microextraction
THF	Tetrahydrofuran
EF	enrichment factor
DLLME-SFOD	solidified organic droplet
ST-DLLME	solvent terminated-DLLME
AA-DLLME	Air assisted-DLLME
VA-DLLME	vortex assisted-DLLME
USA-DLLME	Ultrasound assisted-DLLME
MSA-assisted DLLME	magnetic stirrer assisted DLLME
n-DLLME	normal DLLME
DES	deep eutectic solvent
NADES	Natural deep eutectic solvent
IL	ionic liquid
PIL	polymeric ionic liquid
CAC	critical aggregation concentration
CMC	critical micelle concentration
K	Partition coefficient
LDS	low density solvent
UV	ultraviolet spectrophotometry
MS	mass spectrometry
GC	gas chromatography
HPLC	high performance liquid chromatography
CE	capillary electrophoresis
FAAS	flame atomic absorption spectroscopy
FIA	flow injection analysis
SIA	sequential injection analysis
USAEME	USA emulsification microextraction
LWCC	liquid waveguide capillary cell
ISFME	in situ solvent formation microextraction
DeA	decanoic acid
PAN	2-pyridylazo-2-naphthol

1 Introduction

Liquid–liquid extraction (LLE) is the most commonly used technique of sample preparation. In LLE, a few milliliters of a water immiscible organic solvent is mixed and shaken with the aqueous to allow for analyte partitioning. After that, the extract is left to evaporate under a stream of nitrogen to evade sample oxidation. Then, the residue is dissolved is the least possible amount of a suitable organic solvent to keep the sample highly concentrated. These procedure are not only time consuming, but also health hazardous due to the large volume of organic solvents that either evaporate or disposed after extraction [1]. In addition, The automation of LLE steps is a real challenge [2]. For these reasons, LLE is considered ecologically unfriendly and laborious.

Miniaturization of extraction technologies has grown significantly in recent decades [3] to solve the problems of classical LLE while benefiting from its cost-effectiveness and high efficiency. A tiny amount of a water-immiscible organic solvent is employed in liquid–liquid microextraction (LLME) to extract target analytes from aqueous samples. This smaller variant of LLE offers a more environmentally friendly approach to improving extraction efficiency with fewer chemicals and quicker analytical times. Moreover, LLME was readily automated, which aided in the analytical process and safeguarded workers [4].

DLLME technique was developed in 2006 by Rezaee and colleagues as a modification of the LLME technique [5, 6]. The purpose was to improve the recovery rate of LLME. In DLLME, an immiscible organic solvent is combined with an organic disperser to create an emulsion. Manual shaking is then used to disperse the organic extractant into tiny droplets, resulting in a homogeneous solution. This dispersion process extends the contact surface area between the extractant and the sample, thereby increasing extraction kinetics. Following this, the sample is centrifuged to separate the extractant and break up the emulsion. In 2007, Zanjani et al. proposed a new variation of DLLME that uses low-density solvents of long-chain alcohol. These solvents solidify at low temperatures, allowing for easy phase separation. This method, called DLLME with solidified organic droplet (DLLME-SFOD), has been widely used in various applications [7]. A year later, ultrasound was utilized instead of manual or mechanical shaking to induce dispersion, eliminating the need for a dispersing solvent [8]. In 2010, Chen et al. introduced the solvent-terminated DLLME (ST-DLLME) technique, which avoids the centrifugation step by adding an auxiliary solvent to break the dispersion and induce phase separation [9]. This mode aided in the automation of the method. In 2011, Jafarvand and Shemirani used coacervates and reverse micelles to form supramolecular self-assemblies, resulting in higher extraction efficiency and selectivity for specific analytes [10, 11]. The following year, Farajzadeh and Mogaddam introduced air-assisted DLLME (AA-DLLME) using repeated aspiration/injection cycles to induce dispersion [12]. In 2014, magnetic ionic liquids were employed in DLLME to induce phase separation using a strong magnet, eliminating the need for centrifugation [13]. In 2020, water-immiscible natural deep eutectic solvents (NADES) were utilized in DLLME to extract various analytes, including nine phthalic acid esters [14]. Figure 1 illustrates the milestones of DLLME development over the last years. DLLME is one of the most successful miniaturized sample preparation techniques, due to the high EF, high sensitivity, acceptable precision, accuracy and selectivity according to the acceptance criteria and guidelines of the Food and Drug Administration (FDA). In addition, DLLME is a fast mode of sample preparation in comparison with conventional techniques. The speed of DLLME could be even accelerated by using semiautomated-DLLME or fully automated DLLME [15].

2 Fundamentals

The efficiency of the DLLME technique is governed by the same experimental conditions as LLE. Both extraction and microextraction processes are equilibrium-based and are controlled by the partition coefficient (K), which can be calculated using the following equation:

$$K = \frac{C_{org,Eq}}{C_{aq,Eq}}$$

where, $C_{org,Eq}$ represents the concentration of the analyte in the extracting solvent, and $C_{aq,Eq}$ represents the concentration of the analyte in the aqueous sample, both measured at equilibrium. The main difference between microextraction and extraction lies in the use of tiny amounts of the extractant (microliters) in DLLME, compared to milliliters in conventional LLE. As a result, $C_{org,Eq}$ is substantially higher in DLLME when compared to LLE for two primary reasons. Firstly, the small volumes of organic solvents used in DLLME leads to the analyte being highly concentrated due to the inverse relationship between volumes and concentrations. Secondly, $C_{aq,Eq}$ at equilibrium is very high because only a small amount of the analyte migrates to the small layer of organic extractant. However, K must remain constant, which only occurs if $C_{org,Eq}$ is also very high to maintain the ($C_{org,Eq}/C_{aq,Eq}$) ratio.

In DLLME, the analyte partitioning takes place at the interface between the aqueous sample and the immiscible organic extracting solvent. Increasing this interface enhances the efficacy of partitioning and in turn, the efficiency of microextraction. In DLLME, the organic extract is dispersed in the aqueous sample with the aid of a disperser, mechanical force, or both. This dispersion step increases the contact surface area between the two layers, leading to better extraction and higher efficiency. The efficacy of the process can be assessed by calculating the enrichment factor (EF) using the following formula:

$$EF = \frac{C_{org,Eq}}{C_{aq,int}}$$

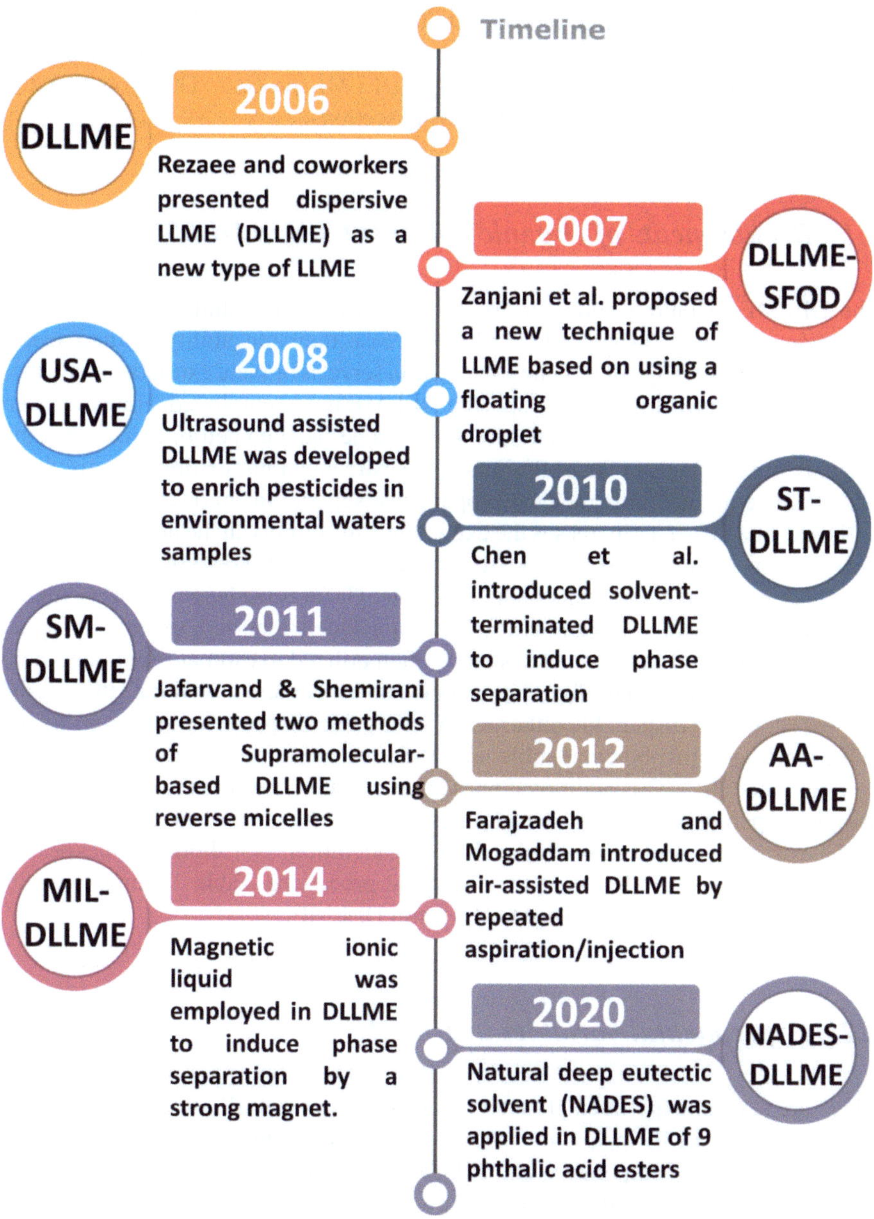

Fig. 1 Timeline of the development in dispersive liquid–liquid microextraction (DLLME)

where, $C_{aq,int}$ denotes the starting concentration of the analyte in the aqueous sample. The EF can be enhanced by selecting appropriate organic solvents and optimizing the experimental settings that influence the DLLME process. These optimization techniques will be discussed in the following sections.

2.1 Requirements of Organic Solvents Used in DLLME

The DLLME technique relies on the utilization of water immiscible solvents as extractants, while a disperser is used to increase their miscibility in the aqueous sample. A diverse range of organic solvents can serve as organic extractants, and their properties are determined by the specific DLLME mode employed. However, there are certain fundamental characteristics that must be satisfied before utilizing organic solvents as extractants in DLLME. Firstly, the extractant must exhibit low miscibility with the aqueous medium to achieve proper phase separation; this is especially critical as the use of a disperser increases the extractant's miscibility in the aqueous medium. Secondly, the extractant should possess the capacity to dissolve the target analyte, with high partition coefficients being desirable. Unfortunately, partition coefficient data for all analytes across various solvents is not widely reported, so the documented Kow value for the octanol/water system is often utilized to estimate the lipophilicity of the target analyte. Thirdly, after manual or instrumental shaking, the organic solvent should be dispersible either using an organic disperser or not. Fourthly, the extractant used must be compatible with the subsequent procedure, or else it must be evaporated first. This additional phase may negatively affect the accuracy of the sample preparation procedure, besides the effort and time involved. Finally, the cost of the extractant should also be taken into consideration, as it should be inexpensive to minimize the overall cost of the analytical procedure. Table 1 summarizes the properties of the most widely used solvents in DLLME.

2.2 Experimental Variables in DLLME

There are several experimental factors that can be optimized to increase extraction efficiency in DLLME including solvents types and volume used in extraction and dispersion, sample temperature and pH, salt addition, extraction duration and stirring rate. The most significant of these factors are proper choices of the kind and amount of disperser and extractant. In traditional DLLME, Halogenated hydrocarbons, including chloroform, are frequently employed as extractants, however, in cases where low density solvents are utilized in DLLME modes, 1-undecanol has emerged as the most prevalent extractant. Typically, maximum extraction efficiency is found at lower extractant quantities (20–100 μL). The type and volume of disperser come next in significance. Acetonitrile (ACN) [17] and methanol [18, 19] are the most often utilized dispersants. A few hundreds of microliters (200–800 μL) are

Table 1 Physicochemical properties of common organic solvents in DLLME [16]

Solventé	Chemical formula	Molecular weight (g/mol)	Boiling point (°C)	Melting point (°C)	Density (g/mL)	Solubility in water (g/L)
Amyl alcohol	$C_5H_{12}O$	88.15	131–132	−78.2	0.809	Slightly soluble
Benzene	C_6H_6	78.11	80.1	5.5	0.88	1.79
Butyl acetate	$C_6H_{12}O_2$	116.16	126.1	−78.8	0.882	1.11
Carbon tetrachloride	CCl_4	153.82	76.7	−22.9	1.586	0.8
Chloroform	$CHCl_3$	119.38	61.2	−63.5	1.49	8.98
Cyclohexane	C_6H_{12}	84.16	80.7	6.5	0.779	2.9
Decanol	$C_{10}H_{22}O$	158.3	232	10-Sep	0.824	Insoluble
Dichloroethane	$C_2H_4Cl_2$	98.96	83.5	−35.7	1.256	Slightly soluble
Diethyl ether	$C_4H_{10}O$	74.12	34.6	−116.3	0.713	6.9
Dichloromethane	CH_2Cl_2	84.93	39.6	−95	1.33	1.33
Diisopropyl ether	$(CH_3)_2CHOCH(CH_3)_2$	102.18	68.2	−111.4	0.725	Insoluble
Ethyl acetate	$C_4H_8O_2$	88.11	77.1	−83.6	0.902	77.68
Hexane	C_6H_{14}	86.18	68.7	−95	0.659	0.0053
Isoamyl alcohol	$C_5H_{12}O$	88.15	131–133	−117.3	0.815	Slightly soluble
Isopropyl ether	$(CH_3)_2CHOCH_2$	102.18	68.2	−117.5	0.725	12.5
Methyl tert-butyl ether (MTBE)	$C_5H_{12}O$	88.15	55	−109.8	0.74	3.9
n-Butyl alcohol	$C_4H_{10}O$	74.12	117.7	−89	0.81	Slightly soluble

(continued)

Table 1 (continued)

Solventé	Chemical formula	Molecular weight (g/mol)	Boiling point (°C)	Melting point (°C)	Density (g/mL)	Solubility in water (g/L)
Propyl acetate	$C_5H_{10}O_2$	102.13	101.3	−95	0.889	0.63
Toluene	C_7H_8	92.14	110.6	−93	0.866	0.52

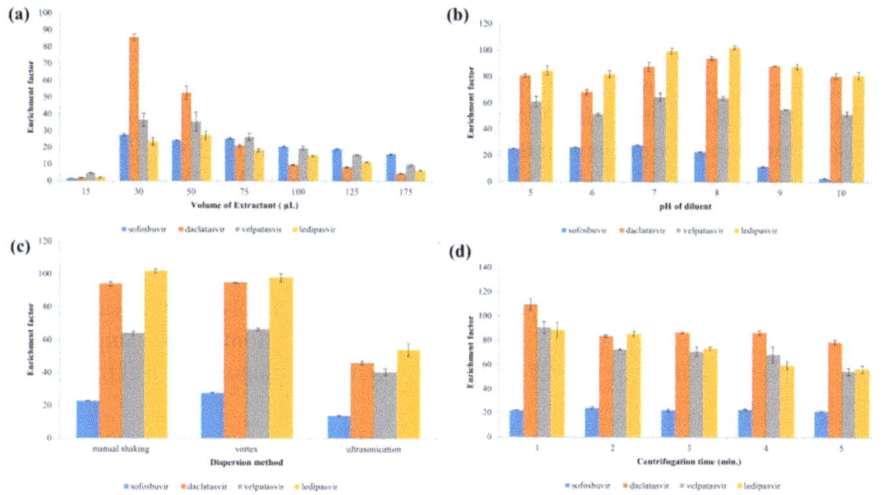

Fig. 2 Optimization of **a** extractant volume, **b** pH of diluent, **c** dispersion technique and **d** centrifugation time ($n = 3$). Reprinted from [21] with permission from Elsevier

frequently sufficient to spread the extracting solvent in the sample. Greater quantities of dispersants are mot recommended due to the unwanted co-solvency, which reduces the efficiency of DLLME [20]. While optimizing the extraction conditions, it is essential to consider the potential interactions between variables. The extraction efficiency of ionizable solutes can be affected by sample pH. The use of acids or bases can potentially alter the ionization process towards the unionized form of the analyte, which is theoretically easier to extract. Similarly, the salting out effect can boost extraction efficiency. Investigating the effects of sample temperature, salt concentration, stirring rate, and extraction duration may aid in achieving the best extraction conditions. Figure 2 shows the effect of extractant volume, pH, dispersion method and centrifugation time on the EF of four different antivirals. As the figure indicated, the most crucial factor was the extractant volume, with markedly higher EFs at lower volumes of the extractant [21]. These factors may be modified at the same time utilizing chemometrics, which can predict the optimal conditions for DLLME with the fewest experimentation, while also predicting variables' interactions [22, 23].

2.3 *Modes of DLLME*

The conventional mode of DLLME (also known as normal DLLME or n-DLLME) employs high density organic solvents as extractants, and manual shaking to facilitate in dispersion [24]. The different modes of DLLME can be classified according to the extractant type or the dispersion technique. As for the extractant type, different solvents have been utilized such as low density solvent (LDS), ionic liquids (ILs), deep eutectic solvent (DES), and supramolecular (SUPRAS). On the other hand, DLLME can be categorized according to the dispersion technique into vortex-assisted (VA), ultrasound-assisted (USA), AA and magnetic stirrer assisted (MSA)-DLLME. Phase separation after dispersion is usually induced by centrifugation [25], although solvent-terminated DLLME has been frequently reported [26]. When a dispersion is subjected to centrifugal force, the tiny droplets within the dispersion experience a radial outward force. This force causes these droplets to move away from the axis of rotation, towards the outer edges of the sample centrifuge tube. The centrifugal force in this case accelerates the phase separation. Denser solvents will settle faster than lighter ones, causing them to migrate towards the bottom of the sample, while lighter solvents float on the top [27]. Figure 3 depicts the categorization of the various DLLME modes.

2.3.1 n-DLLME

The most widely employed method for biological analysis is conventional dispersive liquid–liquid microextraction (n-DLLME), which involves combining a suitable disperser with an extractant that is heavier than water [28]. Upon injection of the combined extractant/disperser solvents into the sample, shaking the solution leads to the formation of an unstable emulsion which can be rapidly disrupted by centrifugation. The bottom layer is then collected using a syringe and supplied to the assay equipment. In this context, n-DLLME has been successfully utilized to determine different classes of drugs including antipsychotics [29], antidepressants [30, 31], antimicrobials [32, 33], immunosuppressants [34], antiarrhythmics [35], and drugs of addiction [36–38]. Chloroform [39–41] is the most commonly used extractant, while methylene chloride [35] and carbon tetrachloride [37] are other often used halogenated hydrocarbons. It should be noted that the disperser needs to possess miscibility with both the sample and the extracting solvent in order to serve as a dispersant. Commonly employed dispersers in n-DLLME include ACN [30], methanol [29], acetone [33], ethanol [37], and tetrahydrofuran (THF) [40]. In some instances, the organic solvent is evaporated before utilizing the analytical tool, and the sample residue is reconstituted in a compatible solvent [41]. n-DLLME has been employed in a range of analytical techniques, including ultraviolet spectrophotometry (UV) [30], mass spectrometry (MS) [34], gas chromatography (GC) [36], high performance liquid chromatography (HPLC) [32], and capillary electrophoresis (CE) [40, 41]. The n-DLLME technique has certain limitations, such as low manual shaking efficiency, high toxicity of organic extractants, and difficulties in automation. Researchers have addressed these challenges by modifying the default processes and developing new modes of DLLME [28].

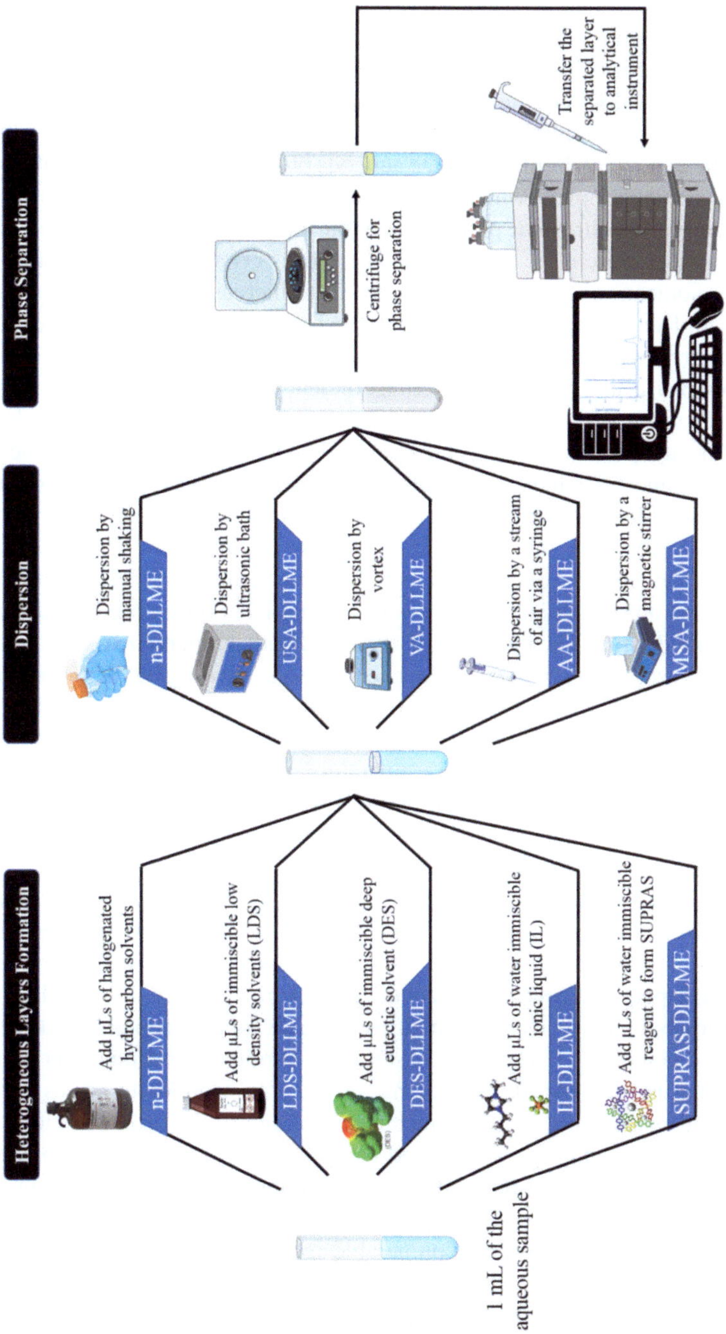

Fig. 3 Different modes of dispersive liquid–liquid microextraction

2.3.2 Ultrasound Assisted DLLME

The most critical step in DLLME is the dispersion. In USA-DLLME, ultrasonic waves are employed to induce better dispersion than manual shaking. This mode has also been termed USA emulsification microextraction (USAEME). The ultrasonic energy was preferentially employed to increase the turbidity of the solution and spread the extractant droplets into the aqueous phase. As a result, the analyte was trapped in these tiny droplets, which make it easily separated from the aqueous solution. Furthermore, the ultrasonic power hastens the transfer of analyte to the extractant phase. To monitor these effects, the ultrasonic settings (temperature, duration, and amplitude of sonication) could be optimized [42] to increase the frequency and rate of material molecular motion, enhance solvent penetration, and thus increase the dispersion degree of extraction solvents. This accelerates the speed of the analytes in the extraction phase, and promote extraction efficiency. Altunay et al. [42] developed USA-DLLME using NADES as extractants for extraction of trace metals from honey by using flame atomic absorption spectroscopy (FAAS). This application studied the effect of ultrasound time and temperature. The effect of sonication period on mass transfer and metal ion recovery was studied from 0 to 20 min at a maximum amplitude of 70%. The recovery values for metal ions were relatively low when ultrasound was not used. The recovery rate for all metal ions rose significantly as the ultrasound period increased up to 10 min, and there was no significant difference in recovery at longer ultrasound times. The influence of ultrasonic bath temperature on the production of NADES droplets with metal ion recovery was also investigated at temperatures ranging from 25 to 60 °C. The best phase separation was attained at 35 °C. No phase separation was accomplished, especially at temperatures over 45 °C. As a result, an ultrasound period of 10 min and a temperature of 35 °C were determined to be appropriate. Generally, ultrasonic bath [43–47] was widely used in USA-DLLME in addition to the ultrasound homogenizer probe, which could be more suitable for limited sample volumes or small extraction vessels [48].

Fernández et al. [49] examined n-DLLME and USA-DLLME for the detection of benzodiazepines in biological fluids; USA-DLLME had greater efficiencies due to the increased dispersion. Moreover, the ultrasonic waves in USA-DLLME obviated the need for a dispersant, reducing solvent usage [50]. Yet, most USA-DLLME applications employ both a disperser and an ultrasonic bath for enhancing the extraction. Fernández and coworkers [51] used USA-DLLME for determination of antidepressants in human plasma by adding 2.5 mL of ACN to samples to act as protein precipitant and dispersant. As an extractant, a volume of 200 μL of chloroform was used. The extracted drugs were tested using UPLC/UV, and the findings were compared to those obtained using traditional LLE connected to GC/MS, there were no substantial differences between the two techniques which indicated that DLLME could enable UV detection to give comparable results to the highly sensitive MS detection. USA-DLLME was also utilized in flow injection analysis (FIA) with inline derivatization through diazotized *p*-sulfanilic acid to determine tetracyclines in egg supplement samples [52].The reaction was carried out at 45 °C in a slightly alkaline media, and the absorbance at 435 nm was measured using a liquid waveguide capillary cell (LWCC). GC–MS was also utilized to detect seven recreational drugs in human plasma, including meperidine, ketamine, methadone, amphetamine, and amphetamine derviatives. The impact of ultrasound application duration ranged from 0 to 5 min. After 2 min, the optimal sonication was reached. Because of the potential demulsification impact, prolonged duration in the ultrasounds application through a bath was undesired. USA-DLLME was used to extract different analytes from different matrices including aqueous samples [43, 48], food [42] and biological samples [53].

2.3.3 VA-DLLME

Instead of shaking or using ultrasonic waves, the sample can be vortexed with an extractant and a dispersant to induce dispersion. A principal role of vortex is to break down any extraction solvent into tiny droplets, increase the contact area between the droplets and water, and speed up extraction equilibrium, which is dependent on vortex speed and duration [54]. As a result, the vortex promotes the equilibration and dispersion of the target analytes in the extractant and aqueous solution, reducing the equilibrium period [55]. So, vortex time should be enough to achieve equilibration between the aqueous sample and the extractant [56]. Usually, the vortex step is performed with the aid of a vortex agitator [56–58]. The multi-tube vortexers were also employed to increase sample throughput. This allows for more applications within the same timeframe and facilitates automation.

Compared with other modes of DLLME, the vortex outperformed the other mechanical agitators in terms of extraction efficiency [59]. The sample/extractant combination is vortexed with or without an organic disperser to generate an emulsion in VA-DLLME. Herrera-Herrera et al. [60] created the VA-DLLME technique for extracting various sulfonamides and quinolones prior to HPLC–UV. As an extractant, chloroform was employed, while acetonitrile was used as a dispersant. Before centrifugation, the ternary mixture was vortexed for 3 min. Interestingly, vortexing had little effect on extraction efficiency, but it dramatically enhanced accuracy. This describes how vortices help to accelerate the equilibrium process. VA-DLLME was widely used for extraction various analytes from diverse sample types including beverages [61], biologicals [62, 63], food [64], and sewage [65]. However, VA-DLLME biological applications are still lower than expected [66].

2.3.4 AA-DLLME

To eliminate the need of equipment in DLLME, an AA-DLLME technique was introduced, in which dispersion was produced simply by aspirating and injecting the extraction mixture with a syringe repeatedly. This approach requires no extra instruments, facilitating the automation process [67]. The principle of AA-DLLME was similar to DLLME in many ways, but there was no requirement for an organic solvent to disperse an extractant into the sample solution. A hydrophobic organic solvent at μL-concentration (extraction solvent) was dispersed into the sample solution using a syringe fitted with a needle and sucking/dispersing cycles numerous times. Despite the absence of a dispersant solvent, this approach considerably enhanced the contact area of the extraction solvent with the sample solution. The investigations revealed that the two primary factors in liquid phase micro extraction (LPME) procedures were extraction solvent viscosity and interfacial tension [68]. These parameters influence both the extractant droplet size and the analyte mass transfer rate. Aspirating-dispersing cycles transform the extraction solvent into extremely small droplets, increasing the contact area of the sample solution with the extraction solvent dramatically. During the aspirating/dispersing cycles in AA-DLLME, there is intense turbulence in the solution, and mass transfer of the analytes is mostly regulated by the convective process [67].

A syringe is used in AA-DLLME to repetitively withdarw and inject the extractant and sample until a hazy solution forms. Farajzadeh et al. [31] used AA-DLLME to preconcentrate NSAIDs in biological fluids. The hazy solution was back-extracted into 10 μL of ammonia buffer (0.1 M, pH 9) after four rounds of recurrent withdrawal and injection before being delivered to an HPLC equipment with UV detection. When the analytical figures of merit were compared to other methods of LPME, it was discovered that AA-DLLME had the highest EF, the maximum sensitivity, and a suitable extraction duration of 10 min. NSAIDs might potentially be chemically derivatized and extracted concurrently with AA-DLLME [69]. The derivatizing agent for GC-FID was butyl chloroformate, while the catalyst was picoline. A syringe was used to aspirate and disseminate the material, which was combined with the chloroform, in presence of the catalyst and the reagents. The number of extraction cycles was investigated; four rounds of repeated injection and withdrawal were found to be adequate to generate dispersion. With five or more extraction cycles, no further improvements in signal intensities were detected. The discovered method's sensitivity was higher than previously reported GC–MS approaches, and it was less expensive and time-saving. The AA-DLLME techniques' simplicity and ease of automation improve their potential in biological applications. Barfi et al. [70] compared the performance of ultrasound-enhanced AA-DLLME (USE-AA-DLLME) with previous DLLME techniques that extracted NSAIDs using an organic disperser. Higher EFs were reported with USE-AA-DLLME, which might be attributable to disperser-induced improved solubility in the aqueous sample. So, because of these advantages AA-DLLME including facilitating of the automation and absence of disperser, AA-DLLME was widely used for determination different analytes from different matrices including water [71–73], beverages [74], food [75–77], biological [78] and biodiesel samples [79].

Rahmani and coworkers [80] compared USA-DLLME, AA-DLLME and VA-DLLME to extract benzene, toluene, ethylbenzene and xylene isomers (BTEX) from water samples. There was no need for a dispersive solvent in any of these procedures, as the extractant is dispersed by air bubbles, vortex, and ultrasound for AA-DLLME, VA-DLLME, and USA-DLLME, respectively. The findings revealed that the three techniques used were highly effective, and the hazy solutions formed were capable of extracting analytes in a relatively short time and with good recoveries (BTEX was used as a simple analyte in these tests). These three procedures were faster, simpler, more sensitive, less costly, and more environmentally friendly than the previous DLLME methods due to the absence of the dispersive solvent. AA-DLLME required a larger volume of extractant and, as a result, required less time to centrifuge. Consequently, AA-DLLME had the shortest analysis time (3 min). The linear dynamic range of AA-DLLME was greater (50–2600 μg /L), although VA-DLLME utilizes less solvent (only 25 μL) and had the best RSD. USA-DLLME has the highest enrichment factor, up to 245-fold.

3 New Developments in DLLME

Since the introduction of DLLME in 2006, tremendous efforts have been exerted to enhance the performance and widen its scope of application. These advances can be categorized into three main trends. The first involves exploring new extractants such as LDS, IL, or DES. The second direction focuses on facilitating the phase separation step such as in DLLME-SFOD. The third direction is geared towards method automation. The following sections discuss these new trends in more details.

3.1 New Extractants in DLLME

3.1.1 Using Low Density Solvents in DLLME

The main obstacles in n-DLLME is the restricted number of extraction solvents and the high toxicity of the halogenated hydrocarbons. These barriers were overcome by utilizing nontoxic organic solvents, with densities lower than water, such as hexane, toluene, xylene, octanol and others in a mode known as low density solvent DLLME (LDS-DLLME). Following dispersion and termination, the floating layer could be delivered to the analytical equipment using a syringe. The LDS-DLLME theory has been studied [81]; nonetheless, applicability in biological fluids are still quite beyond expectations. Ghambari et al. [82] used LDS-DLLME followed by HPLC/UV to extract and evaluate warfarin in plasma. The extraction was carried out in a separate cell. The extraction cell contains deproteinized plasma (pH 2.3 adjusted), an LDS (octanol, 150 μL), a disperser (methanol, 150 μL), and a magnetic stir bar. After the extraction procedure, the extractant collection was facilitated by the long and narrow neck of a special extraction cell. Warfarin extraction recovery was 91%. Applying ultrasonic waves during the dispersion stage boosted LDS-DLLME efficiency. Meng et al. [83] employed SA-LDS-DLLME to determine illicit drugs in plasma. The authors compared their approach to hollow fiber liquid-phase microextraction (HFLPME); USA-LDS-DLLME achieved greater efficiency in shorter extraction times. The fundamental benefit of LDS-DLLME is that the

LDS organic extractants are compatible with routinely used HPLC mobile phases. As a result, there is no need for solvent evaporation or reconstitution prior to sample injection. Unfortunately, LDS-DLLME has intrinsic limitations due to the incomplete phase separation following extraction and the challenging automation of the centrifugation step. Other DLLME modes, such as SFOD [84] and ST-DLLME [9, 85] could overcome these problems.

3.1.2 Using ILs in DLLME

ILs have inspired scientists in a variety of research and industrial fields over the last decade. This is demonstrated by the large number of articles in the area of analytical chemistry pertaining to ionic liquids. ILs can be effectively isolated and reused to greatly decrease application costs [86–89]. Another significant benefit of ionic liquids is the ability to select from a wide range of ions to create an IL with the desired physical and chemical characteristics such as melting point, viscosity, density, and miscibility with water and other solvents. As a result, ILs are frequently referred to as modelling solvents. Ionic liquids' distinctive characteristics piqued the curiosity of scientists and engineers in the field of extraction and separation [90]. For these reasons, ILs and polymeric ionic liquids (PILs) are used in a variety of applications in DLLME [91–93].

The physical and chemical characteristics of ILs are primarily determined by the size, placement, and type of the organic cation and the organic/inorganic anion. The potential of constructing IL structures by selecting the cation and anion that give the necessary physicochemical qualities opens up the possibility of widespread usage of these substances in academic research and industrial applications [94, 95]. ILs are commonly regarded as "green solvents" for their excellent solvation characteristics, low vapour pressure, and low toxicity [92].

Zhou et al. [96] and Baghdadi et al. [97] were the first to use ionic solutions in the DLLME method and to coin the term IL-DLLME [60]. Liu et al. [98] used this IL-DLLME mode for preconcentration and isolation of heterocyclic pesticides in water before HPLC/DAD determination. The IL employed was [C6MIm] [PF6], and the dispersive liquid was methanol [99]. This approach has recently been modified by changing the sample temperature, using ultrasound, microwaves, or more radical modifications such as the in situ IL formation during ME. This in situ IL-DLLME mode was initially suggested by Bahdadi and Shemirani in 2009 and is often referred to as in situ solvent formation microextraction (ISFME) [100]. This in situ IL-DLLME technique works by dissolving hydrophilic IL in an aqueous solution containing the analytes of interest, then adding an ion-exchange reagent to create an insoluble IL. An ion-exchange reagent supports a metathesis reaction, which transforms the hydrophilic IL into a hydrophobic one that settles and preconcentrates the analytes. Yao and Anderson [101] used a similar method for the measurement of aromatic hydrocarbons.

Although ILs have been shown to be good extractants in DLLME, phase separation still needs centrifugation, which is time-consuming and difficult to automate. So, an innovative family of ionic liquids with magnetic properties, known as magnetic ionic liquids (MILs), has been synthesized, which frequently feature an imidazolium [102], choline [103], or phosphonium cation [104] and a paramagnetic metal (Fe, Co, Mn, or Gd) chloride anion. The higher the magnetic susceptibility, the easier the phase separation in the presence of an external magnet. For this reason, Abdelaziz et al. [105] used a hydrophobic gadolinium-based MILfor the first time as extraction solvent in DLLME. In this work, the produced Gd(III)-based MIL demonstrated hydrolysis resilience in aqueous samples as well as a minimal UV noise signal. Furthermore, the suggested MIL's acceptable viscosity promotes analyte partitioning, speeds phase separation, and simplifies extract handling and transfer into the analytical instrument. Furthermore, the introduced Gd-mased MIL showed significantly high magnetic susceptibility, enabling for quicker extraction solvent recovery with a powerful magnet.

3.1.3 Using Deep Eutectic Solvent in DLLME

One of the objectives of implementing the DLLME method is to utilize environmentally friendly green solvents [106]. The critical step in this aspect is to prepare a solvent that is not only green but also offers efficient extraction [107]. Typically, the solvents used in DLLME methods are toxic, which has spurred the development of remarkable and ecologically favorable green solvents [108]. DESs are typically made up of hydrogen bond acceptors (HBAs) and hydrogen bond donors (HBDs). HBAs are frequently quaternary ammonium compounds, whereas HBDs are amines, carboxylic acids, alcohols, polyols, or carbohydrates [109, 110]. Because of the creation of intramolecular hydrogen bonds, DESs have a much lower melting point than their separate components. DESs have low volatility, low vapour pressure, a reasonably broad liquid range, and high heat durability [111]. Furthermore, DESs are readily produced without the need for purification stages, and they are made from low-cost compounds with low or minimal toxicity. DES are also biodegradable and easily reusable. These characteristics make DESs superior to traditional solvents used in extraction processes [112], especially DLLME extraction and isolation of bioactive substances [113, 114]. One important benefit of DESs, for example, is their ability to be tuned to accomplish specific functionality due to the numerous possibilities of beginning components. The selectivity of DESs for extraction and separation can be adjusted by altering the structure and molar ratio of their hydrogen-bonding components [115]. DESs have been categorized into four kinds, as shown in Fig. 4: Type I (metal halide and quaternary salt), Type II (quaternary salt and hydrated metal halide), Type III (quaternary salt and hydrogen bond donor), Type IV (metal halide and HBDs), and Type V (HBD and HBA). This class is particularly important in the microextraction and sample preparation of ionic and highly polar analytes [116–118].

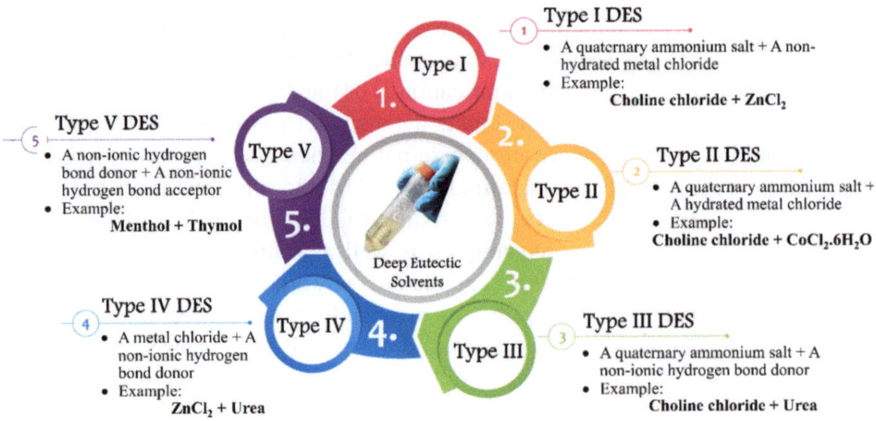

Fig. 4 Types of deep eutectic solvents (DES), with examples

3.1.4 Using SUPRAS in DLLME

SUPRASs are nano-structured liquids produced through self-assembly processes occurring at molecular and nanometer scales from amphiphiles [119]. These solvents have been used in extraction processes for many years under different names such as cloud point technique and coacervates [120, 121], and offer a set of appealing intrinsic properties, including the use of self-assembly based synthetic procedures, widespread availability of amphiphiles, tunability of solvent properties, and excellent solvation properties for various compounds [119]. SUPRASs are formed through consecutive self-assembly processes that occur at molecular and nanometer levels, where amphiphiles form three-dimensional aggregates that separate from the bulk solution as a new liquid phase via coacervation when the critical aggregation concentration is reached (Fig. 5) [122]. Two main types of SUPRAS, vesicle-based and reverse micelle-based, have been developed for analytical extractions, with driving forces for effective solubilization and high extraction efficiency being dispersion forces between hydrocarbon chains and analytes, cation interactions between aromatic rings of complexes and amphiphiles, and hydrogen bonding between nitrogen and oxygen atoms in complexes and carboxylic acids from [123, 124]. Reverse micelle-based SUPRAS using THF has shown greater potential for DLLME compared to vesicle-based SUPRAS [125, 126].

The initial self-assembly process in supramolecular solvent production is the accumulation of amphiphilic molecules in a variety of nanostructures. As a critical aggregation concentration (CAC) is reached, amphiphiles spontaneously aggregate to minimise adverse solvophobic interactions [128, 129]. It becomes energetically advantageous for amphiphiles to interact with one another at the CAC. Colloidal self-assembled structures result from the intricate interplay of solute–solvent and solute–solute interactions.

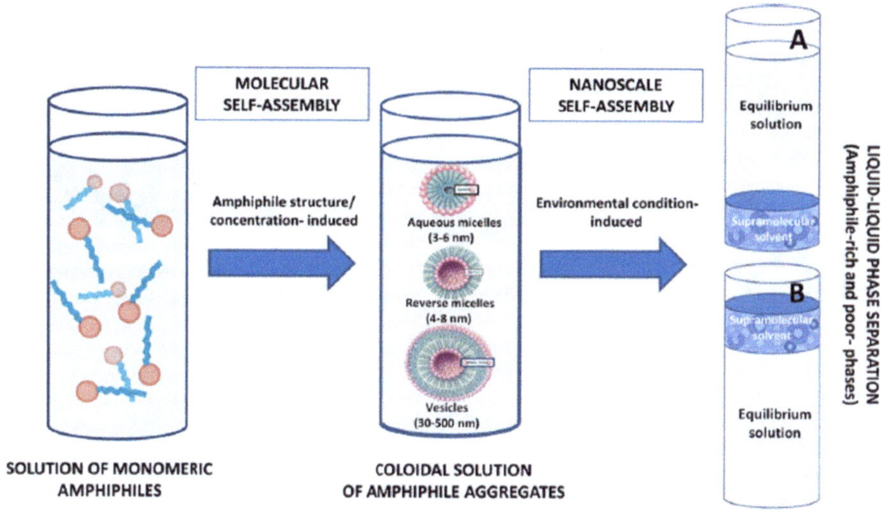

Fig. 5 Self-assembly processes in supramolecular solvent formation. Reprinted from [127] with permission from Elsevier

Seidi et al. [123] developed DLLME-SFOD based on a vesicular SUPRAS of decanoic acid and quaternary ammonium compound for extraction of cadmium, the extraction mechanism of cadmium depended on the SUPRAS structure contains polar and apolar groups, their various interactions with analytes can boost extraction efficiency. The interactions between the vesicular SUPRAS and the Cd(II)-(2-pyridylazo)-2-naphthol (PAN), PAN complex are shown in Fig. 6. The main extraction driving forces appear to be three types of interactions: (1) dispersion forces between the hydrocarbon chains of the amphiphile and the analyte; (2) -cation interactions between the aromatic rings of the Cd(II)-PAN complex and Bu4N+; and (3) hydrogen bonding between the nitrogen and oxygen atoms in the Cd(II)-PAN complex and hydrogen of carboxylic acid. These interactions allow for effective solubilization of Cd (II)-PAN in the SUPRAS as well as high extraction efficiency.

Bendito et al. [124] proposed a novel type of SUPRAS-based extraction constituted of reverse micelles of decanoic acid (DeA) distributed in a water/THF combination in 2007. They demonstrated that polar and non-polar molecules were extracted into SUPRAS using hydrogen bonding and Van der Waals interactions that reverse micelles may generate. A series self-assembly model predicts that the dissolved DeA in THF producing reverse micelles has at least three critical micelle concentration (CMC) points (4.8 $\pm$ 0.2, 7.6 $\pm$ 0.4, and 51 $\pm$ 2 M).When water is added to this combination, the aggregates partially dissolve, facilitating contact and encouraging the formation of larger reverse micelles as an immiscible liquid phase separate from the THF/water bulk solution [125, 126]. It is worth mentioning that using THF based SUPRAS is much more than vesicular SUPRAS in DLLME.

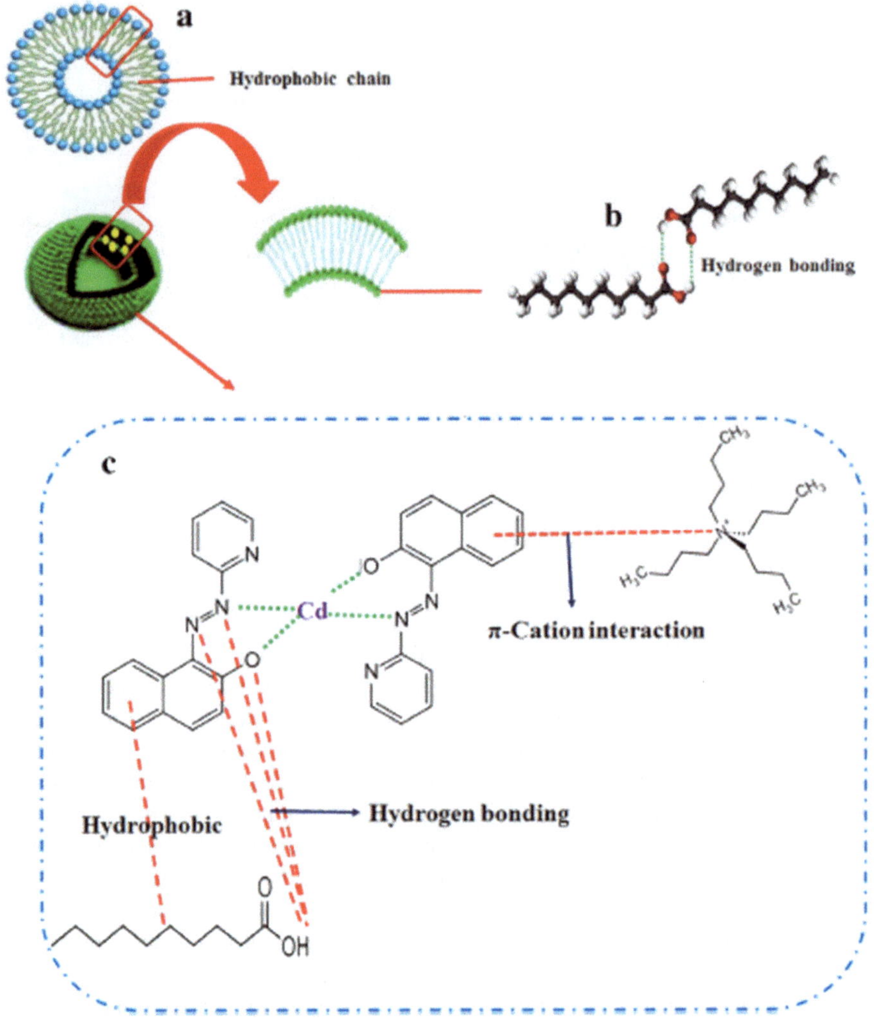

Fig. 6 **a** Chemical interaction can influence vesicle formation and its stability, **b** hydrogen bonding in vesicular formation, and **c** molecular mechanism of microextraction and different interactions between Cd (II)—PAN complex and the vesicle. Reprinted from [123] with permission from Springer Nature

3.2 Phase Separation by SFOD

DLLME-SFOD use low melting point solvents (10–25 °C), such as 1-undecanol and 1-dodecanol [84, 130]. The floating droplet is solidified using an icebox after dispersion and phase separation and then transferred using spatula or forceps. The key benefit of DLLME-SFOD is the ease with which the extracted phase may be separated. In pharmaceutical and biological analysis, DLLME-SFOD is the second most often utilized mode of DLLME. DLLME-SFOD has been used to determine several pharmaceuticals [131–138] and drugs of abuse [139–142] in dosage forms as well as biological fluids such as urine [136], plasma [138, 140], milk [143] and tissues. While being extensively recognised in biomedical analysis, DLLME-SFOD has several drawbacks; to extract the analyte of interest, two organic solvents (the extractant and the disperser) are required. This issue can be avoided by using mechanically-induced dispersion, as in USA-DLLME-SFOD and AA-DLLME-SFOD [144]. The other issue stemmed from the centrifugation stage, which slowed the extraction process and hampered automation. ST-DLLME [145] can be used to solve this problem by adding a demulsifying solvent. To break the emulsion and produce phase separation, a demulsifying solvent is added to the sample/extractant/disperser combination in ST-DLLME. In this situation, the centrifugation stage can be skipped, allowing for process automation and a reduction in overall analytical time. The main barrier in DLLME-SFOD is the limited number of solvents that can solidify at relatively low temperature without causing the whole sample to freeze. Exploring other solvents especially those from botanical origin with relatively low melting point is highly recommended.

3.3 *Automation of DLLME*

Automation is one of the DLLME technique's ongoing problems. Several developments in DLLME have relied on flow analysis methods [146]. Initially, DLLME was automated using the sequential injection analysis (SIA) approach and used to determine metals by flame or electrothermal atomic absorption spectrometry [147–150]. A comparable method was developed utilizing the flow injection analysis methodology, which performed online DLLME using ionic liquids [151–153]. DLLME has also been automated utilizing a dual SIA system, which connects both syringe burettes by a conical tube that serves as the extraction container [154].

SIA rendered the automation of DLLME possible [155], through a multi-axis robotic arm with an integrated phase separator and temperature control. This setup allowed for the automatic solidification of the organic phase, followed by the collection of the organic extract for analyte measurement. The automated DLLME-SFOD of parabens was examined as a proof-of-concept, followed by analyte separation using liquid chromatography. Medina et al. [156] and coworkers developed an automated method in which everything was automated by combining a SIA technology with a custom-made robotic phase separator. Then, phase separation was performed in a 3D printed device incorporating a Peltier cell set and placed on a multi-axis robotic arm. A single software package controls the combined action of the flow system and the robotic arm, allowing for the solidification/melting and collection of the organic phase for subsequent analyte measurement as indicated in Fig. 7.

Another approach for DLLME automation is the completion of the extraction "in-syringe" [15]. In case, syringes are employed as DLLME containers, and the separated extractant droplets can be collected at the top of the syringe, ready to be

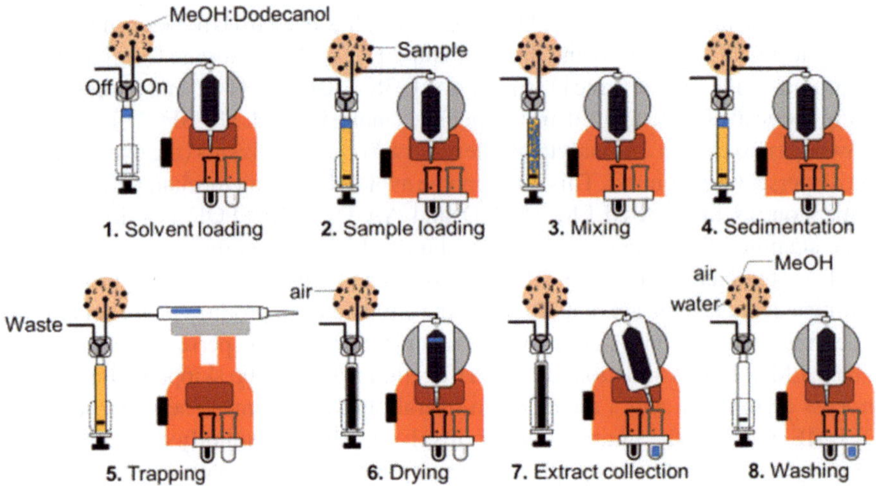

Fig. 7 Major steps of the automated DLLME-SFOD. Reprinted from [156] with permission from Elsevier

automatically injected into the detection system, which is interfaced by an injection valve, utilizing solvents lighter than water [81]. Shishov et al. [157] developed an automated in syringe DLLME for chromium detection in beverages. As indicated in Fig. 8, in the first phase, 0.6 mL of extraction mixture (port a, valve) was aspirated into the syringe via channel 1 by back movement of the syringe pump plunger at a speed of 1.5 mL/ min. The valve was then switched to port b, and 4 mL of sample was aspirated into the syringe at a rate of 10 mL /min. Furthermore, 0.4 mL of air (port c, valve) was sucked to eliminate any leftover sample in channel 1. For 60 s, the sample and extraction liquid were mixed together. The syringe pump and stirrer were turned off for 30 s to allow for extraction and phase separation. Finally, the upper phase was transported into the flow cell of the UV–Vis detector (channel 2), and absorbance was measured under stopped-flow conditions for 5 s at 540 nm before the solution was supplied to trash. After each measurement, the syringe and flow cell were rinsed with 1 mL of isopropyl alcohol (port d). The automation did not compromise the analytical figures of merits, including linearity, selectivity, sensitivity, accuracy and precision. Maya et al. developed a fully automated DLLME for the determination of rhodamine B with integrated spectrophotometric detection [15]. The results indicated that rhodamine B was measured in a working range of 0.023–2 mg/L with a limit of detection of 0.007 mg/L. The method also showed good repeatability for 10 successive extractions, with % RSD values of up to 3.2%. The EF for a 1 mg/L rhodamine B standard was found to be 23, and the method was capable of performing 51 extractions in 1 h.

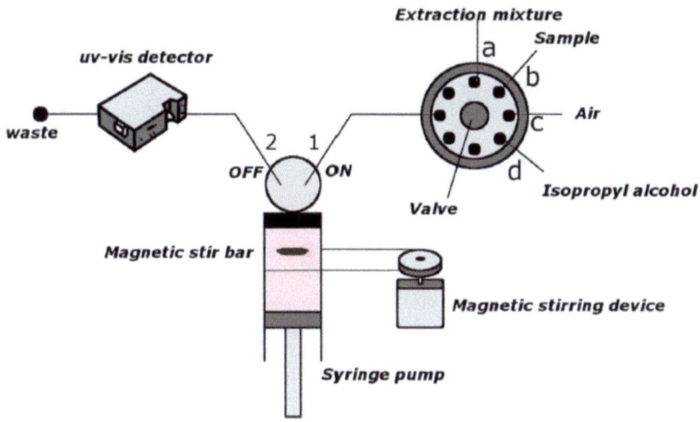

Fig. 8 The manifold of automated procedure for the determination of chromium (VI) in beverages. Reprinted from [157] with permission from Elsevier

4 Application of DLLME

The different modes of DLLME have been extensively used to pre-concentrate analytes of different nature from a variety of samples. Plasma, urine, hair, milk, fruits, vegetables, seafood and water samples were treated by DLLME before analysis. Drugs, toxins, pesticides, preservatives and heavy metals were all enriched with the aid of different modes of DLLME. According to the dispersion technique, n-DLLME and USA-DLLME are the most commonly used modes, followed by VA-DLLME and AA-DLLME. The average sample size is 5–10 mL, but amounts as small as 0.05 mL were also reported. In this case, a dilution step is required before sample preparation to facilitate dispersion. Large sample volumes were also prepared using DLLME, to allow for ultrasensitive determination of heavy metals. The type of extracting solvent depends on the selected DLLME mode, where chloroform is the widely used solvent in n-DLLME, while decanol is very common in DLLME-SFOD. Methanol, ACN and THF are the most popular dispersers due to availability, and high miscibility with both organic solvents and aqueous samples. The volume of disperser is usually less than 1000 μL, and it is highly dependent on the sample size and the extractant volume and type. DLLME has been extensively coupled to HPLC with different detectors including UV, FLD and MS. Application of DLLME before CE was also reported. Both HPLC and CE require minimal sample volumes to be injected into the instrument, which may explain the wide spread of DLLME with these particular analytical techniques. DLLME could also be used before UV/Vis spectrophotometric and spectrofluorometric determinations, if a microcuvette was available. An alternative approach in UV/Vis spectrophotometry was to measure the extracted small sample via a Nanodrop spectrophotometer. Table 2 shows some selected applications of the different modes of DLLME.

Table 2 Application of DLLME in various matrices

Analyte	Sample	Sample volume (mL)	Mode	Extractant	Extractant Volume (μL)	Disperser
Gliflozins	Human plasma	5	USA-DLLLME	1-dodecanol	100 μL	Methanol
Vincristine	Plasma of children	5	VA-DES-DLLME	DES: Methyl tri octyl ammonium chloride (MTOAC) and n-butanol in a molar ratio of 1: 3	80	N/A
Methotrexate	Plasma of children	5	MSA-DLLME	DES-SUPRAS 1-undecanol/ 1-dodecanol; 1:2 v/v)	45	N/A
Favipiravir	Human plasm	2	VA-IL-DLLME	Gadolinium based IL	50 mg	THF
Nateglinide	Human plasma	0.5	VA-DLLME	1-octanol	30	Methanol
Endocrine disrupting compounds	Water	5.0	AA-DES-DLLME-SFOD	DES: nonanoic (C9) acid, decanoic (C10): dodecanoic acid (C12) at ratio 1:1:1	200	N/A
NSAIDS	Milk and water samples	10	USA-DES-DLLLME	DES [TMGH]Cl: thymol at ratio 1:2 1,1,3,3-tetramethylguanidine (TMG)	200	N/A
Sartans	Water	1.2	IL-DLLME	Gadolinium based IL	30 mg	Methanol
Nitrophenol	Water samples	12	VA-DES-DLLME	DES Tetrabutylammonium bromide: thymol: octanoic acid with composition of 1:1:3	70	N/A

(continued)

Table 2 (continued)

Analyte	Sample	Sample volume (mL)	Mode	Extractant	Extractant Volume (μL)	Disperser
Sb(III) and Sb(V)	Water and plasma samples	9	AA-DLLME and narrow bore tube-DLLME	Octanol	100	Ethanol
Phthalic acid esters	Water and beverage samples	20	DES-DLLME	DES Menthol: acetic acid	100	N/A
Neonicotinoid Insecticide Residues	Water, Soil and Egg Yolk Samples	10	VA-DES-DLLME	DES Tetrabutylammonium bromide: decanoic acid	100	ACN
Enrofloxacin	Surface waters	8	USA-DLLME	Chloroform	500	ACN
Sulfonamides antibiotics	Water and Seafood Samples	NA	USA-DLLME	Tetrachloroethane	500	ACN
Chromium	Beverages and vegetables	125	USA-SUPRAS-DLLME	SUPRAS THF, tetrabutylammonium hydroxide and decanol	250	NA
Aucubin	Rat serum	0.05	VA-SUPRAS-DLLME	Pentanol	1000	THF
Carbaryl	Water, fruits and vegetables	15	VA-SUPRAS-DLLME	Heptanol	200	THF
Tricyclic antidepressants	Urine	10	VA-SUPRAS-DLLME	Decanol	50	THF
Traces of maneb	Food and water	10	USA-SUPRAS-DLLME	Decanol	200	THF
Copper	Water and hair	30	USA-SUPRAS-DLLME	Decanol	150	THF
Parabens	Environmental water samples	5	VA-MIL-SA-DLLME	MIL $Co(DMBG)_2NTf_2$	5 mg	ACN

(continued)

Table 2 (continued)

Analyte	Disperser volume (mL)	Instrument of analysis	Linearity (ng/mL)	%RSD	REF
Gliflozins	1	HPLC/DAD	1.1–2500	7.5–9.03	[158]
Vincristine	N/A	HPLC/UV	0.06–300	3.7	[62]
Methotrexate	N/A	HPLC /UV	0.1–150	2.6–4.8	[159]
Favipiravir	0.15	HPLC /UV	25–1.0 $*10^5$	4.07–11.84	[160]
Nateglinide	0.2	HPLC /UV	50–20,000	<6	[161]
Endocrine disrupting compounds	N/A	HPLC/PDA	3–300	<7	[71]
NSAIDS	N/A	HPLC /UV	5–2000	0.05–12.86	[162]
Sartans	70	UHPLC/UV–VIS	250–8000	2.48–4.07	[105]
Nitrophenol	N/A	HPLC /UV	1.0–500	<5.0	[163]
Sb(III) and Sb(V)	300	Spectrophotometric determination	0.03–20	3.5–4.7	[164]
Phthalic acid esters	N/A	HPLC/UV	4–425	<7.5	[14]
Neonicotinoid Insecticide Residues	400	HPLC/UV	3–1000	<5%	[165]
Enrofloxacin	500	HPLC/FLD	10–300	<2%	[166]
Sulfonamides antibiotics	900	HPLC/DAD	5–5000	0.1–8.1%	[167]

(continued)

Table 2 (continued)

Analyte	Disperser volume (mL)	Instrument of analysis	Linearity (ng/mL)	%RSD	REF
Chromium	NA	FAAS	0.1–350	2.1%	[168]
Aucubin	4000	UPLC/MS–MS	3–10,000	0.33–14.27	[169]
Carbaryl	800	UPLC-MS/MS	30–4000	7.1	[170]
Tricyclic antidepressants	200	HPLC/DAD	30–400	1.3–12.9	[171]
Traces of maneb	600	Spectrophotometric	67–1067	4.2	[172]
Copper	600	FAAS	N/A	2.2	[173]
Parabens	500	HPLC/UV	2.8–400	2.1–13.0	[174]

5 Conclusions and Future Trends

DLLME has attracted the interest of the analytical community since its introduction in 2006, owing to its simplicity and strong analytical capabilities. However, traditional DLLME has had one shortcoming from its emergence: the use of high-density halogenated solvents. Despite the high efficiency, these halogenated solvents are very hazardous, and a process utilising this type of solvent cannot be termed green even if the amount required was in microliter units. As a result, scientists have been seeking for solvents that are not only safe for the environment and operators, but also capable of improving the extraction efficiency of DLLME-based procedures. Many approaches have been presented in this regard. The development of the various types of solvents utilized in DLLME over the last 5 years has been examined in this chapter. SUPRAs and DESs offer exceptional qualities for microextraction and some advantages from being termed green solvents. Nonetheless, the field of chemistry is conservative in certain ways, and many DLLME experiments continue to employ traditional halogenated solvents in accordance with the guideline. The benefits of these traditional solvents, such as their ease of use and high density are obvious but we must not overlook the significant impact that these solvents have on health and the environment. As a result, the adoption of newer and greener solvents must be the goal of DLLME in the next years, with an emphasis on tailorable green solvents with high extraction capabilities and simple and safe synthesis.

The authors have declared no conflict of interest.

References

1. Hammad SF, Abdallah IA, Bedair A, Mansour FR (2022) Homogeneous liquid–liquid extraction as an alternative sample preparation technique for biomedical analysis. J Sep Sci 45:185–209. https://doi.org/10.1002/jssc.202100452
2. Lee J, Lee HK, Rasmussen KE, Pedersen-Bjergaard S (2008) Environmental and bioanalytical applications of hollow fiber membrane liquid-phase microextraction: a review. Anal Chim Acta 624:253–268. https://doi.org/10.1016/j.aca.2008.06.050
3. Kannouma RE, Hammad MA, Kamal AH, Mansour FR (2022) Miniaturization of liquid–liquid extraction; the barriers and the enablers. Microchem J 182:107863. https://doi.org/10.1016/j.microc.2022.107863
4. Yamini Y, Rezazadeh M, Seidi S (2019) Liquid-phase microextraction—the different principles and configurations. TrAC—Trends Anal Chem 112:264–272. https://doi.org/10.1016/j.trac.2018.06.010
5. Rezaee M, Assadi Y, Milani Hosseini M-R, Aghaee E, Ahmadi F, Berijani S (2006) Determination of organic compounds in water using dispersive liquid–liquid microextraction. J Chromatogr A 1116:1–9. https://doi.org/10.1016/j.chroma.2006.03.007
6. Mansour FR, Khairy MA (2017) Pharmaceutical and biomedical applications of dispersive liquid–liquid microextraction. J Chromatogr B 1061–1062:382–391. https://doi.org/10.1016/J.JCHROMB.2017.07.055
7. Khalili Zanjani MR, Yamini Y, Shariati S, Jönsson JÅ (2007) A new liquid-phase microextraction method based on solidification of floating organic drop. Anal Chim Acta 585:286–293. https://doi.org/10.1016/j.aca.2006.12.049

8. Regueiro J, Llompart M, Garcia-Jares C, Garcia-Monteagudo JC, Cela R (2008) Ultrasound-assisted emulsification—microextraction of emergent contaminants and pesticides in environmental waters. J Chromatogr A 1190:27–38. https://doi.org/10.1016/j.chroma.2008.02.091
9. Chen H, Chen R, Li S (2010) Low-density extraction solvent-based solvent terminated dispersive liquid–liquid microextraction combined with gas chromatography-tandem mass spectrometry for the determination of carbamate pesticides in water samples. J Chromatogr A 1217:1244–1248. https://doi.org/10.1016/j.chroma.2009.12.062
10. Jafarvand S, Shemirani F (2011) Supramolecular-based dispersive liquid–liquid microextraction: a novel sample preparation technique utilizes coacervates and reverse micelles. J Sep Sci 34:455–461. https://doi.org/10.1002/jssc.201000630
11. Jafarvand S, Shemirani F (2011) Supramolecular—based dispersive liquid–liquid microextraction: a novel sample preparation technique for determination of inorganic species. Microchim Acta 173:353–359. https://doi.org/10.1007/s00604-011-0564-9
12. Farajzadeh MA, Mogaddam MRA (2012) Air-assisted liquid–liquid microextraction method as a novel microextraction technique; Application in extraction and preconcentration of phthalate esters in aqueous sample followed by gas chromatography–flame ionization detection. Anal Chim Acta 728:31–38. https://doi.org/10.1016/j.aca.2012.03.031
13. Wang Y, Sun Y, Xu B, Li X, Jin R, Zhang H, Song D (2014) Magnetic ionic liquid-based dispersive liquid–liquid microextraction for the determination of triazine herbicides in vegetable oils by liquid chromatography. J Chromatogr A 1373:9–16. https://doi.org/10.1016/j.chroma.2014.11.009
14. Ortega-Zamora C, González-Sálamo J, González-Sálamo J, Hernández-Sánchez C, Hernández-Sánchez C, Hernández-Borges J, Hernández-Borges J (2020) Menthol-based deep eutectic solvent dispersive liquid–liquid microextraction: a simple and quick approach for the analysis of phthalic acid esters from water and beverage samples. ACS Sustain Chem Eng 8:8783–8794. https://doi.org/10.1021/acssuschemeng.0c02603
15. Maya F, Estela JM, Cerdà V (2012) Completely automated in-syringe dispersive liquid–liquid microextraction using solvents lighter than water. Anal Bioanal Chem 402:1383–1388. https://doi.org/10.1007/s00216-011-5572-4
16. Dean JA (1999) Lange's handbook of chemistry, fifteenth. McGraw Hill Inc., New York, USA
17. Mahmoudpour M, Mohtadinia J, Mousavi M, Ansarin M, Nemati M (2017) Application of the microwave-assisted extraction and dispersive liquid–liquid microextraction for the analysis of PAHs in smoked rice. Food Anal Methods 10:277–286. https://doi.org/10.1007/s12161-016-0579-2
18. Kalhor H, Hashemipour S, Yaftian MR, Shahdousti P (2016) Determination of carbamazepine in formulation samples using dispersive liquid–liquid microextraction method followed by ion mobility spectrometry. Int J Ion Mobil Spectrom 19:51–56. https://doi.org/10.1007/s12127-015-0184-x
19. Mookantsa SOS, Dube S, Nindi MM (2016) Development and application of a dispersive liquid–liquid microextraction method for the determination of tetracyclines in beef by liquid chromatography mass spectrometry. Talanta 148:321–328. https://doi.org/10.1016/j.talanta.2015.11.006
20. Ahmadi-Jouibari T, Fattahi N, Shamsipur M (2014) Rapid extraction and determination of amphetamines in human urine samples using dispersive liquid–liquid microextraction and solidification of floating organic drop followed by high performance liquid chromatography. J Pharm Biomed Anal 94:145–151. https://doi.org/10.1016/j.jpba.2014.01.044
21. Kannouma RE, Hammad MA, Kamal AH, Mansour FR (2021) A dispersive liquid–liquid microextraction method based on solidification of floating organic droplet for determination of antiviral agents in environmental water using HPLC/UV. Microchem J 171:106790. https://doi.org/10.1016/j.microc.2021.106790
22. Wang R, Qi X, Zhao L, Liu S, Gao S, Ma X, Deng Y (2016) Ionic-liquid-based dispersive liquid–liquid microextraction coupled with high-performance liquid chromatography for the forensic determination of methamphetamine in human urine. J Sep Sci 39:2444–2450. https://doi.org/10.1002/jssc.201600170

23. Monzón CM, Teglia CM, Delfino MR, Goicoechea HC (2016) Chemometric optimization and validation of a novel dispersive liquid–liquid microextraction–HPLC method for gliclazide, glibenclamide and glimepiride quantitation in serum samples. Microchem J 127:113–119. https://doi.org/10.1016/j.microc.2016.02.011
24. Šandrejová J, Campillo N, Viñas P, Andruch V (2016) Classification and terminology in dispersive liquid–liquid microextraction. Microchem J 127:184–186. https://doi.org/10.1016/j.microc.2016.03.007
25. Kamal AH, Hammad MA, Kannouma RE, Mansour FR (2022) Response surface optimization of a vortex-assisted dispersive liquid–liquid microextraction method for highly sensitive determination of repaglinide in environmental water by HPLC/UV. BMC Chem 16:33. https://doi.org/10.1186/s13065-022-00826-w
26. Mansour FR, Danielson ND (2018) Solvent-terminated dispersive liquid–liquid microextraction: a tutorial. Anal Chim Acta 1016:1–11. https://doi.org/10.1016/j.aca.2018.02.005
27. Cölfen H, Rosenberg R, Haffke D, Stemplinger S, Zemb T, Horinek D (2023) Two types of liquid phase separation induced by soft centrifugation in aqueous ethyl acetate using ethanol as cosolvent. Research 6:A67. https://doi.org/10.34133/research.0026
28. Viñas P, Campillo N, López-García I, Hernández-Córdoba M (2014) Dispersive liquid–liquid microextraction in food analysis. A critical review Microextraction Techniques. Anal Bioanal Chem 406:2067–2099. https://doi.org/10.1007/s00216-013-7344-9
29. Fisichella M, Odoardi S, Strano-Rossi S (2015) High-throughput dispersive liquid/liquid microextraction (DLLME) method for the rapid determination of drugs of abuse, benzodiazepines and other psychotropic medications in blood samples by liquid chromatography–tandem mass spectrometry (LC-MS/MS) and app. Microchem J 123:33–41
30. Bulatov A, Medinskaia K, Aseeva D, Garmonov S, Moskvin L (2014) Talanta Determination of antipyrine in saliva using the dispersive liquid–liquid microextraction based on a stepwise injection system. Talanta 133:66–70. https://doi.org/10.1016/j.talanta.2014.05.064
31. Farajzadeh MA, Reza M, Mogaddam A, Bamorowat M (2015) Determination of unconjugated non-steroidal anti-inflammatory drugs in biological fluids using air-assisted liquid–liquid microextraction combined with back extraction followed by high performance liquid chromatography. Anal Methods 7:1372–1379. https://doi.org/10.1039/c4ay01996b
32. Deng K-J, Lan X-H, Sun G, Ji LY, Zheng X-L (2016) Determination of sulfonamide residues in chicken liver using high-performance liquid chromatography. Food Anal Methods 9:3337–3344. https://doi.org/10.1007/s12161-016-0514-6
33. Lin SY, Chen PS, Chang SY (2015) Detection of posaconazole by surface-assisted laser desorption/ionization mass spectrometry with dispersive liquid–liquid microextraction. J Am Soc Mass Spectrom 26:530–533. https://doi.org/10.1007/s13361-014-1061-2
34. Chen P-S, Cheng Y-H, Lin S-Y, Chang SY (2016) Determination of immunosuppressive drugs in human urine and serum by surface-assisted laser desorption/ionization mass spectrometry with dispersive liquid–liquid microextraction. Anal Bioanal Chem 408:629–637
35. Jouyban A, Sorouraddin MH, Farajzadeh MA, Somi MH, Fazeli-Bakhtiyari R (2015) Determination of five antiarrhythmic drugs in human plasma by dispersive liquid–liquid microextraction and high-performance liquid chromatography. Talanta 134:681–689. https://doi.org/10.1016/j.talanta.2014.12.008
36. Cunha RL, Lopes WA, Pereira PAP (2016) Determination of free (unconjugated) amphetamine-type stimulants in urine samples by dispersive liquid–liquid microextraction and gas chromatography coupled to mass spectrometry (DLLME-GC-MS). Microchem J 125:230–235. https://doi.org/10.1016/j.microc.2015.11.017
37. Habibollahi S, Tavakkoli N, Nasirian V, Khani H (2015) Determination of tramadol by dispersive liquid–liquid microextraction combined with GC-MS. J Chromatogr Sci 53:655–661. https://doi.org/10.1093/chromsci/bmu118
38. Gardner MA, Sampsel S, Jenkins WW, Owens JE (2015) Analysis of fentanyl in urine by DLLME-GC-MS. J Anal Toxicol 39:118–125. https://doi.org/10.1093/jat/bku136
39. Shamsipur M, Mirmohammadi M (2014) High performance liquid chromatographic determination of ultra traces of two tricyclic antidepressant drugs imipramine and trimipramine

in urine samples after their dispersive liquid–liquid microextraction coupled with response surface optimization. J Pharm Biomed Anal 100:271–278. https://doi.org/10.1016/j.jpba.2014.08.008

40. Hsieh M, Chang SY (2015) Sensitive detection of piperazinyl phenothiazine drugs by field-amplified sample stacking in capillary electrophoresis with dispersive liquid–liquid microextraction. Electrophoresis. https://doi.org/10.1002/elps.201500260
41. de Albuquerque NCP, de Gaitani CM, de Oliveira ARM (2015) A new and fast DLLME-CE method for the enantioselective analysis of zopiclone and its active metabolite after fungal biotransformation. J Pharm Biomed Anal 109:192–201. https://doi.org/10.1016/j.jpba.2015.02.039
42. Altunay N, Elik A, Gürkan R (2019) Monitoring of some trace metals in honeys by flame atomic absorption spectrometry after ultrasound assisted-dispersive liquid liquid microextraction using natural deep eutectic solvent. Microchem J 147:49–59. https://doi.org/10.1016/j.microc.2019.03.003
43. Ali J, Tuzen M, Citak D, Uluozlu OD, Mendil D, Kazi TG, Afridi HI (2019) Separation and preconcentration of trivalent chromium in environmental waters by using deep eutectic solvent with ultrasound-assisted based dispersive liquid–liquid microextraction method. J Mol Liq 291:111299. https://doi.org/10.1016/j.molliq.2019.111299
44. Campone L, Celano R, Piccinelli AL, Pagano I, Cicero N, Di SR, Carabetta S, Russo M, Rastrelli L (2019) Ultrasound assisted dispersive liquid–liquid microextraction for fast and accurate analysis of chloramphenicol in honey. Food Res Int 115:572–579. https://doi.org/10.1016/j.foodres.2018.09.006
45. Altunay N (2021) Optimization of ultrasound-assisted dispersive liquid–liquid microextraction of niacinamide in pharmaceutical and cosmetic samples using experimental design. Microchem J 170:106659. https://doi.org/10.1016/j.microc.2021.106659
46. Custodio-Mendoza JA, Caamaño-Fernandez C, Lage MA, Almeida PJ, Lorenzo RA, Carro AM (2022) GC–MS determination of malondialdehyde, acrolein, and 4-hydroxy-2-nonenal by ultrasound-assisted dispersive liquid–liquid microextraction in beverages. Food Chem 384:132530. https://doi.org/10.1016/j.foodchem.2022.132530
47. Hsieh CZ, Chung WH, Ding WH (2021) Experimental design approaches to optimize ultrasound-assisted simultaneous-silylation dispersive liquid–liquid microextraction for the rapid determination of parabens in water samples. RSC Adv 11:23607–23615. https://doi.org/10.1039/d1ra04195a
48. Werner J (2020) Ligandless, deep eutectic solvent-based ultrasound-assisted dispersive liquid–liquid microextraction with solidification of the aqueous phase for preconcentration of lead, cadmium, cobalt and nickel in water samples. J Sep Sci 43:1297–1305. https://doi.org/10.1002/jssc.201901184
49. Fernández P, González C, Pena MT, Carro AM, Lorenzo RA (2013) A rapid ultrasound-assisted dispersive liquid–liquid microextraction followed by ultra-performance liquid chromatography for the simultaneous determination of seven benzodiazepines in human plasma samples. Anal Chim Acta 767:88–96. https://doi.org/10.1016/j.aca.2013.01.016
50. Sajedi-Amin S, Assadpour-Zeynali K, Panahi-Azar V, Kebriaeezadeh A, Khoubnasabjafari M, Ansarin K, Jouyban-Gharamaleki V, Jouyban A (2015) Spectroscopic analysis of bosentan in biological samples after a liquid–liquid microextraction. Bioimpacts 5:191–197. https://doi.org/10.15171/bi.2015.28
51. Fernández P, Taboada V, Regenjo M, Morales L, Alvarez I, Carro AM, Lorenzo RA (2016) Optimization of ultrasound assisted dispersive liquid–liquid microextraction of six antidepressants in human plasma using experimental design. J Pharm Biomed Anal 124:189–197. https://doi.org/10.1016/j.jpba.2016.02.041
52. Rodríguez MP, Pezza HR, Pezza L (2016) Ultrasound-assisted dispersive liquid–liquid microextraction of tetracycline drugs from egg supplements before flow injection analysis coupled to a liquid waveguide capillary cell. Anal Bioanal Chem 408:6201–6211. https://doi.org/10.1007/s00216-016-9732-4

53. Zhou GS, Yuan YC, Yin Y, Tang YP, Xu RJ, Liu Y, Chen PD, Yin L, Duan JA (2020) Hydrophilic interaction chromatography combined with ultrasound-assisted ionic liquid dispersive liquid–liquid microextraction for determination of underivatized neurotransmitters in dementia patients' urine samples. Anal Chim Acta 1107:74–84. https://doi.org/10.1016/j.aca.2020.02.027
54. Wang Q, Fan J (2019) A hydrophobic deep eutectic solvent-based vortex-assisted dispersive liquid–liquid microextraction combined with HPLC for the determination of nitrite in water and biological samples. J Sep Sci 42(2):574–581. https://doi.org/10.1002/jssc.201800921
55. Li T, Song Y, Li J, Zhang M, Shi Y, Fan J (2019) New low viscous hydrophobic deep eutectic solvents in vortex-assisted liquid–liquid microextraction for the determination of phthalate esters from food-contacted plastics. Food Chem 309:125752. https://doi.org/10.1016/j.foodchem.2019.125752
56. Gholami Z, Hossein M, Yousefinejad S (2021) Vortex-assisted dispersive liquid–liquid microextraction based on hydrophobic deep eutectic solvent for the simultaneous identification of eight synthetic dyes in jellies and drinks using HPLC-PDA. Microchem J 170:106671. https://doi.org/10.1016/j.microc.2021.106671
57. Abdallah IA, Hammad SF, Bedair A, Mansour FR (2022) Menthol-assisted homogenous liquid–liquid microextraction for HPLC/UV determination of favipiravir as an antiviral for COVID-19 in human plasma. J Chromatogr B 1189:123087. https://doi.org/10.1016/j.jchromb.2021.123087
58. Abdallah IA, Hammad SF, Bedair A, Elshafeey AH, Mansour FR (2022) Determination of favipiravir in human plasma using homogeneous liquid–liquid microextraction followed by HPLC/UV. Bioanalysis 14:205–216. https://doi.org/10.4155/bio-2021-0219
59. Guiñez M, Martinez LD, Fernandez L, Cerutti S (2017) Dispersive liquid–liquid microextraction based on solidification of floating organic drop and fluorescence detection for the determination of nitrated polycyclic aromatic hydrocarbons in aqueous samples. Microchem J 131:1–8. https://doi.org/10.1016/j.microc.2016.10.020
60. Herrera-herrera AV, Hernández-borges J, Borges-miquel TM, Rodríguez-delgado MÁ (2013) Dispersive liquid–liquid microextraction combined with ultra-high performance liquid chromatography for the simultaneous determination of 25 sulfonamide and quinolone antibiotics in water samples. J Pharm Biomed Anal 75:130–137. https://doi.org/10.1016/j.jpba.2012.11.026
61. Santana-mayor A, Herrera-herrera AV, Rodr R, Miguel A (2021) Development of a green alternative vortex-assisted dispersive liquid−liquid microextraction based on natural hydrophobic deep eutectic solvents for the analysis of phthalate esters in soft drinks. ACS Sustain Chem Eng 9(5):2161–2170. https://doi.org/10.1021/acssuschemeng.0c07686
62. Golpayegani MR, Akramipour R, Gheini S, Amini MV, Fattahi F, Mohebbi A, Fattahi N (2022) Sensitive determination of vincristine in plasma of children with leukaemia using vortex-assisted dispersive liquid–liquid microextraction based on hydrophobic deep eutectic solvent. RSC Adv 12:3611–3617. https://doi.org/10.1039/d1ra07981f
63. Bocato MZ, Cesila CA, Lataro BF, de Oliveira ARM, Campíglia AD, Barbosa F (2020) A fast-multiclass method for the determination of 21 endocrine disruptors in human urine by using vortex-assisted dispersive liquid–liquid microextraction (VADLLME) and LC-MS/MS. Environ Res 189:109883. https://doi.org/10.1016/j.envres.2020.109883
64. Elencovan V, Joseph J, Yahaya N, Abdul N, Raoov M, Lim V, Nadhirah N, Zain M (2022) Exploring a novel deep eutectic solvents combined with vortex assisted dispersive liquid–liquid microextraction and its toxicity for organophosphorus pesticides analysis from honey and fruit samples. Food Chem 368:130835. https://doi.org/10.1016/j.foodchem.2021.130835
65. Wu Y, Chen M, Wang X, Zhou Y, Xu M, Zhang Z (2021) Development and validation of vortex-assisted dispersive liquid–liquid microextraction method based on solidification of floating hydrophobic deep eutectic solvent for the determination of endocrine disrupting chemicals in sewage. Microchem J 163:105915. https://doi.org/10.1016/j.microc.2020.105915
66. Bosch Ojeda C, Sánchez Rojas F (2014) Vortex-assisted liquid–liquid microextraction (VALLME): applications. Chromatographia 77:745–754. https://doi.org/10.1007/s10337-014-2669-x

67. Mogaddam A (2019) Air–assisted liquid–liquid microextraction; principles and applications with analytical instruments. Trends Anal Chem 112:115734. https://doi.org/10.1016/j.trac.2019.115734
68. Pena-pereira F, Lavilla I, Bendicho C (2010) Liquid-phase microextraction techniques within the framework of green chemistry. Trends Anal Chem 29:617–628. https://doi.org/10.1016/j.trac.2010.02.016
69. Farajzadeh MA, Nasrollahpour H, Mogaddam MRA, Khoshmaram L (2016) Determination of widely used non-steroidal anti-inflammatory drugs in biological fluids using simultaneous derivatization and air-assisted liquid–liquid microextraction followed by gas chromatography–flame ionization detection. J Iran Chem Soc 13:289–298. https://doi.org/10.1007/s13738-015-0736-3
70. Barfi B, Asghari A, Rajabi M, Goochani Moghadam A, Mirkhani N, Ahmadi F (2015) Comparison of ultrasound-enhanced air-assisted liquid–liquid microextraction and low-density solvent-based dispersive liquid–liquid microextraction methods for determination of nonsteroidal anti-inflammatory drugs in human urine samples. J Pharm Biomed Anal 111:297–305. https://doi.org/10.1016/j.jpba.2015.03.034
71. El-Deen AK, Shimizu K (2020) A green air assisted-dispersive liquid–liquid microextraction based on solidification of a novel low viscous ternary deep eutectic solvent for the enrichment of endocrine disrupting compounds from water. J Chromatogr A 1629:461498. https://doi.org/10.1016/j.chroma.2020.461498
72. Ge D, Zhang Y, Dai Y, Yang S (2018) Air-assisted dispersive liquid–liquid microextraction based on a new hydrophobic deep eutectic solvent for the preconcentration of benzophenone-type UV filters from aqueous samples. J Sep Sci 41:1635–1643. https://doi.org/10.1002/jssc.201701282
73. Jia L, Yang J, Zhao W, Jing X (2019) Microextraction based on solidification of the aqueous phase for the determination of triazole fungicides in water samples by high-performance. RSC Adv 9(63):36664–36669. https://doi.org/10.1039/c9ra07348e
74. Chaikhan P, Udnan Y, Ampiah-bonney RJ, Chuachuad W (2021) Air-assisted solvent terminated dispersive liquid–liquid microextraction (AA-ST-DLLME) for the determination of lead in water and beverage samples by graphite furnace atomic absorption spectrometry. Microchem J 162:105828. https://doi.org/10.1016/j.microc.2020.105828
75. Rahmani M, Ghasemi E, Sasani M (2016) Application of response surface methodology for air assisted-dispersive liquid–liquid microextraction of deoxynivalenol in rice samples prior to HPLC—DAD analysis and comparison with solid phase extraction cleanup. Talanta 165:27–32. https://doi.org/10.1016/j.talanta.2016.12.031
76. Teglia CM, Gonzalo L, Culzoni MJ, Goicoechea HC (2017) Determination of six veterinary pharmaceuticals in egg by liquid chromatography: chemometric optimization of a novel air assisted-dispersive liquid–liquid microextraction by solid floating organic drop Abbreviated running title: Determination of six vet. Food Chem 273:194–202. https://doi.org/10.1016/j.foodchem.2017.08.034
77. Karami-osboo R, Maham M (2018) Pre-concentration and extraction of aflatoxins from rice using air-assisted dispersive liquid–liquid microextraction. Food Anal Methods 11(10):2816–2821
78. Ferrone AV, Cotellese R, Di ML, Carlucci G (2017) Air assisted dispersive liquid–liquid microextraction with solidification of the floating organic droplets (AA-DLLME-SFO) and UHPLC-PDA method: Application to antibiotics analysis in human plasma of hospital acquired pneumonia patients. J Pharm Biomed Anal 151:266–273. https://doi.org/10.1016/j.jpba.2017.12.039
79. Shishov A, Trufanov I, Nechaeva D, Bulatov A (2019) A reversed-phase air-assisted dispersive liquid–liquid microextraction coupled with colorimetric paper-based analytical device for the determination of glycerol, calcium and magnesium in biodiesel samples. Microchem J 150:104134. https://doi.org/10.1016/j.microc.2019.104134

80. Rahmani M, Kaykhaii M, Safari Z, Ansari O (2017) Comparison of air—assisted, vortex—assisted and ultrasound—assisted dispersive liquid–liquid microextraction for the determination of BTEX compounds in water samples prior to GC–FID analysis. Chromatographia 80:109–117. https://doi.org/10.1007/s10337-016-3216-8
81. Kocúrová L, Balogh IS, Šandrejová J, Andruch V (2012) Recent advances in dispersive liquid–liquid microextraction using organic solvents lighter than water: a review. Microchem J 102:11–17. https://doi.org/10.1016/j.microc.2011.12.002
82. Ghambari H, Hadjmohammadi M (2012) Low-density solvent-based dispersive liquid–liquid microextraction followed by high performance liquid chromatography for determination of warfarin in human plasma. J Chromatogr B 899:66–71. https://doi.org/10.1016/j.jchromb.2012.04.035
83. Meng L, Zhang W, Meng P, Zhu B, Zheng K (2015) Comparison of hollow fiber liquid-phase microextraction and ultrasound-assisted low-density solvent dispersive liquid–liquid microextraction for the determination of drugs of abuse in biological samples by gas chromatography–mass spectrometry. J Chromatogr B 989:46–53. https://doi.org/10.1016/j.jchromb.2015.02.039
84. Mansour FR, Danielson ND (2017) Solidification of floating organic droplet in dispersive liquid–liquid microextraction as a green analytical tool. Talanta 170:22–35. https://doi.org/10.1016/j.talanta.2017.03.084
85. Tolcha T, Merdassa Y, Megersa N (2013) Low-density extraction solvent based solvent-terminated dispersive liquid–liquid microextraction for quantitative determination of ionizable pesticides in environmental waters. J Sep Sci 36:1119–1127. https://doi.org/10.1002/jssc.201200849
86. Carda-broch S, Berthod A, Angel MJR (2008) Ionic liquids in separation techniques. J Chromatogr A 1184:6–18. https://doi.org/10.1016/j.chroma.2007.11.109
87. Nerín C, Salafranca J, Aznar M, Batlle R (2009) Critical review on recent developments in solventless techniques for extraction of analytes. Anal Bioanal Chem 393:809–833. https://doi.org/10.1007/s00216-008-2437-6
88. Sun P, Armstrong DW (2010) Analytica Chimica Acta Ionic liquids in analytical chemistry. Analytica Chimica Acta 661:1–16. https://doi.org/10.1016/j.aca.2009.12.007
89. Liu R, Liu J, Yin Y, Hu X (2009) Ionic liquids in sample preparation. Anal Bioanal Chem 393:871–883. https://doi.org/10.1007/s00216-008-2445-6
90. Danielson ND, Mansour FR, Zhou L, Connell CV, Dotlich EM, Gibler JN, Norman BE, Grossman S, Wei W, Zhang Y (2018) Liquid chromatography with alkylammonium formate ionic liquid mobile phases and fluorescence detection. J Chromatogr A 1559:128–135. https://doi.org/10.1016/j.chroma.2018.03.020
91. Borrull F, Marce RM (2012) Ionic liquids in solid-phase extraction. Trends Anal Chem 41:15–26. https://doi.org/10.1016/j.trac.2012.08.010
92. Rykowska I, Ziemblińska J, Nowak I (2018) Modern approaches in dispersive liquid–liquid microextraction (DLLME) based on ionic liquids: a review. J Mol Liq 259:319–339. https://doi.org/10.1016/j.molliq.2018.03.043
93. Abdelaziz MA, Saleh AM, Mansour FR, Danielson ND (2022) A gadolinium-based magnetic ionic liquid for dispersive liquid–liquid microextraction of ivermectin from environmental water. J Chromatogr Sci. https://doi.org/10.1093/chromsci/bmac101
94. Kokosa JM (2013) Advances in solvent—microextraction techniques. Trends Anal Chem 43:2–13. https://doi.org/10.1016/j.trac.2012.09.020
95. Plechkova NV, Seddon KR (2008) Applications of ionic liquids in the chemical industry. Chem Soc Rev 37(1):123–150. https://doi.org/10.1039/b006677j
96. Zhou Q, Bai H, Xie G, Xiao J (2008) Trace determination of organophosphorus pesticides in environmental samples by temperature-controlled ionic liquid dispersive liquid-phase microextraction. J Chromatogr A 1188:148–153. https://doi.org/10.1016/j.chroma.2008.02.094
97. Baghdadi M, Shemirani F (2008) Cold-induced aggregation microextraction: a novel sample preparation technique based on ionic liquids. Anal Chim Acta 613:56–63. https://doi.org/10.1016/j.aca.2008.02.057

98. Liu Y, Zhao E, Zhu W, Gao H, Zhou Z (2009) Determination of four heterocyclic insecticides by ionic liquid dispersive liquid–liquid microextraction in water samples. J Chromatogr A 1216:885–891. https://doi.org/10.1016/j.chroma.2008.11.076
99. Soylak M (2016) Latest trends, green aspects, and innovations in liquid-phase—Based microextraction techniques: a review. Turkish J Chem 40. https://doi.org/10.3906/kim-160 5-26
100. Baghdadi M, Shemirani F (2009) In situ solvent formation microextraction based on ionic liquids: a novel sample preparation technique for determination of inorganic species in saline solutions. Anal Chim Acta 634:186–191. https://doi.org/10.1016/j.aca.2008.12.017
101. Yao C, Anderson JL (2009) Dispersive liquid–liquid microextraction using an in situ metathesis reaction to form an ionic liquid extraction phase for the preconcentration of aromatic compounds from water. Anal Bioanal Chem 395:1491–1502. https://doi.org/10.1007/s00216-009-3078-0
102. Del Sesto RE, McCleskey TM, Burrell AK, Baker GA, Thompson JD, Scott BL, Wilkes JS, Williams P (2002) Structure and magnetic behavior of transition metal based ionic liquids. Chem Commun 8:447–449. https://doi.org/10.1039/b711189d
103. Rosatella AA, Siopa F, Frade RFM, Afonso CAM (2016) New low viscous cholinium-based magnetic ionic liquids. New J Chem 40:3124–3129. https://doi.org/10.1039/C5NJ03165F
104. Santos E, Albo J, Rosatella A, Afonso CAM, Irabien Á (2014) Synthesis and characterization of Magnetic Ionic Liquids (MILs) for CO2 separation. J Chem Technol Biotechnol 89:866–871. https://doi.org/10.1002/jctb.4323
105. Abdelaziz MA, Mansour FR, Danielson ND (2021) A gadolinium-based magnetic ionic liquid for dispersive liquid–liquid microextraction. Anal Bioanal Chem 413:205–214. https://doi.org/10.1007/s00216-020-02992-z
106. Mousavi L, Tamiji Z, Khoshayand MR (2018) Applications and opportunities of experimental design for the dispersive liquid–liquid microextraction method—a review. Talanta 190:335–356. https://doi.org/10.1016/j.talanta.2018.08.002
107. Sajid M, Alhooshani K (2018) Dispersive liquid–liquid microextraction based binary extraction techniques prior to chromatographic analysis: a review. TrAC—Trends Anal Chem 108:167–182. https://doi.org/10.1016/j.trac.2018.08.016
108. Paquin F, Rivnay J, Salleo A, Stingelin N, Silva C (2015) Multi-phase semicrystalline microstructures drive exciton dissociation in neat plastic semiconductors. J Mater Chem C 3:10715–10722. https://doi.org/10.1039/b000000x
109. Cvjetko Bubalo M, Ćurko N, Tomašević M, Kovačević Ganić K, Radojcic Redovnikovic I (2016) Green extraction of grape skin phenolics by using deep eutectic solvents. Food Chem 200:159–166. https://doi.org/10.1016/j.foodchem.2016.01.040
110. Khandelwal S, Tailor YK, Kumar M (2016) Deep eutectic solvents (DESs) as eco-friendly and sustainable solvent/catalyst systems in organic transformations. J Mol Liq 215:345–386. https://doi.org/10.1016/j.molliq.2015.12.015
111. Florindo C, Branco LC, Marrucho IM (2017) Development of hydrophobic deep eutectic solvents for extraction of pesticides from aqueous environments. Fluid Phase Equilib 448:135–142. https://doi.org/10.1016/j.fluid.2017.04.002
112. Duan L, Dou LL, Guo L, Li P, Liu EH (2016) Comprehensive evaluation of deep eutectic solvents in extraction of bioactive natural products. ACS Sustain Chem Eng 4:2405–2411. https://doi.org/10.1021/acssuschemeng.6b00091
113. Zeng Q, Wang Y, Huang Y, Ding X, Chen J, Xu K (2014) Deep eutectic solvents as novel extraction media for protein partitioning. Analyst 139:2565–2573. https://doi.org/10.1039/c3an02235h
114. Nam MW, Zhao J, Lee MS, Jeong JH, Lee J (2015) Enhanced extraction of bioactive natural products using tailor-made deep eutectic solvents: application to flavonoid extraction from Flos sophorae. Green Chem 17:1718–1727. https://doi.org/10.1039/c4gc01556h
115. Lasarte-Aragonés G, Lucena R, Cárdenas S, Valcárcel M (2015) Use of switchable hydrophilicity solvents for the homogeneous liquid–liquid microextraction of triazine herbicides from environmental water samples. J Sep Sci 38:990–995. https://doi.org/10.1002/jssc.201401224

116. Hamad A, Elshahawy M, Negm A, Mansour FR (2020) Analytical methods for determination of glutathione and glutathione disulfide in pharmaceuticals and biological fluids. Rev Anal Chem 38:20190019. https://doi.org/10.1515/revac-2019-0019
117. Mansour FR, Danielson ND (2012) Separation methods for captopril in pharmaceuticals and biological fluids. J Sep Sci 35:1213–1226. https://doi.org/10.1002/jssc.201200057
118. Mansour FR, Wei W, Danielson ND (2013) Separation of carnitine and acylcarnitines in biological samples: a review. Biomed Chromatogr 27:1339–1353. https://doi.org/10.1002/bmc.2995
119. Ballesteros-Gómez A, Rubio S, Pérez-Bendito D (2009) Potential of supramolecular solvents for the extraction of contaminants in liquid foods. J Chromatogr A 1216:530–539. https://doi.org/10.1016/j.chroma.2008.06.029
120. Carabias-Martínez R, Rodríguez-Gonzalo E, Moreno-Cordero B, Pérez-Pavón JL, García-Pinto C, Fernández Laespada E (2000) Surfactant cloud point extraction and preconcentration of organic compounds prior to chromatography and capillary electrophoresis. J Chromatogr A 902:251–265. https://doi.org/10.1016/S0021-9673(00)00837-2
121. Cantero M, Rubio S, Pérez-Bendito D (2004) Determination of non-ionic polyethoxylated surfactants in sewage sludge by coacervative extraction and ion trap liquid chromatography-mass spectrometry. J Chromatogr A 1046:147–153. https://doi.org/10.1016/j.chroma.2004.06.073
122. Menger FM, Sykes BM (1998) Anatomy of a coacervate. Langmuir 14:4131–4137. https://doi.org/10.1021/la980208m
123. Seidi S, Alavi L, Jabbari A (2018) Trace determination of cadmium in rice samples using solidified floating organic drop microextraction based on vesicular supramolecular solvent followed by flow-injection analysis–flame atomic absorption spectrometry. J Iran Chem Soc 15:2083–2092. https://doi.org/10.1007/s13738-018-1401-4
124. Ruiz FJ, Rubio S, Pérez-Bendito D (2007) Water-induced coacervation of alkyl carboxylic acid reverse micelles: Phenomenon description and potential for the extraction of organic compounds. Anal Chem 79:7473–7484. https://doi.org/10.1021/ac0708644
125. García-Prieto A, Lunar L, Rubio S, Pérez-Bendito D (2008) Decanoic acid reverse micelle-based coacervates for the microextraction of bisphenol A from canned vegetables and fruits. Anal Chim Acta 617:51–58. https://doi.org/10.1016/j.aca.2008.01.061
126. Tayyebi M, Yamini Y, Moradi M (2012) Reverse micelle-mediated dispersive liquid–liquid microextraction of 2,4-dichlorophenoxyacetic acid and 4-chloro-2-methylphenoxyacetic acid. J Sep Sci 35:2491–2498. https://doi.org/10.1002/jssc.201200150
127. Ballesteros-Gómez A, Sicilia MD, Rubio S (2010) Supramolecular solvents in the extraction of organic compounds: a review. Anal Chim Acta 677:108–130. https://doi.org/10.1016/j.aca.2010.07.027
128. Mabrouk MM, Hamed NA, Mansour FR (2022) Physicochemical and electrochemical methods for determination of critical micelle concentrations of surfactants: a comprehensive review. Monatshefte für Chemie—Chem Mon 153:125–138. https://doi.org/10.1007/s00706-022-02891-2
129. Mabrouk MM, Hamed NA, Mansour FR (2023) Spectroscopic methods for determination of critical micelle concentrations of surfactants; a comprehensive review. Appl Spectrosc Rev 58:206–234. https://doi.org/10.1080/05704928.2021.1955702
130. Viñas P, Campillo N, Andruch V (2015) Recent achievements in solidified floating organic drop microextraction. Trends Anal Chem 68:48–77. https://doi.org/10.1016/j.trac.2015.02.005
131. Farhadi K, Hatami M, Forough M, Molaei R (2013) Dispersive liquid–liquid microextraction of propranolol enantiomers from human plasma based on the solidification of a floating organic droplet. Bioanalysis 5:701–710. https://doi.org/10.4155/bio.13.23
132. Suh JH, Lee YY, Lee HJ, Kang M, Hur Y, Lee SN, Yang D-H, Han SB (2013) Dispersive liquid–liquid microextraction based on solidification of floating organic droplets followed by high performance liquid chromatography for the determination of duloxetine in human plasma. J Pharm Biomed Anal 75:214–219. https://doi.org/10.1016/j.jpba.2012.11.041

133. Jia S, Li J, Park S-R, Ryu Y, Park IH, Park JH, Hong S-S, Kwon SW, Lee J (2013) Combined application of dispersive liquid–liquid microextraction based on the solidification of floating organic droplets and charged aerosol detection for the simple and sensitive quantification of macrolide antibiotics in human urine. J Pharm Biomed Anal 86:204–213. https://doi.org/10.1016/j.jpba.2013.07.024
134. Taheri S, Jalali F, Fattahi N, Bahrami G (2015) Sensitive determination of atorvastatin in human plasma by dispersive liquid–liquid microextraction and solidification of floating organic drop followed by high-performance liquid chromatography. J Sep Sci 38:309–315. https://doi.org/10.1002/jssc.201401115
135. Zhao L, Zhao P, Wang L, Ma X, Hou X, Li F (2014) A dispersive liquid–liquid microextraction method based on the solidification of a floating organic drop combined with HPLC for the determination of lovastatin and simvastatin in rat urine. Biomed Chromatogr 28:895–900. https://doi.org/10.1002/bmc.3205
136. Shukri DSM, Sanagi MM, Ibrahim WAW, Abidin NNZ, Aboul-Enein HY (2015) Liquid chromatographic determination of NSAIDs in urine after dispersive liquid–liquid microextraction based on solidification of floating organic droplets. Chromatographia 78:987–994. https://doi.org/10.1007/s10337-015-2920-0
137. Pebdani AA, Shabani AMH, Dadfarnia S, Talebianpoor MS, Khodadoust S (2016) Preconcentration of valsartan by dispersive liquid–liquid microextraction based on solidification of floating organic drop and its determination in urine sample: central composite design. J Sep Sci 39:1935–1944. https://doi.org/10.1002/jssc.201600010
138. Zamani-Kalajahi M, Fazeli-Bakhtiyari R, Amiri M, Golmohammadi A, Afrasiabi A, Khoubnasabjafari M, Jouyban A (2013) Dispersive liquid–liquid microextraction based on solidification of floating organic droplet followed by spectrofluorimetry for determination of carvedilol in human plasma. Bioanalysis 5:437–448. https://doi.org/10.4155/bio.12.326
139. Yamini Y, Rezaee M, Khanchi A, Faraji M, Saleh A (2010) Dispersive liquid–liquid microextraction based on the solidification of floating organic drop followed by inductively coupled plasma-optical emission spectrometry as a fast technique for the simultaneous determination of heavy metals. J Chromatogr A 1217:2358–2364. https://doi.org/10.1016/j.chroma.2009.11.046
140. Ahmadi-Jouibari T, Fattahi N, Shamsipur M, Pirsaheb M (2013) Dispersive liquid–liquid microextraction followed by high-performance liquid chromatography-ultraviolet detection to determination of opium alkaloids in human plasma. J Pharm Biomed Anal 85:14–20. https://doi.org/10.1016/j.jpba.2013.06.030
141. Taheri S, Jalali F, Fattahi N, Jalili R, Bahrami G (2015) Sensitive determination of methadone in human serum and urine by dispersive liquid–liquid microextraction based on the solidification of a floating organic droplet followed by HPLC-UV. J Sep Sci 38:3545–3551. https://doi.org/10.1002/jssc.201500636
142. Akramipour R, Fattahi N, Pirsaheb M, Gheini S (2016) Combination of counter current salting-out homogenous liquid–liquid extraction and dispersive liquid–liquid microextraction as a novel microextraction of drugs in urine samples. J Chromatogr B 1012–1013:162–168. https://doi.org/10.1016/j.jchromb.2016.01.031
143. Asadi M, Dadfarnia S, Haji Shabani AM (2016) Simultaneous extraction and determination of albendazole and triclabendazole by a novel syringe to syringe dispersive liquid phase microextraction-solidified floating organic drop combined with high performance liquid chromatography. Anal Chim Acta 932:22–28. https://doi.org/10.1016/j.aca.2016.05.014
144. Hemmati M, Rajabi M, Asghari A (2017) Efficient and clean pre-concentration of ultra-trace calcium channel blockers from biological matrices via a hyphenated procedure of two sequential dispersive solid/liquid phase microextractions. Anal Chim Acta 960:138–150. https://doi.org/10.1016/j.aca.2017.01.001
145. Guo L, Lee HK (2014) Automated dispersive liquid−liquid microextraction−gas chromatography−mass spectrometry. Anal Chem
146. Silvestre CIC, Santos JLM, Lima JLFC, Zagatto EAG (2009) Liquid–liquid extraction in flow analysis: a critical review. Anal Chim Acta 652:54–65. https://doi.org/10.1016/j.aca.2009.05.042

147. Lenehan CE, Barnett NW, Lewis SW (2002) Sequential injection analysis. Analyst 127:997–1020. https://doi.org/10.1039/b106791p
148. Anthemidis AN, Ioannou K-IG (2009) On-line sequential injection dispersive liquid–liquid microextraction system for flame atomic absorption spectrometric determination of copper and lead in water samples. Talanta 79:86–91. https://doi.org/10.1016/j.talanta.2009.03.005
149. Anthemidis AN, Ioannou KIG (2010) Development of a sequential injection dispersive liquid–liquid microextraction system for electrothermal atomic absorption spectrometry by using a hydrophobic sorbent material: Determination of lead and cadmium in natural waters. Anal Chim Acta 668:35–40. https://doi.org/10.1016/j.aca.2009.10.063
150. Anthemidis AN, Ioannou KIG (2011) Sequential injection dispersive liquid–liquid microextraction based on fatty alcohols and poly(etheretherketone)-turnings for metal determination by flame atomic absorption spectrometry. Talanta 84:1215–1220. https://doi.org/10.1016/j.talanta.2010.12.017
151. Berton P, Martinis EM, Wuilloud RG (2010) Development of an on-line temperature-assisted ionic liquid dispersive microextraction system for sensitive determination of vanadium in environmental and biological samples. J Hazard Mater 176:721–728. https://doi.org/10.1016/J.JHAZMAT.2009.11.094
152. Berton P, Wuilloud RG (2011) An online ionic liquid-based microextraction system coupled to electrothermal atomic absorption spectrometry for cobalt determination in environmental samples and pharmaceutical formulations. Anal Methods 3:664–672. https://doi.org/10.1039/c0ay00616e
153. Martinis EM, Escudero LB, Berton P, Monasterio RP, Filippini MF, Wuilloud RG (2011) Determination of inorganic selenium species in water and garlic samples with on-line ionic liquid dispersive microextraction and electrothermal atomic absorption spectrometry. Talanta 85:2182–2188. https://doi.org/10.1016/J.TALANTA.2011.07.065
154. Andruch V, Acebal CC, Škrlíková J, Sklenářová H, Solich P, Balogh IS, Billes F, Kocúrová L (2012) Automated on-line dispersive liquid–liquid microextraction based on a sequential injection system. Microchem J 100:77–82. https://doi.org/10.1016/j.microc.2011.09.006
155. Ruzicka J, Marshall GD (1990) Sequential injection: a new concept for chemical sensors, process analysis and laboratory assays. Anal Chim Acta 237:329–343. https://doi.org/10.1016/S0003-2670(00)83937-9
156. Medina DAV, Santos-Neto ÁJ, Cerdà V, Maya F (2018) Automated dispersive liquid–liquid microextraction based on the solidification of the organic phase. Talanta 189:241–248. https://doi.org/10.1016/J.TALANTA.2018.06.081
157. Shishov A, Terno P, Moskvin L, Bulatov A (2020) In-syringe dispersive liquid–liquid microextraction using deep eutectic solvent as disperser: Determination of chromium (VI) in beverages. Talanta 206:120209. https://doi.org/10.1016/j.talanta.2019.120209
158. Mabrouk MM, Soliman SM, El-Agizy HM, Mansour FR (2020) Ultrasound-assisted dispersive liquid–liquid microextraction for determination of three gliflozins in human plasma by HPLC/DAD. J Chromatogr B 1136:121932. https://doi.org/10.1016/j.jchromb.2019.121932
159. Akramipour R, Fattahi N, Golpayegani MR (2021) Sensitive determination of methotrexate in plasma of children with acute leukemia using double-solvent supramolecular systemas a novel extractant for dispersive liquid–liquid microextraction. J Chromatogr B Anal Technol Biomed Life Sci 1171:122628. https://doi.org/10.1016/j.jchromb.2021.122628
160. Abdallah IA, Hammad SF, Bedair A, Abdelaziz MA, Danielson ND, Elshafeey AH, Mansour FR (2022) A gadolinium-based magnetic ionic liquid for supramolecular dispersive liquid–liquid microextraction followed by HPLC/UV for the determination of favipiravir in human plasma. Biomed Chromatogr 36:1–10. https://doi.org/10.1002/bmc.5365
161. Hammad MA, Kamal AH, Kannouma RE, Mansour FR (2021) Vortex-assisted dispersive liquid–liquid microextraction coupled with deproteinization for determination of Nateglinide in human plasma using HPLC/UV. J Chromatogr Sci 59:297–304. https://doi.org/10.1093/chromsci/bmaa096

162. Qiao L, Sun R, Yu C, Tao Y, Yan Y (2021) Novel hydrophobic deep eutectic solvents for ultrasound-assisted dispersive liquid–liquid microextraction of trace non-steroidal anti-inflammatory drugs in water and milk samples. Microchem J 170:106686. https://doi.org/10.1016/j.microc.2021.106686
163. Faraji M, Noormohammadi F, Adeli M (2020) Preparation of a ternary deep eutectic solvent as extraction solvent for dispersive liquid–liquid microextraction of nitrophenols in water samples. J Environ Chem Eng 8:103948. https://doi.org/10.1016/j.jece.2020.103948
164. Moradi M, Zarabi S, Heydari R (2021) Spectrophotometric determination of trace amounts of Sb(III) and Sb(V) in water and biological samples by in-tube dispersive liquid–liquid microextraction and air-assisted liquid–liquid microextraction. Chem Pap 75:6499–6508. https://doi.org/10.1007/s11696-021-01818-1
165. Kachangoon R, Vichapong J, Santaladchaiyakit Y, Burakham R, Srijaranai S (2020) An eco-friendly hydrophobic deep eutectic solvent-based dispersive liquid–liquid microextraction for the determination of neonicotinoid insecticide residues in water, soil and egg yolk samples. Molecules 25(12):2785. https://doi.org/10.3390/molecules25122785
166. Dias RAS, Sousa ER, Silva GS, Silva LK, Freitas AS, Lima DLD, Sousa ÉML (2021) Ultrasound-assisted dispersive liquid–liquid microextraction for determination of enrofloxacin in surface waters. Microchem J 160:105633. https://doi.org/10.1016/j.microc.2020.105633
167. Wang Y, Li J, Ji L, Chen L (2022) Simultaneous determination of sulfonamides antibiotics in environmental water and seafood samples using ultrasonic-assisted dispersive liquid–liquid microextraction coupled with high performance liquid chromatography. Molecules 27(7):2160. https://doi.org/10.3390/molecules27072160
168. Tuzen M, Elik A, Altunay N (2021) Ultrasound-assisted supramolecular solvent dispersive liquid–liquid microextraction for preconcentration and determination of Cr(VI) in waters and total chromium in beverages and vegetables. J Mol Liq 329:115556. https://doi.org/10.1016/j.molliq.2021.115556
169. Lian X, Wang N, Ma L, Jiang H, Bai D, Xue H, Ma Q (2020) Determination of aucubin by supramolecular solvent-based dispersive liquid–liquid microextraction and UPLC-MS/MS: application to a pharmacokinetic study in rats with type 1 diabetes. J Pharm Biomed Anal 186:113301. https://doi.org/10.1016/j.jpba.2020.113301
170. ALOthman ZA, Yilmaz E, Habila MA, Alhenaki B, Soylak M, Ahmed AYBH, Alabdullkarem EA (2022) Development of combined-supramolecular microextraction with ultra-performance liquid chromatography-tandem mass spectrometry procedures for ultra-trace analysis of carbaryl in water, fruits and vegetables. Int J Environ Anal Chem 102:1491–1501. https://doi.org/10.1080/03067319.2020.1738419
171. de Oliveira FM, Scheel GL, Augusti R, Tarley CRT, Nascentes CC (2020) Supramolecular microextraction combined with paper spray ionization mass spectrometry for sensitive determination of tricyclic antidepressants in urine. Anal Chim Acta 1106:52–60. https://doi.org/10.1016/j.aca.2020.01.061
172. Soylak M, Agirbas M, Yilmaz E (2021) A new strategy for the combination of supramolecular liquid phase microextraction and UV–Vis spectrophotometric determination for traces of maneb in food and water samples. Food Chem 338:128068. https://doi.org/10.1016/j.foodchem.2020.128068
173. Aydin F, Soylak M (2015) RSC Advances liquid microextraction of copper from water and hair samples. RSC Adv 5:40422–40428. https://doi.org/10.1039/C4RA17116K
174. Tao Y, Jia L, Qin H, Niu R, Qiao L (2022) A new magnetic ionic liquid based salting-out assisted dispersive liquid–liquid microextraction for the determination of parabens in environmental water samples. Anal Methods 14:4775–4783. https://doi.org/10.1039/d2ay01403c

Homogeneous Liquid–Liquid Microextraction

Alaa Bedair and Fotouh R. Mansour

Abstract The development of liquid phase microextraction methods has led to significant progress in extraction processes by overcoming several challenges associated with conventional liquid–liquid extraction techniques. Liquid phase microextraction is a more cost-effective and eco-friendly alternative that is easier to implement compared to the traditional method. However, the use of water immiscible solvents as extractants in both liquid–liquid and dispersive liquid–liquid microextraction methods posed a challenge in extracting polar drugs. To address this limitation, homogeneous liquid–liquid microextraction emerged as the preferred mode for extracting polar analytes from complex matrices. HLLME uses hydrophilic, water-miscible solvents as extractants, leading to the formation of a homogeneous phase between the extractant and aqueous media. Because there is no interface between the sample and the extractant, HLLME provides superior extraction efficiency compared to other modes of liquid phase microextraction. Phase separation can be achieved by adding chemicals such as salt or sugar or manipulating the extractant's physicochemical properties, such as temperature or pH. In this chapter, we provide a detailed discussion of different homogeneous liquid–liquid microextraction modes, with emphasis on the fundamentals, the new developments and the applications.

Keywords Homogeneous liquid–liquid microextraction · Sample preparation · Green analytical chemistry · Salting-out · Sugaring-out

---

A. Bedair
Department of Analytical Chemistry, Faculty of Pharmacy, University of Sadat City, Sadat City 32897, Monufia, Egypt
e-mail: alaa.mostafa@fop.usc.edu.eg

F. R. Mansour (✉)
Department of Pharmaceutical Analytical Chemistry, Faculty of Pharmacy, Tanta University, Tanta 31111, Egypt
e-mail: Fotouhrashed@pharm.tanta.edu.eg

M. Á. Rodríguez-Delgado et al. (eds.), *Microextraction Techniques*, Integrated Analytical Systems, https://doi.org/10.1007/978-3-031-50527-0_10

Abbreviations

ACN	Acetonitrile
CHCl	Choline chloride
DES	Deep eutectic solvent
DLLME	Dispersive liquid–liquid microextraction
EHLLME	Emulsification induced homogeneous liquid–liquid microextraction
HBA	Hydrogen bond acceptor
HBD	Hydrogen bond donor
HLLME	Homogeneous liquid–liquid microextraction
LLE	Liquid–liquid extraction
LPME	Liquid phase microextraction
LLME	Liquid–liquid microextraction
PSA	Phase separating agent
SALLME	Salting-out induced liquid–liquid microextraction
SHS	Switchable hydrophilic solvent
SULLME	Sugaring-out induced liquid–liquid microextraction
THF	Tetrahydrofuran
DMSO	Dimethyl sulfoxide
IPA	Isopropyl alcohol
SIPTE	Solvent induced phase transition extraction
SULLE	Sugaring-out induced liquid–liquid extraction
EF	Enrichment factor
ISFME	In situ Solvent formation microextraction
HLLE	Homogeneous liquid–liquid extraction
IL	Ionic liquid
SA-HLLME	Surfactant-assisted HLLME
MA-IL-HLLME	Microwave assisted-IL-HLLME
NPs	Nanoparticles
APA	Analytical process automation
DEHPA	Bis(2-ethylhexyl) phosphoric acid
HSLLME	Hydrophobic substance induced HLLME

1 Introduction

Nowadays, there is a growing need for novel analytical techniques to monitor drugs, residues, and pollutants in complex matrices. To meet this demand, technological advancements have been made in the field of analytical chemistry. Analysing analytes at very low concentrations involves various processes, including sample preparation and data processing, each of which can impact analytical performance. Despite recent developments in analytical chemistry instruments, a sample treatment step is necessary before instrumental analysis in most cases to process raw complex matrices. This step aims to remove matrix interferents, clean up the sample, and preconcentrate target compounds before injection into the instrument, which can improve analyte response [1, 2].

Sample preparation is typically considered the most error-prone and time-consuming phase in the analytical workflow. Traditional extraction procedures, such as liquid–liquid extraction (LLE), were invented decades ago and are still used in sample processing [3, 4]. LLE involves using large volumes of water immiscible organic solvents (e.g. chloroform, ether, or ethyl acetate) combined with the aqueous sample, which are then separated, evaporated, and reconstituted in the lowest amount of a suitable solvent.

However, these classic sample preparation procedures have significant limitations, including high solvent consumption, high production of waste, a tedious routine, a common source of sample contamination, and analytical mistakes arising from the operator's handling required to conduct these processes [3, 4]. Liquid–liquid microextraction (LLME) is a contemporary trend that aims to address these issues by saving solvents, preserving the environment, and enhancing sensitivity [5, 6]. In LLME, small volumes of a water immiscible organic solvent are used as extractants, which makes the analyte highly concentrated in the organic phase.

LLME can be categorised into homogeneous or heterogeneous modes, based on the miscibility of the extractant in the aqueous sample [7–9]. Heterogeneous LLME relies on using a water immiscible organic solvent similar to conventional LLE, but the volume used is much less (30–200 μL versus 3–10 mL). This kind of heterogeneous LLME is known as liquid phase microextraction (LPME) and results in a significant increase in analytical method sensitivity due to the tiny volume of the extractant, which makes the analyte extremely concentrated. Dispersive liquid–liquid microextraction (DLLME) is a kind of heterogeneous LLME that uses a dispersant (such as methanol or acetonitrile (ACN) with the water immiscible extractant to increase the contact surface with the aqueous phase and improve analyte accessibility [10–12]. However, both modes show low extraction efficiency for polar analytes due to their dependence on hydrophobic, water immiscible solvents as extractants.

Homogeneous liquid–liquid extraction (HLLE) depends on the creation of a homogeneous phase between the extractant and the aqueous sample, which enormously increases the accessibility of the extractant to the target analyte. The system can be composed of water-miscible organic solvents, water-immiscible solvents/cosolvents, surfactants, or smart polymer to create the homogeneous phase. When water-miscible solvents are used, homogeneous phases emerge spontaneously. In this case, the extraction solution is a binary combination of the water-miscible solvent and the aqueous sample at room temperature. Despite the expanded number of water miscible solvents available, only a handful were used in HLLE. This could be explained by the difficulties of initiating phase separation following homogeneous phase development in most common solvents. Phase separation is achieved by the addition of chemicals such as salt, sugar, buffer, hydrophobic substances, or by changing the environment of the system (pH, temperature).

The principle of HLLE was first introduced by Matkovich and Christian using salt as a phase separating agent (PSA) to extract polar analytes [13]. The first miniaturized form of HLLE was developed by Verma's group in 2009 to extract carbonyl compounds, before HPLC/UV analysis [14]. In the same year, Baghdadi and Shemirani developed in situ solvent formation microextraction (ISFME) as modified form of homogeneous liquid–liquid microextraction (HLLME) [15]. Since then, HLLME was employed to overcome the challenge of extracting polar analytes from aqueous samples by using a few microlitres of water-miscible organic solvents such as acetonitrile, ethanol, acetone, or tetrahydrofuran (THF). In 2010, solvent induced phase transition extraction (SIPTE) was tried for the first time to extract three structurally diverse drugs from human plasma [16]. Later on, new solvents were tested in HLLME such as ionic liquid (IL) [17] and deep eutectic solvents (DES) [18]. In 2021, sugaring-out assisted HLLME (SULLME) was employed in the extraction of 14 drugs from three theraputic classes, and was found more efficient than salting-out assisted homogenous liquid–liquid microextraction (SALLME) and DLLME [5]. Figure 10.1 shows the milestones of HLLME development over the last decades.

2 Fundamentals

HLLME is a technique of sample preparation that involves the formation of a homogeneous phase between an aqueous sample and a small amount of a water-miscible extractant, such as acetonitrile, acetone or tetrahydrofuran. The separation of phases is achieved using a PSA, which may be a salt, sugar, or hydrophobic substance. Depending on the type of PSA used, HLLME can be classified into three categories: SALLME, SULLME, and hydrophobic substance-assisted HLLME.

The contact surface between the aqueous phase and the extractant in HLLME is enormous, which enhances the accessibility of the target analyte to the extractant. Consequently, HLLME facilitates higher extraction efficiency than other microextraction techniques such as LPME [19] and DLLME [20, 21]. In addition, HLLME is a greener techniques of sample preparation because the water-miscible solvents used are typically safer and more environmentally friendly. It is worth mentioning that water-miscible solvents like THF, ACN, IPA are greener than water immiscible solvents such as chloroform (hepatotoxic) and ethers which are harmful to the lung and the other organs. HLLME is particularly suitable for extracting very hydrophilic drugs due to the high dielectric constants of the water-miscible extractants employed, as indicated in Table 10.1.

The various types of homogeneous liquid–liquid extraction (HLLE) are categorized based on the dmethod used to create a uniform phase and the process of separation. The creation of a uniform phase can be achieved through the use of a water-miscible organic solvent (in HLLME), an ionic liquid (in IL-HLLME), a deep eutectic solvent (in DES-HLLME) a surfactant (in SA-HLLME), or a smart polymer (in switchable hydrophilic solvent (SHS-HLLME). Additionally, the separation process can be initiated by adding certain chemicals or altering the physical

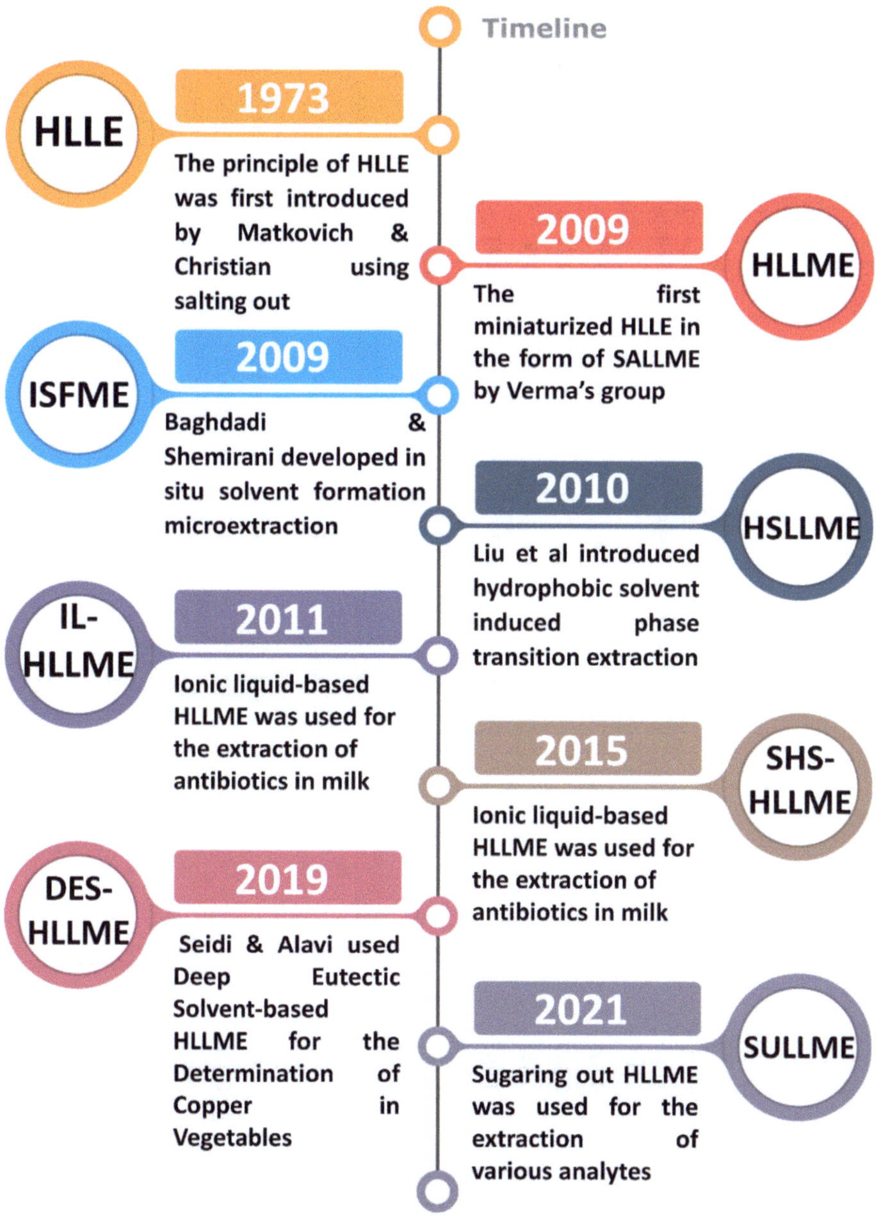

Fig. 10.1 Timeline of the development in Homogeneous liquid–liquid microextraction (HLLME)

Table 10.1 Properties of common organic solvents in HLLME [22]

Solvent	UV cut off (nm)[4]	Viscosity (mPa s)[4]	BP (°C)[4]	Polarity (P')[2]	Refractive Index[3] n^D_{20}	Dipole[2,5] π*	Acidity[2] α	Basicity[2] β	Density at 20 °C
Acetone	330	0.306	56.2	5.1	1.359	0.56	0.06	0.38	0.791
Acetonitrile	190	0.369	81.6	5.8	1.344	0.60	0.15	0.25	0.787
Chloroform	245	0.542	61.2	4.1	1.443	0.57	0.43	0.00	1.4832
Dimethysulfoxide (DMSO)	265	1.987	189	7.2	1.478	0.57	0.00	0.43	1.101
Dioxane	215	1.177	101.2	4.8	1.420	0.60	0.00	0.40	1.036
Ethanol	210	1.074	78.3	5.2	1.361	0.25	0.39	0.36	0.79
Hexane	210	0.300	68.7	0.1	1.357	–	–	–	0.659
Isopropyl alcohol (IPA)	210	2.04	82.4	3.9	1.377	0.24	0.36	0.40	0.785
Methanol	210	0.544	64.7	5.1	1.328	0.28	0.43	0.29	0.792
THF	220	0.456	66.0	4.0	1.407	0.51	0.00	0.49	0.888
Water	191	0.89	100	10.2	1.333	0.45	0.43	0.18	1.000

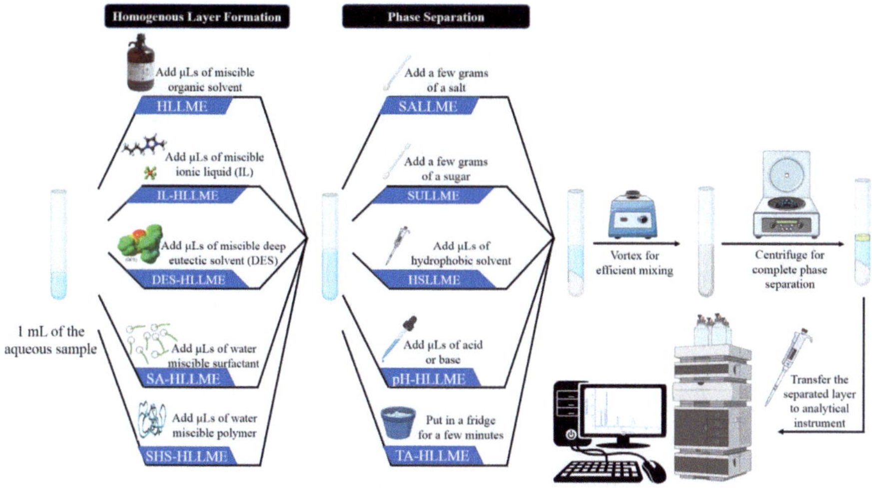

Fig. 10.2 Different modes of homogeneous liquid–liquid microextraction

conditions of the extraction media. Chemical-induced phase separations are the most common form of HLLE, where an extractant is removed from the extraction media by adding a more water-soluble substance such as salts (in SALLME), sugars (in SULLME), or a hydrophobic solvent (in HSLLME). Phase separation can also be induced by changing the pH in case of smart polymers or by changing the temperature. Figure 10.2 illustrates the different modes of HLLME.

HLLME is an efficient, cost-effective, and environmentally friendly approach for extracting biomolecules, such as enzymes, hormones, and proteins. HLLME is compatible with liquid chromatography-mass spectrometry (LC–MS) for metabolomics and proteome research. The effectiveness of HLLME depends on several factors, including the type and amount of extractant, the type of PSA, the volume of the aqueous sample and extractant, and the pH of the aqueous sample.

The Hydrophilicity of the extractants plays a significant role in the extraction process, as it follows the like-dissolves-like principle. Therefore, the polarity of the extractant should be matched with the target analyte's polarity. Similarly, the type of PSA used affects the efficacy of the extraction process. Not all sugars and salts can achieve phase separation, and not all water-miscible solvents can be separated.

Although SULLME has weaker phase-separation ability than SALLME, it has more preconcentration capacity in extraction because it achieves a lower retrieval volume. It is important to use an appropriate amount of PSA to achieve reproducible phase separation. However, excessive PSA can reduce preconcentration capacity by dilution effects. The hydration hypothesis may explain the salting out phenomenon [23]. It should be noted that PSA should have considerably higher solubility in aqueous samples than the extractant to induce phase separation.

Generally, sample volume directly correlates with preconcentration capacity, while extractant volume inversely correlates with extraction efficiency. Finally, pH plays a major role in extraction efficiency, as it influences analyte solubility. The analyte should be in a non-ionized form for better extraction.

The preconcentration capacity of HLLME is assessed using the enrichment factor (EF) which can be calculated from the ratio between the analyte concentration in the extractant (C_{org}) and the concentration in the sample (C_{aq}):

$$\mathrm{EF} = \mathrm{C_{org}} / \mathrm{C_{aq}} \tag{10.1}$$

The extraction recovery can also be calculated based on the ratio between the analyte amount in the extractant (n_{org}) and the initial amount in the aqueous sample (n_{aq}) as follows:

$$\%ER = \frac{n_{org}}{n_{aq}} \times 100 = \frac{C_{org}V_{org}}{C_{aq}V_{aq}} \times 100 = \%EF \times \frac{V_{org}}{V_{aq}} \times 100 \tag{10.2}$$

where V_{org} and V_{aq} are the volumes of the separated extractant and of the aqueous sample, respectively.

One of the most interesting aspects of HLLME is that it fits under the umbrella of green analytical chemistry (GAC). Historically, GAC is a branch of green chemistry that focuses on the role of analytical chemists in making laboratory operations more environmentally friendly, and it has attracted the interest of chemists [24–26]. The 12 principles of green chemistry were developed by Anastas and Warner in 1998 [27]. However, only a few of these principles can be directly applied to analytical chemistry because they were originally designed for synthetic chemistry. Gałuszka et al. [28] reported 12 principles of GAC that are better suited for analytical methods. From the perspective of environmentalists who also consider the economic aspects of analytical methodologies, special attention should be paid to the inherent risks of certain sample types and solvents used, the energy consumption associated with advanced instrumentation, and, of course, the resulting laboratory wastes and emissions from the numerous steps of analytical methodologies. In this regard, HLLME is a miniaturized sample preparation technique that reduces reagent consumption, minimizes waste production, and utilizes greener extractants such ILs and DESs. Moreover, HLLME is time-saving and cheaper compared to conventional sample preparation techniques, and it has a high potential for automation. The green characteristics of HLLEM and the high efficiency in extracting highly polar analytes may explain the increasing numbers of publications every year.

3 Novel Developments

HLLME modes are classified based on the approach utilized to produce a homogenous phase or the technique used in the separation of phases. Recent developments in HLLME research have led to three emerging trends being pursued by researchers. The first trend involves exploring novel materials that can form a homogeneous layer with water. This approach is accomplished through the use of IL [17, 29], DES [30], surfactants [31] or switchable solvents [32] to facilitate the formation of a stable homogenous liquid phase, leading to a more efficient extraction process. The second trend focuses on developing advanced techniques for the separation of phases. In particular, the use of gas flotation [33], liquid nitrogen [34], magnetic fields [35], and ultrasound [36] has gained considerable attention due to their effectiveness in separating the two liquid phases. These techniques also enable better control over the extraction process, allowing for enhanced precision in the determination of target analytes.

The third trend is the automation of HLLME processes. Automation offers several advantages, including increased efficiency, reproducibility, and reduced risk of human error. The use of automated systems also facilitates the handling of large sample volumes, which is particularly useful in industrial settings. These emerging trends demonstrate the ongoing efforts to optimize HLLME techniques, improve their efficiency, and expand their applications. In the following sections, we will delve into each trend in greater detail to provide a more comprehensive understanding of their potential contributions to the advancement of HLLME.

3.1 New Solvents for HLLME

3.1.1 Using Ionic Liquids in HLLME

ILs have emerged as a promising class of solvents due to their unique combination of properties that offer advantages over common organic solvents. These solvents are a type of green solvents that are nonmolecular, ionic and have melting temperatures below 100 °C [37–39]. They possess unique physicochemical properties such as low vapor pressure, high solubility in both organic and inorganic compounds, and excellent thermal stability [40–42]. Due to the customizable structure of ionic liquids, they are often referred to as designer solvents [43, 44]. By altering the cation/anion combination or incorporating specific functional groups into their structure, the characteristics of ionic liquids can be easily modified [45]. As a result, ionic liquids exhibit specific solvation characteristics that provide selectivity and a range of separation mechanisms. Consequently, they have gained acceptance in various liquid-phase microextraction techniques, including HLLME [46]. Wang et al. [47] performed microwave-assisted IL-HLLME (MA-IL-HLLME) for the determination of anthraquinones in *Rheum palmatum L.* by HPLC–DAD. The extractant used was 1-octyl-3-methylimidazolium tetrafluoroborate ($[C_8MIM][BF_4]$), an ionic liquid dissociated into its ions in aqueous media and converted into a hydrophilic form, as illustrated in Fig. 10.3. Following the extraction, ammonium hexafluorophosphate (NH_4PF_6) was utilized as an ion pairing agent. This produced $[C_8MIM][PF_6]$, a hydrophobic compound that promoted phase separation. The MA-IL-HLLME approach demonstrated high sensitivity, with extraction recoveries for anthraquinones ranging from 81.13% to 93.07%, indicating its effectiveness

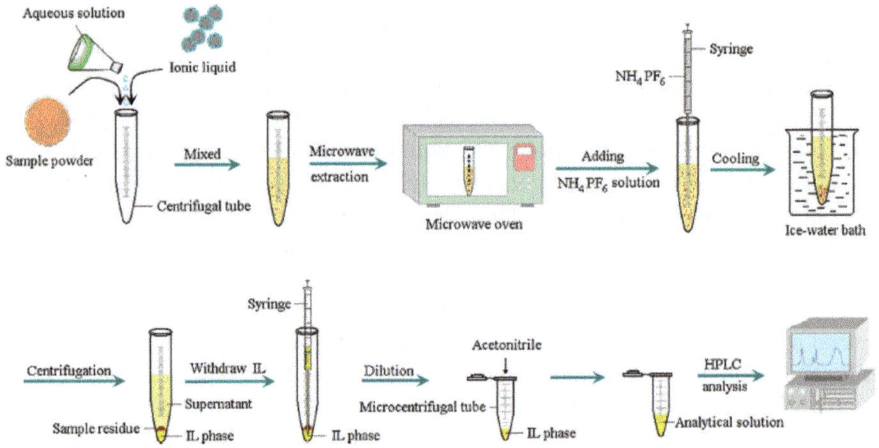

Fig. 10.3 Schematic procedure of MA-IL-HLLME. Reprinted from [47] with permission from Elsevier

in extracting these compounds from plant sources [47]. These applications and advantages collectively make ionic liquids attractive alternatives to traditional solvents in HLLME. The major challenge in using IL-HLLME is to develop a hydrophilic IL that can be miscible with water, and then to find the proper PSA to induce phase separation. This area of research is still overlooked, and there are a lot to expect in terms of future developments and progress.

3.1.2 Using DES in HLLME

The continuous effort to promote sustainable chemistry has resulted in the development of new materials that achieve the principles of green chemistry. In this context, DESs have risen as one of the most promising alternatives to the use of toxic organic solvents. Their unique properties have led to a massive development of these materials and a sharp increase in their applications in analytical chemistry in recent years [48]. DES is considered a natural extension of ILs due to their similar preparation and properties. However, DES offer several advantages over ILs, as they can be prepared from readily available and renewable starting materials, which makes DES cost-effective, environmentally friendly, and less toxic, than ILs. Many recent researches on the environmental effect of ILs have revealed that, despite their unique features and evident benefits in a growing number of applications and processes, ILs are not inherently green. DESs were created in the hunt for biodegradable and low toxicity ILs [49].

DESs have distinct and adjustable features including easy tunability by components variations. Furthermore, they may be made from a wide range of readily accessible components using simple and low-cost synthesis processes. In addition, DESs also have a low vapor pressure and remain liquids over a wide temperature range.

Furthermore, eutectic mixtures intended for use as solvents must meet basic requirements in order to be considered as such. In this regard, it should be noted that the many features changed depending on the components, although there are three main criteria in general: (i) the eutectic point must be significantly lower than the melting points of its individual components and far lower than the projected melting point of an ideal liquid combination; (ii) the eutectic point is dependent on the molar composition of the mixture; and (iii) DESs resemble liquids at ambient temperature. Towards the development greener analytical method, natural deep eutectic solvents (NADESs) were developed depending on natural components in synthesis of NADES including sugars, certain amino acids, choline salts, and organic acids such as malic acid, citric acid, lactic acid, and succinic acid [48].

In 1884, the term "eutectic" was coined by British chemist Frederick Guthrie to describe metal alloys with lower melting temperatures than their basic components [50]. The term "eutectic mixture" now refers to a combination of two or more compounds with a certain molar ratio indicating a minimum melting point on the relevant phase diagram. This position is known as the eutectic point in phase diagrams [51]. There are five types of DESs: type I, formed by combining a quaternary ammonium salt and a non-hydrated metal chloride; type II, formed by combining a quaternary ammonium salt and a hydrated metal chloride; type III, formed by combining a quaternary ammonium salt as a hydrogen bond acceptor compound (HBA) and a hydrogen bond donor compound (HBD); type IV, formed by combining metal chloride and HBD., while type V DES composed of non-ionic chemicals [52].

One of the enticing properties of DESs is the ease of their production methods, which include heating, freeze-drying, and grinding. The most common method for preparing DESs is heating, which involves stirring and heating the mixture until a homogenous and clear solution develops. The physicochemical properties of DESs may be altered by adjusting the synthesis temperature, the kind or molar ratio of the constituents, and the addition of a certain amount of water. Florindo et al. [49] offered adjusted densities, viscosities, and refractive indices for DESs made using choline chloride as the hydrogen bond acceptor and different carboxylic acids as the hydrogen bond donors (levulinic, glutaric, malonic, oxalic, and glycolic). The thermophysical characteristics of the produced DESs were evaluated using two separate synthetic techniques, heating and grinding. A range of eutectic combinations were synthesized in this work using a reasonably simple, cost-effective, and ecologically friendly approach. DESs were created by combining cholinium chloride with several carboxylic acids that served as hydrogen bond donors. Due to differences in the experimental thermophysical properties, particularly viscosity, two different synthetic methods were used, and the formation of an ester during the heating method led to the conclusion that when carboxylic acids are used as HBD in combination with cholinium chloride, the grinding method should be preferred to prepare DESs. The comparison of the thermophysical parameters of the created DESs with the comparable ILs revealed that DESs had identical densities but substantially lower viscosities, making mass transfer procedures easier. Furthermore, DESs may be made with varied molar ratios of HBA:HBD, providing further tunability.

DESs have recently attracted attention as an ecologically friendly alternative to hydrophilic organic solvents commonly used as extraction solvents, particularly in the conventional HLLME approach. Using DES in HLLME can be achieved by using an aprotic solvent as a phase separation agent, a mode known as emulsification-induced HLLME (EHLLME). The suggested approach was used to successfully extract several organic chemical components from water samples.

Switchable solvents are liquids that may be transformed reversibly between hydrophilic and hydrophobic forms by altering the pH or temperature of the system. DES can be used as switchable solvents by changing the pH or the temperature of the medium. Both pH-induced HLLME and temperature-induced HLLME were used to enrich the target analytes, with the latter having two types based on the mechanism of phase separation: (1) using DESs with a low melting point and inducing phase separation by cooling the homogeneous solution and freezing the DESs; and (2) using temperature switchable DESs and inducing phase separation by temperature adjustment. At different temperatures, the temperature-switchable DESs have varying water solubility. Table 10.2 summarizes the most recent applications of DES in HLLME.

3.2 New Techniques of Phase Separation

A crucial aspect of HLLME development is the phase separation technique. Proper phase separation is essential for achieving accurate and efficient extraction of analytes from complex matrices. Traditionally, centrifugation has been used to accomplish this task. However, centrifugation can be time-consuming and requires skilled technician. To overcome these limitations, novel techniques have been developed that eliminate the need for centrifugation.

One such technique is flotation-assisted HLLME (FA-HLLME), pioneered by Hosseini et al. [33, 74]. To perform this technique, a special microextraction cell (Fig. 10.4) was constructed, and organic solvent was transported to the conical section of the cell using N_2 or air flotation. This approach eliminated the need for centrifugation and made the procedure faster and more automated. The technique was successfully applied to extract polyaromatic hydrocarbons from soil and water samples followed by GC/FID. Rezaee et al. [75] developed a simple home-designed extraction cell for extracting malathion from water samples using GC/FID. Another technique involves ultrasound-assisted HLLME, as demonstrated by Xu et al. [76]. In this method, sonication was used to float a water-miscible organic phase such as ACN in a Pasteur pipette, with salt added to promote the separation of the phases. The technique was successfully applied to extract triazole pesticides from aqueous samples. Hosseini et al. [33] applied FA-HLLME for the extraction of polycyclic aromatic hydrocarbons (PAHs) from soil samples. The homemade extraction cell was created to allow for the collection of the low-density extraction solvent without the need of centrifugation. PAHs were extracted from soil samples into methanol and water (1:1, v/v) in two phases using ultrasound, followed by filtering as a clean-up

Table 10.2 Application of DES in HLLME

Analyte	Sample	Sample volume (mL)	DES composition	DES Volume (µL)	PSA	Type of HLLME	Technique of analysis	Linearity ng /mL	LOD ng/mL	%RSD	REF
Sulfonamides	Water	1.5	CHCl: glycol	100	THF	EHLLME	HPLC/UV	N/A	1.2–2.3	≤4.15	[53]
Malachite Green	Farmed and ornamental aquarium fish water samples	10	CHCl: Phenol	500	THF	EHLLME	UV–VIS spectrophotometry	45–900	3.6	2.7	[54]
Aluminum	Water	25	CHCl: Phenol	500	THF	EHLLME	ETAAS	0.05–20.0	0.032	3.3	[55]
Anti-depressants	Human plasma and pharmaceutical waste water sample	6	CHCl: Phenol	200	THF	EHLLME	HPLC/UV	10–8000	3.0–4.5	3.6–5.7	[56]
Vanadium	Food	25	CHCl: Phenol	1000	THF	EHLLME	ETAAS	0.5–5.0	0.025	3.4	[57]
Selenium species	Water and food samples	25	CHCl: Phenol	500	THF	EHLLME	ETAAS	0.2–8	0.00461	4.1	[58]
Curcumin	Food and Herbal Tea Samples	10	CHCl: Phenol	400	THF	EHLLME	UV–VIS spectrophotometry HPLC/DAD	9–920	2.9	1.8	[59]
Cadmium	Food and water sample	50	CHCl: Phenol	500	THF	EHLLME	ETAAS	0.005–0.15	0.000023	3.1	[60]
Lead	Food and water sample	30	CHCl: Phenol	600	THF	EHLLME	GFAAS	0.12–2.5	0.0006	2.9	[61]
Phthalates	Beverages	10	CHCl: Phenol	440	THF	EHLLME	HPLC/DAD	170–4490	5.3–17.8	<11	[62]
Thiophenols	Water samples	1.5	CHCl: p-cresol	50	Acetone	EHLLME	GC/FID	20–100,000	10–15	≤3.3	[63]
Arsenic and selenium	Rice samples	1.5	Proline: malic acid	500	THF	EHLLME	AAS	0.005–0.45	0.003	≤2.6	[64]
Methyl-mercury and total mercury	Fish and environmental waters	25	Betaine hydro chloride:sorbitol:water	600	ACN	EHLLME	UV–VIS spectrophotometry	0.7–340	0.25	≤3.2	[65]

(continued)

Table 10.2 (continued)

Analyte	Sample	Sample volume (mL)	DES composition	DES Volume (μL)	PSA	Type of HLLME	Technique of analysis	Linearity ng /mL	LOD ng/mL	%RSD	REF
Mercury	Water and biological samples	10	CHCl: Phenol	500	THF	EHLLME	ETAAS	0.1–10.0	0.073, 0.091	≤4.05	[66]
Pesticides	Traditional Chinese medicine	10	CHCl: Phenol	650	THF	EHLLME	HPLC/DAD	54–107,000	20–200	<4.7	[67]
Carbamazepine	Plasma	1	CHCl: Phenol	314	THF	EHLLME	HPLC/UV	10–1500	1.17	≤6.85	[68]
Curcumin	Tea and honey species	5	CHCl: maltose	762.5	THF	EHLLME	UV–VIS spectrophotometry	0.4–120	1.2×10^{-5}	≤4.3	[69]
Benzotriazole and benzothiazole derivatives	Surface water samples	5	CHCl: Phenol	1000	THF	EHLLME	UHPLC- MS	5–200	0.02–0.15	≤8	[70]
Copper	Vegetables	20	Benzyl triphenyl phosphonium bromide:ethylene glycol	80 mg	Room temperature	Temperature induced HLLME	FAAS	5.0–250	0.13	≤2.6	[18]
Pesticides	Fruit juice and vegetable samples	5	CHCl:*p*chlorophenol	142	Ice bath (0 °C	Temperature induced HLLME	GC/FID	0.45–5000	0.13–0.31	≤9	[71]
UV-filters	Surface and bathing waters	10	Tetrapropylammonium bromide:octanoic acid	200	Sulfuric acid	pH assisted HLLME	HPLC/DAD	0.139–500	0.021–0.336	≤6.3	[72]
Daclatasvir and sofosbuvir	Urine samples	5	Tetra butyl ammonium chloride:Pamino phenol	105	Ammonia	pH assisted HLLME	HPLC/DAD	1.6–250	1–1.3	≤9.3	[73]

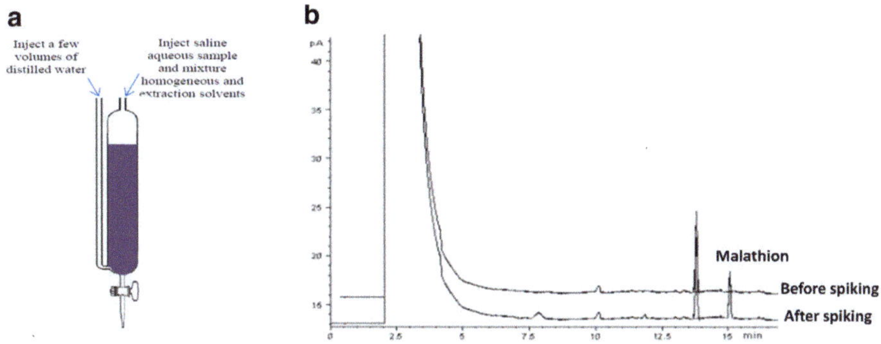

Fig. 10.4 **A** Schematic of the home-designed extraction cell, **B** GC-FID chromatograms of malathion in river water, before spiking (upper) and after spiking with 5.0 μg/L of Malathion (lower). Reprinted from [75] with permission from AJOL

step. The filtrate was mixed with 1.0 mL methanol (homogenous solvent) and 150.0 μL toluene (extraction solvent) in a home-made extraction cell. Using N_2 flotation, the dispersed extraction solvent was transported to the mixture's surface and collected with a micro-syringe. The recovered organic solvent was then fed into the GC-FID for further examination. Gas flotation can improve the HLLME process by enhancing the separation efficiency and reducing the extraction time. It also lowers solvent consumption and reduces emulsion formation during ME.

A novel approach to HLLME without the need for a centrifugation step is magnetic retrieval of SHS-HLLME, developed by Çabuk et al. [35]. In this approach, bis(2-ethylhexyl) phosphoric acid (DEHPA) was employed as SHS, and iron oxide nanoparticles (Fe_3O_4 NPs) were used for magnetic retrieval. After extraction, DEHPA containing the target analytes was separated and collected from the sample solution using Fe_3O_4 NPs, eliminating the need for centrifugation or special extraction devices. This approach was successful due to DEHPA's strong binding affinity for Fe_3O_4 NPs. These methods have been applied to extract various analytes from complex matrices, showing potential for future applications in analytical chemistry. Trying other nanoparticles with higher magnetic susceptibility will make phase separation easier and faster. Magnetic ILs is another area of development in this regard [77, 78].

Liquid nitrogen offers several advantages for enhancing the performance of HLLME. Liquid nitrogen creates a cryogenic environment that prevents thermal degradation or loss of thermally labile compounds during the extraction process, thereby improving their recovery. Additionally, the use of liquid nitrogen accelerates the phase separation process in HLLME, enabling faster formation of distinct phases between the extraction solvent and sample matrix. This accelerates the separation and collection of desired analytes. Lastly, liquid nitrogen cooling enhances precision and reproducibility by minimizing temperature fluctuations during extraction, ensuring consistent extraction conditions and yielding more reliable and repeatable results. However, all precautions must be taken when working with liquid nitrogen due to its extremely low temperature (−196 °C), necessitating adherence to appropriate personal protective equipment and safe handling procedures. Okhravi et al. [34] developed liquid nitrogen HLLME for extraction of Co(II) and Ni(II) ions prior to measurement by FAAS. The proposed method presented a simple and fast sample preparation procedure. More progresses is expected to be made in the future regarding the application of liquid nitrogen-enhanced HLLME, with a multitude of potential advancements on the horizon.

Ultrasound has emerged as a promising green technique for enhancing the efficiency and effectiveness of ME [79]. This method utilizes ultrasound waves to accelerate mass transfer between immiscible liquid phases, thereby accelerating the extraction process. Additionally, ultrasound promotes increased contact area and contact time between the sample and the extraction solvent through its agitation effects. Consequently, ultrasound-assisted HLLME offers reduced extraction time, improved extraction efficiency, and lower solvent consumption, leading to enhanced overall performance and sustainability of the extraction process [36].

3.3 *Automation of HLLME*

Analytical Process Automation (APA) is a recent development in analytical chemistry, which involves the automation and downsizing of various analytical procedures. Flow-based technologies have been shown to be suitable for this purpose [80–82]. Pochivalov and colleagues [83] have designed and developed a fully automated microextraction technique using a SHS, DEHPA, contained within a syringe. The process involved the dissociation of the extractant in an alkaline sample solution leading to a homogenous solution, followed by the development of an organic phase by acidification and subsequent separation. This microextraction process was utilized in determining antimicrobial medications, sulfamethoxazole and sulfamethazine, in human urine samples using HPLC with UV detection. The automation process consisted of aspirating 0.175 mL of DEHPA (port b, 10 mL/min) and 1.5 mL of sample solution (containing 0.4 M NaOH, port a, 0.7 mL/min) into the syringe pump in the first stage (Fig. 10.5A). The sample solution was then transferred using 0.2 mL of air (port c, 15 mL/min) sucked into the syringe pump. The creation of water-soluble sodium bis (2-ethylhexyl) phosphate led to the achievement of a homogenous solution after 15 s of magnetic stirring. In the second stage, 0.210 mL of 3 M H_2SO_4 (port d, 2 mL/min) and 1.7 mL of air (port c, 15 mL/min) were fed into the syringe pump successively, and organic phase formation and analyte extraction were carried out with magnetic stirring. After the cessation of magnetic stirring, phase separation occurred in 15 s. The aqueous phase was removed to waste, while the organic phase was transferred to a 0.5 mL Eppendorf tube containing 0.2 mL of methanol. The process demonstrated good linearity, sensitivity, and a sampling rate of 12 samples per hour.

Vakh and colleagues [84] have devised an automated process of effervescence-assisted switchable solvent-based liquid phase microextraction for the determination

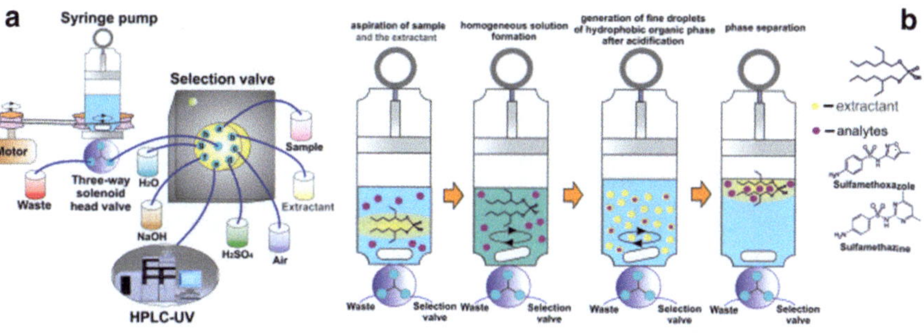

Fig. 10.5 Flow system set-up (**A**) and extraction mechanism (**B**) for in-syringe switchable hydrophilicity solvent-based microextraction method. Reprinted from [83] with permission from Elsevier

of ofloxacin in human urine utilizing a syringe pump-based system. In this extraction process, medium-chain saturated fatty acids were investigated as switchable hydrophilic solvents. In the presence of sodium carbonate, the fatty acids were transformed into a hydrophilic state. The introduction of sulfuric acid into the solution reduced the pH value, leading to the creation of microdroplets of the fatty acid. The in-situ production of carbon dioxide bubbles facilitated the extraction process and the eventual phase separation. The automation of the microextraction procedures was achieved as shown in Fig. 10.6. In the first step, the syringe pump consecutively aspirated 350 μL of 2 M Na_2CO_3 (port 1, valve 1), 1 mL of diluted urine sample (port 2, valve 1), and 50 μL of hexanoic acid (port 3, valve 1) into the mixing coil. A homogeneous solution of sodium hexanoate, ofloxacin, and excess carbonate ions was developed while the reagents were passing through the mixing coil. Subsequently, this mixture was then supplied (through port 4, valve 1) into the mixing chamber by the syringe pump. The mixing coil (MC) was rinsed twice with water (port 5, valve 1) to eliminate sodium carbonate residues. Then, 260 μL of 2.5 M H_2SO_4 was drawn into the MC and fed into the mixing chamber (port 6, valve 1). This led to the transformation of water-soluble hexanoate ions into hydrophobic hexanoic acid, subsequently accompanied by effervescence-assisted microextraction of ofloxacin and phase separation. The aqueous phase was discarded (port 7, valve 1), while the organic phase remained in the mixing chamber. The peristaltic pump then added a mixture of 450 μL of phosphate buffer (pH 6.4) and methanol (1:1, v/v) (port 11, valve 2) into the mixing chamber for dissolving hexanoic acid containing the extracted analyte for HPLC analysis. Air bubbles (port 12, valve 2) stirred the mixture for 20 s. Using a peristaltic pump, the resulting solution was aspirated (port 14, valve 2) into a chromatographic vial and analyzed by HPLC/FLD.

In their study, Cherkashina et al. [85] developed an automated SALLME method for the determination of tetracyclines in urine samples by HPLC/UV. In this approach, 1-octylamine was investigated as a new SALLME extractant. The process involved aspirating 1-octylamine and sample solution into a mixing chamber of a flow system, followed by air-bubble mixing to produce an isotropic solution. A solution of a

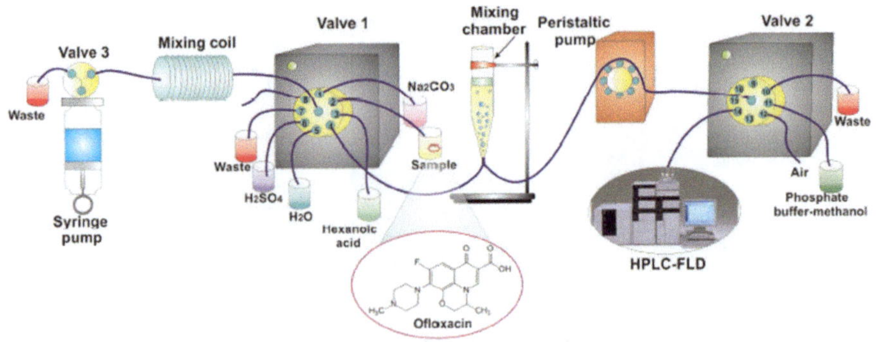

Fig. 10.6 The manifold for automated EA-SS-LPME system coupled with HPLC-FLD for the determination of ofloxacin in human urine samples. Reprinted from [84] with permission from Elsevier

salting-out agent was added to the mixing chamber to enable phase separation. The micellar 1-octylamine phase containing analyte was combined with methanol and delivered to an HPLC/UV system after phase separation.

During the first stage of the process, 50 μL of 1-octylamine (port 1, valve 1) and 1 mL of sample solution (port 2, valve 1) were aspirated using a syringe pump and supplied (port 4, valve 1) to the (MC). Next, the resultant mixture was injected with 300 μL of 20% NaCl (port 5, valve 1) into the MC. The contents of the MC were mixed with air bubbles generated by the peristaltic pump (port 12, valve 2), and phase separation occurred. The aqueous phase was then transferred to the trash (port 6, valve 1), while 50 μL of methanol (port 11, valve 2) was introduced into the MC and mixed for 60 s by air-bubbling. The resulting solution was transferred (through port 14, valve 2) to a chromatographic vial using a peristaltic pump and evaluated by HPLC/UV. The proposed method was found to be cost-effective, simple, and quick, making it a promising approach for the determination of tetracyclines in urine samples. The set-up of this study is presented in Fig. 10.7.

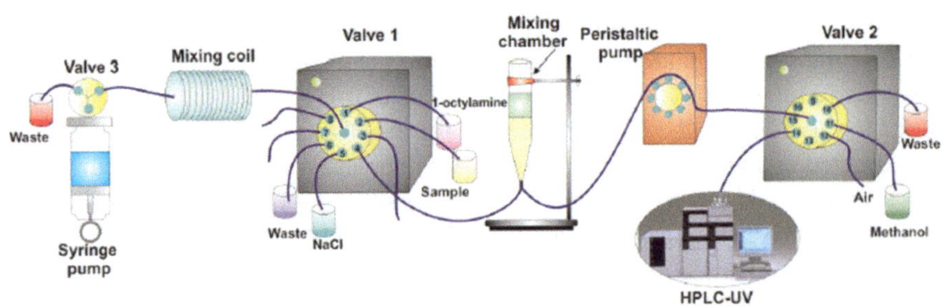

Fig. 10.7 The manifold of the automated salting-out assisted liquid–liquid microextraction procedure for the determination of tetracycline in urine. Reprinted from [85] with permission from Elsevier

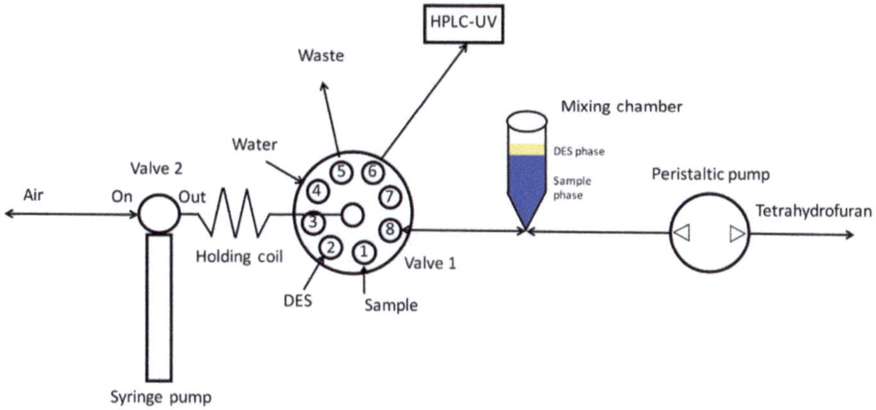

Fig. 10.8 The manifold of the Automated—HLLME procedure. Reprinted from [86] with permission from Elsevier

Shishov et al. [86] developed an automated HLLME method based on DES for the determination of caffeine followed by HPLC/UV. To automate this process, the researchers used a multiport valve, a syringe pump, and a peristaltic pump, as described in Fig. 10.8. The automated method began with the backward movement of the syringe pump plunger to aspirate 1 mL of sample (1) and 50 μL of DES (choline chloride with phenol at a 1:3 molar ratio) (2) into the holding coil via the multiport valve. The resultant mixture was then delivered into the mixing chamber by forward movement of the syringe pump from the holding coil. To dissolve the DES in the sample, the syringe pump was switched to the airport (On), and 10 mL of air was sucked into it before being delivered into the mixing chamber at a flow rate of 10 mL/min through port 8. The peristaltic pump then introduced 50 μL of THF into the mixing chamber leading to DES phase separation and analyte microextraction. The developed method was found to have the potential to be used as an alternative to traditional methods for obtaining organic analytes from aqueous samples. Additionally, it could be combined with other instrumental techniques, such as chromatography, and different detection systems, like UV–VIS spectrophotometry detectors" could be used.

4 New Phase Separation Agents in HLLME

HLLME is a preferred method for extracting hydrophilic analytes from aqueous matrices due to the special hydrophilic characteristics of extractants like ACN, acetone or THF. HLLME also achieves efficient extraction of more hydrophobic analytes by forming a ternary homogeneous system, where the addition of a few microliters of a water-immiscible solvent such as chloroform or toluene to the binary system modifies the hydrophilic nature of the extractant. Phase separation is induced by adding salt (SALLME) or sugar (SULLME). The PSA should be more soluble in aqueous media than the extractant to expel it from the extraction media. The following section provides detailed information on different modes of HLLME, according to the phase separating agent or condition.

4.1 Salting-Out Homogeneous Liquid–Liquid Micro Extraction (SALLME)

SALLME utilizes salt as a PSA to induce phase separation between the aqueous sample and a few microliters of a water-miscible organic extractant. It is important to note that phase separation is achieved through a salting-out phenomenon. When salt is introduced into homogeneous water/organic solutions, the solubility of the water-miscible solvents diminishes, thereby resulting in the formation of a discernible phase either at the top or at the bottom of the aqueous sample. The positioning of this phase is contingent upon the density of the organic solvent. Typically, most water-miscible organic solvents are lighter than water, resulting in their flotation on top of the sample alongside other organic solutes and analytes. In contrast, salting-in [87] refers to increasing the solubility of a nonelectrolyte in water by raising the salt concentration. It has been observed in proteins and hydrophobic solutes due to the counterion impact of salts, which raises the charge on the solute's surface and promotes solubility. However, salting-out is more prevalent and is the focus of this section.

Salting-out is a phenomenon that can be controlled by various factors, such as the type of salt used. The selected salt should have high water solubility, low extractant solvent miscibility, and robust salting-out ability. In general, the anion effect is more significant than the cation effect [88]. Water is a strong electron acceptor solvent and therefore has a stronger ability to solvate anions more than cations. Moreover, anions have a significantly larger ionic radius than cations, resulting in weaker hydration. Anions also have larger and more polarisable electron clouds, making them more prone to selective interactions with nearby cations, which may polarise the anions in their vicinity [89].

The hydration hypothesis may explain salting-out: as the process of dissolution occurs, the interaction between the solvent and ions becomes more pronounced compared to the interaction between ions themselves, resulting in an expansion of the distance between the ions. This interaction of solvation causes the immobilization of water molecules, effectively rendering them unable to function as a solvent. In the case of binary systems involving water, a water-miscible organic solvent, and salt as a precipitating salt agent (PSA), there is a competition between the water-miscible organic solvent and the salt ions for the water molecules. Once the concentration of ions surpasses a specified threshold due to the strong electrostatic contact between salts and water molecules, the interaction between water and organic solvents is significantly diminished. The hydration Gibbs free energy (Ghyd) of salts plays a vital role in determining their ability to induce salting-out. Ions with lower values of Ghyd exhibit a more potent salting-out ability [90]. In addition to Ghyd, other factors that must be taken into consideration when selecting a salt include the charge density of both anions and cations, the chemical properties of the analytes, the type of solvent used for extraction, and the analytical equipment employed. Ammonium acetate is considered the ideal salt for mass spectrometry, whereas non-volatile salts like magnesium chloride cannot be employed [4].

Aside from the salt type, additional parameters that may influence SALLME efficiency include the the nature and quantity of the solvent, the amount of salt used, and the pH of the aqueous sample. The major challenge in selecting water-miscible solvents is determining how to remove them from the aqueous phase. The salting-out phenomenon cannot separate all water-miscible solvents [88]. This explains why acetonitrile was the most commonly used extractant in SALLME applications due to its ease of separation compared to other water-miscible organic solvents, as indicated in Table 10.2. The quantity of salt is also crucial in SALLE optimisation since higher salt concentrations have been linked to better phase separation [91].

Finally, pH plays a critical role in SALLME if the analytes are weak acids or bases. pH adjustment in SALLME is crucial because the optimal extraction conditions are obtained when the drug is in the non-ionised form. The pH of the aqueous media should be two pH units higher than the pKa of the analyte and on the same side (i.e., acidic for acidic drugs and basic for basic compounds). One of the advantages of SALLME is that the chromatograms of plasma samples obtained after SALLME are superior to traditional protein precipitation due to its intense deproteinization by acetonitrile and salting-out. As indicated in Table 10.3. SALLME was employed to extract a diverse array of polarities of analytes from various matrices, encompassing food, water, and biological fluids.

4.2 Sugaring-Out Homogeneous Liquid–liquid Microextraction (SULLME)

SULLME is a miniaturised version of sugaring-out assisted liquid–liquid extraction (SULLE), which relies on using sugars as a PSA. SULLE is an extraction mode that can serve as an alternative to SALLE in bioanalysis, given its eco-friendliness, compatibility with MS detection and inertness, and less likelihood of modifying the pH of the medium compared to SALLE [106]. SULLE can enrich analytes with varying polarities by utilising the optimal sugar-to-aqueous phase ratio. It is worth mentioning that both SULLME and SALLME are compatible with RP-LC, the most prevalent mode in HPLC [107].

In SULLME, the separation of the phases is achieved by breaking the hydrogen bond between water and extractant, with the type of sugar used being a critical factor in extraction efficiency in both modes. Glucose is the most commonly used sugar in SULLE due to its low cost and availability [107–113]. However, glucose

Table 10.3 Application of SALLME

Sample	Analyte	Sample volume (mL)	Extractant	Extractant volume (mL)	Salt type	Salt amount (g)	Technique of analysis	Linearity (ng/mL)	LOD ng/mL	%RSD	REF
Waste water	Beta-naphthol, naphthalene and anthracene	10	ACN	0.8	Sodium carbonate	2.5	HPLC/UV	1–1000	0.22–8.34	<7.2%	[92]
Water samples	Triazole pesticides	3	ACN	0.65	Sodium chloride	1.3	GC/MS	10–1000	0.4–14.4	0.4–8.1	[76]
Human plasma	Alogliptin benzoate	1	ACN	0.5	Sodium chloride	0.250	HPLC/UV	100–5000	19	≤4.92	[93]
Fruit juice	Pesticides	5	IPA	0.2	Sodium sulfate	2	GC/FID	0.73–5000	0.22–0.48	**≤7**	[94]
Waste water	Azole compounds	5	IL	0.075	Sodium chloride	0.56	HPLC/DAD	25–2000	1.25–5	≤11.1	[95]
Human urine samples	Tetracycline	5	1-octylamine	0.05	Sodium chloride	1	HPLC/UV	500–20,000	170	≤8	[85]
Human serum and plasma	Tricyclic antidepressant	5	ACN	0.5	Ammonium sulfate	0.5	HPLC/UV	2–20,000	0.46–0.58	≤7.9	[96]
Water, food and bioligical fluids	Sulfonamides	0.5	ACN	0.15	Sodium chloride	N/A	HPLC/UV	10–10,000	1.4–4.5	≤8.7	[97]

(continued)

Table 10.3 (continued)

Sample	Analyte	Sample volume (mL)	Extractant	Extractant volume (mL)	Salt type	Salt amount (g)	Technique of analysis	Linearity (ng/mL)	LOD ng/mL	%RSD	REF
Cosmetics and personal care product	Triclosan	5	IPA	0.18	Ammonium sulfate	4	HPLC/UV	0.4–100	0.09	≤5.3	[98]
Water samples	Lead	5	IL	20 mg	Sodium sulfate	N/A	FAAS	5.0–500.0	0.1	1.3	[99]
Aqueous humor	Dorzolomide and timolol	0.15	ACN	0.09	Sodium sulfate	0.11	HPLC/UV	9–500	2.89–7.76	≤1.2	[100]
Human plasma	Daclatasvir	1	ACN	0.5	Ammonium acetate	0.2	Spectrophotometry	500–5000	130	≤8.323	[101]
Water, food and biological samples	Sulfanilamide	0.5	ACN	0.25	Sodium chloride	0.25	HPLC/UV	1–10,000	0.3	1.55	[102]
Water, food and biological samples	Fluoroquinolones	5	ACN	1	Magnesium sulfate	2	HPLC/FLD	2–100	0.07–5.58	≤6.8	[103]
Water samples	Carbonyl compounds	3	ACN	0.5	Ammonium sulfate	0.5	HPLC/UV	7–15,000	0.58–3.2	≤6	[14]
Table salt	Iodine	5	Ethanol	0.8	Ammonium sulfate	3–3.5	HPLC/DAD	10–10,000	3.7	≤12.4	[104]
Food grade salt	Iodate	4.0	2 propanol	1	Ammonium sulfate	2.6–3.5	Spectrophotometry	80–10,000	16	≤12.6	[105]

cannot achieve phase separation in SULLME due to its inability to separate a few microlitres of water-miscible solvent. Sucrose was found to be the most efficient PSA in SULLE for honokiol and magnolol extraction [114], while polyols like glycerol, sorbitol, xylitol, maltitol, and erythritol have also been studied for their ability to induce phase separation [23]. THF and fructose were successfully used in SULLE for diuron pesticide trace detection in water [115].

A previous study investigated different modes of LLME on various drug classes [5]. The results indicated that ternary SALLME had less preconcentration capacity in sample enrichment due to the high volume of separated phase caused by the large amount of acetonitrile required for homogeneous phase formation. NaCl used in SALLME not only separated the extractant (decanol), but it also separated some of the co-solvent (ACN) from the aqueous sample, which diluted the analyte in the extract and compromised the enrichment. In contrast, SULLME was found to be the most efficient mode for extraction of antivirals, antidiabetics, and β-blockers as shown in Fig. 10.9. Although SALLME and SULLME are both homogeneous modes, SULLME had more preconcentration capacity due to the smaller layer of acetonitrile separated on the top of the aqueous sample, making the sample more concentrated. This could be attributed to the electrostatic force of the salt, which makes salting-out more efficient in phase separation but reduces the analyte concentration in the extractant.

Chromatographic separation of the extracted analytes showed higher peak areas and a more stable baseline in SULLME compared to SALLME, improving the signal-to-noise ratio and method sensitivity. Additionally, SULLME uses only acetonitrile as an organic extractant, minimizing the use of other solvents and reducing potential risks associated with lipophilic solvents such as chloroform or long chain alcohols. Acetonitrile is also more compatible with conventional mobile phases because of its low UV cutoff point.

There are challenges associated with the use of SULLME, including the limited ability of some sugars to achieve phase separation, a limited number of extractants that can be separated, and the limited volume of aqueous samples that can be used. Nonetheless, SULLME is suitable for bioanalysis applications, including analysing plasma, urine, and aqueous humour samples [116]. For example, SULLME was successfully used to determine favipiravir in human plasma with comparable or superior sensitivity than the LC–MS/MS approach [116]. The inert nature of sugars makes them less likely to impact the pH of the sample or stability of the analyte. Combining SULLME with sensitive methods like LC–MS/MS shows promise for polar pharmaceuticals that are inefficiently extracted using standard solvents [117–119]. Additionally, self-assembly was formed using THF and fructose, resulting in enhanced extraction efficiency in the self-assembly core [116].

Recently, matrix-induced SULLME has been developed as a novel extraction method that relies on using the sample itself as a PSA source [120]. Alkan et al. developed matrix-induced SULLME for determination of pesticides in jams. The procedures involved weighing 1 g of pre-homogenised jam sample into 2 ml safe-lock Eppendorf microtubes and adding 600 μL of ACN/water combination (50/50%, v/v). The mixture was then shaken at 2500 rpm for 0.5 min before being centrifuged

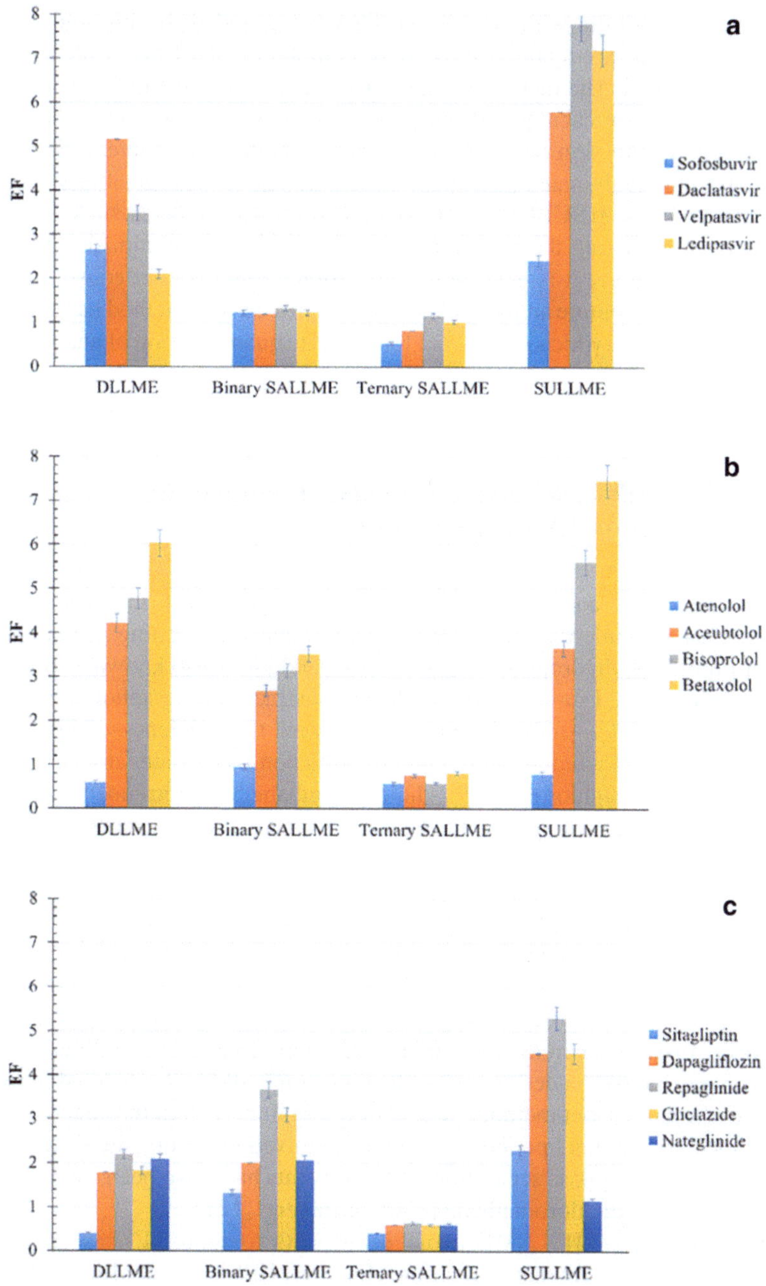

Fig. 10.9 Extraction efficiency of different microextraction modes for the antivirals (**A**), the β-blockers (**B**), and the antidiabetics (**C**). Reprinted from [5] with permission from with Wiley

at 6000 rpm for 4 min, with no external PSA being added. In this study, a quick and simple sample pretreatment procedure combining SULLME with HPLC/UV has been developed. The method is inexpensive and environmentally friendly since ACN is the only chemical required, and phase separation from homogeneous solution occurs through induction of the sample matrix's high sugar content. Satisfactory recoveries from manipulated jam samples indicated good reproducibility.

The matrix-induced SULLME method offers several advantages, such as requiring fewer steps, eliminating the need for an external PSA, and minimising the use of solvents. Furthermore, it is environmentally friendly and cost-effective, making it highly suitable for routine analysis of complex matrices. The method has also shown good reproducibility, which is essential for accurate quantification of analytes in real samples.

4.3 Hydrophobic Substance Induced Homogenous Liquid–Liquid Microextraction

In hydrophobic substance induced HLLME (HSLLME), the homogenous layer is formed as usual, while the phase separation is induced by adding a small amount of water immiscible hydrophobic solvent. This mode was also known as SIPTE and it was developed by Liu et al. [16] in 2010 to extract andrographolide, sildenafil, and finasteride. In this study, the authors examined the efficiency of SIPTE using different modifiers, including six commonly used non-oxygenated organic solvents, i.e. dichloromethane, chloroform, 1,2-didichloroethane, 1,2-dibromoethane, toluene, fluorobenzene and four oxygenated organic solvents i.e. ethyl acetate, ethyl ether, n-hexanol and n-octanol.

The results showed that non-oxygenated solvents were effective as modifiers, whereas oxygenated solvents were less efficient. Specifically, high recoveries of all test substances were obtained using non-oxygenated modifiers, while much lower extraction recoveries were observed with oxygenated solvents. Furthermore, at least 0.3 mL of an oxygenated modifier was required to separate 2 mL of the plasma-acetonitrile mixture, whereas only 0.05 mL of a non-oxygenated modifier was sufficient. The mechanism underlying this phenomenon could be considered a reversed process to the salting-out method. Instead of dissolving a salt in the aqueous sample to expel the extractant, a lipophilic solvent is dissolved in the organic to decrease its polarity, resulting in phase separation. Other solvents have also been used as PSA in HSLLME, such as dichloromethane [106], or toluene [121].

This technique of HSLLME avoid using salts to induced phase separation, which may interfere with mass detection. However, HSLLME employs highly toxic solvents. To overcome this problem, Abdallah et al. [122] developed a menthol-assisted HSLLE method to determine favipiravir in human plasma samples via HPLC/UV detection. This approach is less expensive, simpler, and more environmentally friendly than traditional sample preparation and other HSLLME methods.

Menthol has surfactant-like characteristics and can form micelles, which could be used to extract hydrophilic analytes like favipiravir. The menthol-assisted HSLLME method demonstrated equivalent or even better sensitivity than the LC–MS/MS method and performed well in extracting favipiravir from actual human plasma samples collected during a bioequivalence study examining favipiravir as a potential COVID-19 antiviral medication. This method does not require the use of expensive or complicated instruments, making it a promising sample preparation approach for polar drugs [122].

4.4 pH Induced Homogeneous Liquid–liquid Microextraction

SHSs are a type of environmentally friendly solvents that are increasingly replacing traditional organic solvents in microextraction [123]. These solvents can be categorised into various types based on different criteria, such as miscibility, ionic strength, polarity, surface activity, and properties like fluorescence, solubility, aggregation, hydrophilicity, and charge (CO_2-responsive polymers). It is important to note that the term "switchable solvents" typically refers to tertiary or secondary amines that undergo structural changes in solutions with varying pH values, resulting in altered hydrophilicity and solubility.

$$B + nH^+ \leftrightarrow BHn^{n+} \qquad \text{Equilibrium (1)}$$

In an acidic environment, the equilibrium (1) favours the formation of ionic forms which have high water solubility and can easily form homogeneous mixtures in aqueous solutions. As the pH increases, the equilibrium shifts to the left indicating that molecular compounds are predominant, exhibiting limited solubility in aqueous fluids, and resulting in biphasic systems. It is noteworthy that CO_2 can be used to facilitate such conversions, as has been reported in previous studies [124, 125].

$$B + nH_2O + nCO_2 \leftrightarrow BHn^{n+} + nHCO_3^- \qquad \text{Equilibrium (2)}$$

In the presence of CO_2, acidic conditions lead to a shift in equilibrium (2) towards the prevalence of quaternary ammonium cations that are water-soluble [125]. To obtain ionic forms from amines, small amounts of dry ice are typically added to a water-amine mixture [126]. The process for obtaining SHS is achieved by maintaining constant stirring until a homogeneous solution is obtained [64]. Prior to adding CO_2 to the aqueous phase, appropriate conditions such as reagent concentration, pH, and ionic strength must be established [125].

There are alternative methods for altering solvent polarity, such as sparging with carbon dioxide, and using carbonate salts, sulfuric acid or perchloric acid, but these are less common [125]. It should be noted that to establish a two-phase system and concentrate analytes, CO_2 must be removed or the amine deprotonated. This can be achieved by adding concentrated alkali solutions or by other methods such as sparging with nitrogen, argon, or air, which can be carried out by heating a homogenous mixture [126].

Although different amines are commonly used for the formation of SHS, not all amines and their derivatives have the ability to transform homogeneous aqueous-organic phases to heterogeneous biphasic ones depending on pH [125]. Some amines, such as diethylamine, diisopropylamine, butylethylamine, and triethanolamine, have high water solubility and cannot form two-phase extraction systems. Conversely, some amines, such as dihexylamine, butylisopropylamine, and trioctylamine, have poor solubility even in acidic conditions, resulting in an inability to produce homogeneous water-organic combinations [125].

When selecting an appropriate SHS, stability, volatility, toxicity, and bioaccumulation are all key considerations. Stability is particularly critical for solvent reuse. From an environmental standpoint, the recommended amines should be as non-toxic and non-volatile as possible. Therefore, high molecular weight amines, which can be synthesized using functional groups such as alcohols, esters, ketones, acetals, and aromatic rings, are preferred [125]. Functional amines are often preferred due to their low toxicity, volatility, flammability, and potential for eutrophication. In addition to amines, other solvents such as saturated fatty acids can also be used as SHS. For instance, Vakh et al. and coworkers [84] used hexanoic acid as a switchable solvent to detect ofloxacin in urine. By interacting with Na_2CO_3 and H_2SO_4, the SHS's hydrophilicity was altered.

The use of SHSs in HLLME depends on solubilizing the solvent into the aqueous phase prior to extraction. This results in a stable homogeneous SHS/water combination that can then be used as the extractant phase. It is common to first dissolve the solvent in the aqueous phase with the aid of dry ice. This produces a stable and uniform mixture of SHS and water, which can then be utilized as the extractant phase. By manipulating the pH level through the addition of an acid or base, it is possible to alter the solubility of the SHS and induce a separation of the phases. The method described here employs a water-soluble extractant phase and utilizes a pH shift as a trigger for phase separation [127]. As indicated in Table 10.4, pH induced HLLME using SHS was widely used for extraction different analytes including antidepressants [128], antioxidants [129], metals [130] fungicide [131], herbicides [127, 132], benzophenone-type UV filters [35], anxiolytics [133] and toxicants [134] from different matrices including water [35], beverages [135], biological [136] and food samples [134].

The pH induced HLLME method offers an appealing advantage over other modes of HLLME, such as SALLME and SULLME. Specifically, this method allows for the use of a large volume of aqueous sample in combination with a very small amount of extractant. This is made possible by taking advantage of the pH shift-induced phase separation, which enables efficient extraction of the target compounds from the sample using a minimal amount of extractant. The most common SHSs used in pH induced HLLME are triethylamine [137], *N, N*-dimethyl cyclohexylamine

Table 10.4 Application of pH induced HLLME

Analyte	Sample	Sample volume (mL)	SHS composition	Extractant volume µL	Phase trigger	Technique of analysis	Linearity (ng/mL)	LOD ng/mL	% RSD	REF
Antidepressants	Waste water, human serum and breast milk	7.0	Dipropyl amine	50	NaOH	GC/FID	5.0–1000	1.0	≤8.7	[128]
Permethrin and deltamethrin	Water samples	10	Triethylamine	500	NaOH	GC/FID	0.5–2500	0.03–0.16	<5%	[137]
Chlorophenols	Water samples	5	bis(2-ethylhexyl) phosphate	60	HCl	HPLC/UV	5–500	1.4–2.7	≤4.7	[140]
Synthetic Antioxidants	Vinegar samples	5	Di-(2-ethylhexyl) phosphoric acid	10	Vinegar	HPLC/UV	10–500	3.2–5.5	<7.8	[129]
Copper	Oil samples	30	Triethyl amine	500	Nitric acid	FAAS	23.0–1000	6.9	≤9.4	[130]
Quercetin	Food samples	30	N, N-dimethyl cyclohexylamine	1000	NaOH	Spectrophotometry	29.9–500	9.0–11.9	≤8.9	[139]
Triazole fungicide	Water samples	10	N, N-dimethyl cyclohexylamine	400	NaOH	GC/MS	5–500	0.46–0.99	≤13.9	[131]
Herbicides	Water samples	10	N, N-dimethyl cyclohexylamine	125	NaOH	GC/MS	N/A	0.1–0.37	≤12.5	[127]
Paraquat	Biological and environmental samples	10	Triethyl amine	500	NaOH	HPLC/UV–VIS	0.5–500	0.2	<5	[132]
Benzophenone-type UV filters	Water samples	8	Di-(2-ethylhexyl) phosphoric acid	40	HCl	HPLC/UV	2.5–1000	0.7–0.8	<6	[35]

(continued)

Table 10.4 (continued)

Analyte	Sample	Sample volume (mL)	SHS composition	Extractant volume μL	Phase trigger	Technique of analysis	Linearity (ng/mL)	LOD ng/mL	% RSD	REF
NSAID	Biological fluids	9.5	N, N-dimethyl cyclohexylamine	500	NaOH	HPLC/DAD	130–100,000	40–180	≤7.7	[136]
Pyrethroid insecticides	Fresh fruits and fruit juices	10	Pivalic acid	260	HCl	GC/MS	0.023–500	0.006–0.038	<9	[135]
Nickel	Food and cigarette samples	10	1-ethyl piperidine	800	NaOH	FAAS	17–500	5.2	6.0	[138]
Nitrazepam	Urine samples	4	N, N-dipropyl amine	100	NaOH	DPV	0.03–20 and 20–450	0.009	7.4	[133]
Methamphetamine	Urine samples	4	Dipropyl amine	100	NaOH	GC/MS	5–1500	1.5	≤7.8	[141]
Bisphenols	Foods and Drinks	5	N, N Dimethyl cyclohexylamine	782	NaOH	HPLC/UV	0.27–0.67	0.17–0.67	≤5.7	[134]

[136], dipropyl amine [128], pivalic acid [135], 1-ethyl piperidine [138] and di-(2-ethylhexyl) phosphoric acid [35]. It is worth mentioning that the most common phase trigger agent was sodium hydroxide [136], hydrochloric acid [135], vinegar [129] and nitric acid [130]. This mode is compatible with wide range of analytical instruments including spectrophotometers [139], HPLC [129], GC [128, 131], FAAS [130] and DPV [133].

5 Conclusions and Future Trends

HLLME is very attractive among other ME modes owing to a lot of merits, including a huge surface area between the aqueous sample and miscible extractants like ACN, acetone, THF, and hydrophilic DESs. Phase separation could be achieved by the addition of chemicals such as salt (SALLME), sugar (SULLME), or by buffer. Additionally, manipulation of the properties of the system may also result in phase separation, such as pH and temperature. One of the important advantages of HLLME is its suitability for automation. HLLME has been successfully applied to extract different analytes from various matrices, including biological fluids and aqueous samples. The various applications and reports have proven that HLLME is a sensitive and effective technique for enriching desired analytes from various matrices. However, it has certain limitations, such as the use of hydrophilic solvents to extract analytes from complex matrices. Conventional HLLME procedures still employ organic solvents as extractants, which pose significant environmental risks due to their toxicity. To address this issue, new green solvents such as supercritical fluids, ILs, and DESs have been developed.

Among these, ILs have gained popularity due to their unique physicochemical properties that make them effective extraction solvents for a wide range of analytes during the HLLME mode. However, some methods of synthesizing ILs involve hazardous halogenated hydrocarbons and extended reaction periods, leading to potential environmental contamination. To overcome these limitations, DESs were developed as a greener alternative to ILs. DESs share similar physicochemical properties with ILs but have additional advantages such as low vapor pressure, nonflammability, low density, and low melting points. By using DESs instead of ILs, researchers can reduce the risk of environmental poisoning and secondary contamination from toxic byproducts produced during high-temperature processes.

Innovative approaches could also include the use of green solvents like propylene glycol and glycerol as extractants in HLLME. These solvents are considered safer alternatives to conventional solvents, and further research should be conducted to determine their efficacy in this application. The development of green solvents such as NADESs and the exploration of new green solvents hold promise for improving the sustainability and safety of HLLME procedures.

The Authors Have Declared no Conflict of Interest.

References

1. Xu L, Qi X, Li X, Bai Y, Liu H (2016) Recent advances in applications of nanomaterials for sample preparation. Talanta 146:714–726. https://doi.org/10.1016/j.talanta.2015.06.036
2. Niu Z, Zhang W, Yu C, Zhang J, Wen Y (2018) Recent advances in biological sample preparation methods coupled with chromatography, spectrometry and electrochemistry analysis techniques. TrAC - Trends Anal Chem 102:123–146. https://doi.org/10.1016/j.trac.2018.02.005
3. Filippou O, Bitas D, Samanidou V (2017) Green approaches in sample preparation of bioanalytical samples prior to chromatographic analysis. J Chromatogr B Anal Technol Biomed Life Sci 1043:44–62. https://doi.org/10.1016/j.jchromb.2016.08.040
4. Hammad SF, Abdallah IA, Bedair A, Mansour FR (2022) Homogeneous liquid–liquid extraction as an alternative sample preparation technique for biomedical analysis. J Sep Sci 45:185–209. https://doi.org/10.1002/jssc.202100452
5. Abdallah IA, Hammad SF, Bedair A, Mansour FR (2021) Sugaring-out induced homogeneous liquid-liquid microextraction as an alternative mode for biological sample preparation: a comparative study. J Sep Sci 44:3117–3125. https://doi.org/10.1002/jssc.202100255
6. da Silva S, Burato J, Vargas Medina DA, de Toffoli AL, Vasconcelos Soares Maciel E, Mauro Lanças F (2020) Recent advances and trends in miniaturized sample preparation techniques. J Sep Sci 43:202–225
7. Altunay N, Elik A, Unal Y, Kaya S (2021) Optimization of an ultrasound-assisted alcohol-based deep eutectic solvent dispersive liquid-phase microextraction for separation and preconcentration of quercetin in wine and food samples with response surface methodology. J Sep Sci 44:1998–2005. https://doi.org/10.1002/jssc.202100048
8. Farajzadeh MA, Sohrabi H, Mohebbi A, Mogaddam MRA (2019) Combination of a modified quick, easy, cheap, efficient, rugged, and safe extraction method with a deep eutectic solvent based microwave-assisted dispersive liquid–liquid microextraction: application in extraction and preconcentration of multiclass pesticide residues in tomato samples. J Sep Sci 42:1273–1280. https://doi.org/10.1002/jssc.201801107
9. Asadollahi T, Dadfarnia S, Haji Shabani AM, Amirkavei M (2015) Separation/preconcentration and determination of quercetin in food samples by dispersive liquid–liquid microextraction based on solidification of floating organic drop -flow injection spectrophotometry. J Food Sci Technol 52:1103–1109. https://doi.org/10.1007/s13197-013-1077-9
10. Mansour FR, Khairy MA (2017) Pharmaceutical and biomedical applications of dispersive liquid–liquid microextraction. J Chromatogr B 1061–1062:382–391. https://doi.org/10.1016/J.JCHROMB.2017.07.055
11. Mansour FR, Danielson ND (2017) Solidification of floating organic droplet in dispersive liquid-liquid microextraction as a green analytical tool. Talanta 170:22–35. https://doi.org/10.1016/j.talanta.2017.03.084
12. Hammad MA, Kamal AH, Kannouma RE, Mansour FR (2021) Vortex-assisted dispersive liquid-liquid microextraction coupled with deproteinization for determination of nateglinide in human plasma using HPLC/UV. J Chromatogr Sci 59:297–304. https://doi.org/10.1093/chromsci/bmaa096
13. Matkovich CE, Christian GD (1973) Salting-out of acetone from water. Basis of a new solvent extraction system. Anal Chem 45:1915–1921. https://doi.org/10.1021/ac60333a023
14. Gupta M, Jain A, Verma KK (2009) Salt-assisted liquid-liquid microextraction with water-miscible organic solvents for the determination of carbonyl compounds by high-performance liquid chromatography. Talanta 80:526–531. https://doi.org/10.1016/j.talanta.2009.07.021
15. Baghdadi M, Shemirani F (2009) In situ solvent formation microextraction based on ionic liquids: a novel sample preparation technique for determination of inorganic species in saline solutions. Anal Chim Acta 634:186–191. https://doi.org/10.1016/j.aca.2008.12.017

16. Liu G, Zhou N, Zhang M, Li S, Tian Q, Chen J, Chen B, Wu Y, Yao S (2010) Hydrophobic solvent induced phase transition extraction to extract drugs from plasma for high performance liquid chromatography–mass spectrometric analysis. J Chromatogr A 1217:243–249. https://doi.org/10.1016/J.CHROMA.2009.11.037
17. Gao S, Jin H, You J, Ding Y, Zhang N, Wang Y, Ren R, Zhang R, Zhang H (2011) Ionic liquid-based homogeneous liquid-liquid microextraction for the determination of antibiotics in milk by high-performance liquid chromatography. J Chromatogr A 1218:7254–7263. https://doi.org/10.1016/j.chroma.2011.08.063
18. Seidi S, Alavi L (2019) Novel and rapid deep eutectic solvent (DES) homogeneous liquid-liquid microextraction (HLLME) with flame atomic absorption spectrometry (FAAS) detection for the determination of copper in vegetables. Anal Lett 52:2092–2106. https://doi.org/10.1080/00032719.2019.1598425
19. Kannouma RE, Hammad MA, Kamal AH, Mansour FR (2022) Miniaturization of Liquid-Liquid extraction; the barriers and the enablers. Microchem J 182:107863. https://doi.org/10.1016/j.microc.2022.107863
20. Kamal AH, Hammad MA, Kannouma RE, Mansour FR (2022) Response surface optimization of a vortex-assisted dispersive liquid–liquid microextraction method for highly sensitive determination of repaglinide in environmental water by HPLC/UV. BMC Chem 16:33. https://doi.org/10.1186/s13065-022-00826-w
21. Kannouma RE, Hammad MA, Kamal AH, Mansour FR (2021) A dispersive liquid–liquid microextraction method based on solidification of floating organic droplet for determination of antiviral agents in environmental water using HPLC/UV. Microchem J 171:106790. https://doi.org/10.1016/j.microc.2021.106790
22. Dean JA (1999) Lange's handbook of chemistry, 15th edn. McGraw Hill Inc., New YorK, USA
23. De Brito CG, Souza IN, Mourão T, Freire MG, Soares CMF, Lima ÁS (2014) Novel aqueous two-phase systems composed of acetonitrile and polyols: phase diagrams and extractive performance. Sep Purif Technol 124:54–60. https://doi.org/10.1016/j.seppur.2014.01.004
24. Armenta S, Garrigues S, de la Guardia M (2008) Green Analytical chemistry. TrAC Trends Anal Chem 27:497–511. https://doi.org/10.1016/j.trac.2008.05.003
25. Koel M, Kaljurand M (2006) Application of the principles of green chemistry in analytical chemistry. Pure Appl Chem 78:1993–2002. https://doi.org/10.1351/pac200678111993
26. Tobiszewski M, Mechlińska A, Namieśnik J (2010) Green analytical chemistry—theory and practice. Chem Soc Rev 39:2869. https://doi.org/10.1039/b926439f
27. Anastas PT, Warner JC (1998) Green chemistry: theory and practice. Oxford University Press, Oxford
28. Gałuszka A, Migaszewski Z, Namieśnik J (2013) The 12 principles of green analytical chemistry and the SIGNIFICANCE mnemonic of green analytical practices. TrAC Trends Anal Chem 50:78–84. https://doi.org/10.1016/j.trac.2013.04.010
29. Zhang L, Yu R, Yu Y, Wang C, Zhang D (2019) Determination of four acetanilide herbicides in brown rice juice by ionic liquid / ionic liquid-homogeneous liquid-liquid micro-extraction high performance liquid chromatography. Microchem J 146:115–120. https://doi.org/10.1016/j.microc.2018.12.062
30. Khezeli T, Daneshfar A, Sahraei R (2015) Emulsification liquid–liquid microextraction based on deep eutectic solvent: an extraction method for the determination of benzene, toluene, ethylbenzene and seven polycyclic aromatic hydrocarbons from water samples. J Chromatogr A 1425:25–33. https://doi.org/10.1016/j.chroma.2015.11.007
31. Pochivalov A, Davletbaeva P, Cherkashina K, Lezov A, Vakh C, Bulatov A (2018) Surfactant-mediated microextraction approach using switchable hydrophilicity solvent: HPLC-UV determination of Sudan dyes in solid food samples. J Mol Liq 271:807–814. https://doi.org/10.1016/j.molliq.2018.09.072
32. Ullah N, Tuzen M (2023) A comprehensive review on recent developments and future perspectives of switchable solvents and their applications in sample preparation techniques. Green Chem 25:1729–1748. https://doi.org/10.1039/D3GC00020F

33. Hosseini MH, Rezaee M, Mashayekhi HA, Akbarian S, Mizani F, Pourjavid MR (2012) Determination of polycyclic aromatic hydrocarbons in soil samples using flotation-assisted homogeneous liquid–liquid microextraction. J Chromatogr A 1265:52–56. https://doi.org/10.1016/j.chroma.2012.09.099
34. Okhravi T, Sorouraddin SM, Farajzadeh MA, Mohebbi A (2020) Development of a liquid-nitrogen-induced homogeneous liquid–liquid microextraction of Co(II) and Ni(II) from water and fruit juice samples followed by atomic absorption spectrometry detection. Anal Bioanal Chem 412:1675–1684. https://doi.org/10.1007/s00216-020-02406-0
35. Çabuk H, Kavaracı O (2022) Magnetic retrieval of a switchable hydrophilicity solvent: fast homogeneous liquid–liquid microextraction for the determination of benzophenone-type UV filters in environmental waters. Int J Environ Anal Chem 102:2569–2585. https://doi.org/10.1080/03067319.2020.1757088
36. Xu X-Y, Ye J, Nie J, Li Z-G, Lee M-R (2015) A new liquid–liquid microextraction method by ultrasound assisted salting-out for determination of triazole pesticides in water samples coupled by gas chromatography-mass spectrometry. Anal Methods 7:1194–1199. https://doi.org/10.1039/C4AY02448F
37. de Souza FLA, de Souza Ramos TJ, Montenegro MCBSM, Pinto L, Cassol TM, Paim APS (2021) Chemometric cleanup to eliminate ionic liquid interferences and enable its application on in-situ IL-DLLME using HPLC-DAD detection. J Mol Liq 330:115627. https://doi.org/10.1016/j.molliq.2021.115627
38. Hassan AA, Tanimu A, Alhooshani K (2021) Iron and cobalt-containing magnetic ionic liquids for dispersive micro-solid phase extraction coupled with HPLC-DAD for the preconcentration and quantification of carbamazepine drug in urine and environmental water samples. J Mol Liq 336:116370. https://doi.org/10.1016/j.molliq.2021.116370
39. Varona M, Eor P, Ferreira Neto LC, Merib J, Anderson JL (2021) Metal-containing and magnetic ionic liquids in analytical extractions and gas separations. TrAC Trends Anal Chem 140:116275. https://doi.org/10.1016/j.trac.2021.116275
40. Rykowska I, Ziemblińska J, Nowak I (2018) Modern approaches in dispersive liquid-liquid microextraction (DLLME) based on ionic liquids: a review. J Mol Liq 259:319–339. https://doi.org/10.1016/j.molliq.2018.03.043
41. Abdelaziz MA, Saleh AM, Mansour FR, Danielson ND (2022) A gadolinium-based magnetic ionic liquid for dispersive liquid-liquid microextraction of ivermectin from environmental water. J Chromatogr Sci. https://doi.org/10.1093/chromsci/bmac101
42. Danielson ND, Mansour FR, Zhou L, Connell CV, Dotlich EM, Gibler JN, Norman BE, Grossman S, Wei W, Zhang Y (2018) Liquid chromatography with alkylammonium formate ionic liquid mobile phases and fluorescence detection. J Chromatogr A 1559:128–135. https://doi.org/10.1016/j.chroma.2018.03.020
43. Ashouri V, Adib K, Fariman GA, Ganjali MR, Rahimi-Nasrabadi M (2021) Determination of arsenic species using functionalized ionic liquid by in situ dispersive liquid-liquid microextraction followed by atomic absorption spectrometry. Food Chem 349:129115. https://doi.org/10.1016/j.foodchem.2021.129115
44. Fan C, Li N, Cao X (2015) Determination of chlorophenols in honey samples using in-situ ionic liquid-dispersive liquid-liquid microextraction as a pretreatment method followed by high-performance liquid chromatography. Food Chem 174:446–451. https://doi.org/10.1016/j.foodchem.2014.11.050
45. Xu W, Li J, Feng J, Wang Z, Zhang H (2021) In-syringe temperature-controlled liquid–liquid microextraction based on solidified floating ionic liquid for the simultaneous determination of triazine and phenylurea pesticide in vegetable protein drinks. J Chromatogr B Anal Technol Biomed Life Sci 1174:122721. https://doi.org/10.1016/j.jchromb.2021.122721
46. Wang H, Hu L, Li W, Yang X, Lu R, Zhang S, Zhou W, Gao H, Li J (2017) In-syringe dispersive liquid-liquid microextraction based on the solidification of ionic liquids for the determination of benzoylurea insecticides in water and tea beverage samples. Talanta 162:625–633. https://doi.org/10.1016/j.talanta.2016.10.035

47. Wang Z, Hu J, Du H, He S, Li Q, Zhang H (2016) Microwave-assisted ionic liquid homogeneous liquid-liquid microextraction coupled with high performance liquid chromatography for the determination of anthraquinones in Rheum palmatum L. J Pharm Biomed Anal 125:178–185. https://doi.org/10.1016/j.jpba.2016.03.046
48. Santana-Mayor Á, Rodríguez-Ramos R, Herrera-Herrera AV, Socas-Rodríguez B, Rodríguez-Delgado MÁ (2021) Deep eutectic solvents: the new generation of green solvents in analytical chemistry. TrAC - Trends Anal Chem 134:116108. https://doi.org/10.1016/j.trac.2020.116108
49. Florindo C, Oliveira FS, Rebelo LPN, Fernandes AM, Marrucho IM (2014) Insights into the synthesis and properties of deep eutectic solvents based on cholinium chloride and carboxylic acids. ACS Sustain Chem Eng 2:2416–2425. https://doi.org/10.1021/sc500439w
50. Guthrie F (1884) On Eutexia. Proc Phys Soc London 6:124–146. https://doi.org/10.1088/1478-7814/6/1/312
51. Gill I, Vulfson E (1994) Enzymic catalysis in heterogeneous eutectic mixtures of substrates. Trends Biotechnol 12:118–122
52. Abranches DO, Martins MAR, Silva LP, Schaeffer N, Pinho SP, Coutinho JAP (2019) Phenolic hydrogen bond donors in the formation of non-ionic deep eutectic solvents: the quest for type V DES. Chem Commun 55:10253–10256. https://doi.org/10.1039/C9CC04846D
53. Liu L, Zhu T (2017) Emulsification liquid–liquid microextraction based on deep eutectic solvents: an extraction method for the determination of sulfonamides in water samples. Anal Methods 9:4747–4753. https://doi.org/10.1039/C7AY01332A
54. Aydin F, Yilmaz E, Soylak M (2017) A simple and novel deep eutectic solvent based ultrasound-assisted emulsification liquid phase microextraction method for malachite green in farmed and ornamental aquarium fish water samples. Microchem J 132:280–285. https://doi.org/10.1016/j.microc.2017.02.014
55. Panhwar AH, Tuzen M, Kazi TG (2018) Deep eutectic solvent based advance microextraction method for determination of aluminum in water and food samples: multivariate study. Talanta 178:588–593. https://doi.org/10.1016/j.talanta.2017.09.079
56. Moghadam AG, Rajabi M, Asghari A (2018) Efficient and relatively safe emulsification microextraction using a deep eutectic solvent for influential enrichment of trace main antidepressant drugs from complicated samples. J Chromatogr B 1072:50–59. https://doi.org/10.1016/j.jchromb.2017.09.042
57. Zounr RA, Tuzen M, Khuhawar MY (2018) Novel ultrasonic-assisted deep eutectic solvent-based dispersive liquid–liquid microextraction for determination of vanadium in food samples by electrothermal atomic absorption spectrometry: a multivariate study. Appl Organomet Chem 32:e4144. https://doi.org/10.1002/aoc.4144
58. Haleem A, Tuzen M, Gul T (2017) Talanta Ultrasonic assisted dispersive liquid-liquid microextraction method based on deep eutectic solvent for speciation, preconcentration and determination of selenium species (IV) and (VI) in water and food samples. Talanta 175:352–358. https://doi.org/10.1016/j.talanta.2017.07.063
59. Aydin F, Yilmaz E, Soylak M (2018) Vortex assisted deep eutectic solvent (DES)-emulsification liquid-liquid microextraction of trace curcumin in food and herbal tea samples. Food Chem 243:442–447. https://doi.org/10.1016/j.foodchem.2017.09.154
60. Ali R, Tuzen M, Deligonul N, Yar M (2018) A highly selective and sensitive ultrasonic assisted dispersive liquid phase microextraction based on deep eutectic solvent for determination of cadmium in food and water samples prior to electrothermal atomic absorption spectrometry. Food Chem 253:277–283. https://doi.org/10.1016/j.foodchem.2018.01.167
61. Zounr RA, Tuzen M, Khuhawar MY (2018) A simple and green deep eutectic solvent based air assisted liquid phase microextraction for separation, preconcentration and determination of lead in water and food samples by graphite furnace atomic absorption spectrometry. J Mol Liq 259:220–226. https://doi.org/10.1016/j.molliq.2018.03.034
62. Santana-Mayor Á, Socas-Rodríguez B, Rodríguez-Ramos R, Rodríguez-Delgado MÁ (2020) A green and simple procedure based on deep eutectic solvents for the extraction of phthalates from beverages. Food Chem 312:125798. https://doi.org/10.1016/j.foodchem.2019.125798

63. Babaee S, Daneshfar A, Sahraei R (2019) Deep eutectic solvent-based emulsification liquid–liquid microextraction coupled with gas chromatography for the determination of thiophenols in water samples. Anal Methods 11:1663–1670. https://doi.org/10.1039/C8AY02719F
64. Elik A, Demirbas A, Altunay N (2019) Developing a new and simple natural deep eutectic solvent based ultrasonic-assisted microextraction procedure for determination and preconcentration of As and Se from rice samples. Anal Methods 11:3429–3438. https://doi.org/10.1039/C9AY00916G
65. Altunay N, Elik A, Gürkan R (2019) Natural deep eutectic solvent-based ultrasound- assisted-microextraction for extraction, pre-concentration and analysis of methylmercury and total mercury in fish and environmental waters by spectrophotometry. Food Addit Contam Part A 36:1079–1097. https://doi.org/10.1080/19440049.2019.1619939
66. Thongsaw A, Udnan Y, Ross GM, Chaiyasith WC (2019) Talanta Speciation of mercury in water and biological samples by eco-friendly ultrasound-assisted deep eutectic solvent based on liquid phase microextraction with electrothermal atomic absorption spectrometry. Talanta 197:310–318. https://doi.org/10.1016/j.talanta.2019.01.018
67. Li X, Wang M, Zhao J, Guo H, Gao X, Xiong Z, Zhao L (2019) Ultrasound-assisted emulsification liquid phase microextraction method based on deep eutectic solvent as extraction solvent for determination of five pesticides in traditional Chinese medicine. J Pharm Biomed Anal 166:213–221. https://doi.org/10.1016/j.jpba.2019.01.018
68. Heidari H, Yari B (2020) Multivariate optimization of an ultrasound-Assisted Deep Eutectic Solvent-Based Liquid-Phase Microextraction Method for HPLC–UV Analysis of Carbamazepine in Plasma. Chromatographia 83:1467–1475. https://doi.org/10.1007/s10337-020-03966-0
69. Altunay N, Unal Y, Elik A (2020) Towards green analysis of curcumin from tea, honey and spices: extraction by deep eutectic solvent assisted emulsification liquid-liquid microextraction method based on response surface design. Food Addit Contam Part A 37:869–881. https://doi.org/10.1080/19440049.2020.1748233
70. Ao Y, Chen Y, Ding W (2021) Deep eutectic solvent-based ultrasound-assisted emulsification microextraction for the rapid determination of benzotriazole and benzothiazole derivatives in surface water samples. J Hazard Mater 401:123383. https://doi.org/10.1016/j.jhazmat.2020.123383
71. Farajzadeh MA, Shahedi Hojghan A, Afshar Mogaddam MR (2018) Development of a new temperature-controlled liquid phase microextraction using deep eutectic solvent for extraction and preconcentration of diazinon, metalaxyl, bromopropylate, oxadiazon, and fenazaquin pesticides from fruit juice and vegetable samples fo. J Food Compos Anal 66:90–97. https://doi.org/10.1016/j.jfca.2017.12.007
72. Lu Y, Wang X, Gu H, Gao M (2021) Morphological transformation assisted switchable deep eutectic solvents combined with HPLC-DAD for the detection of six UV-filters in surface and bathing waters. Microchem J 169:106626. https://doi.org/10.1016/j.microc.2021.106626
73. Jouyban A, Farajzadeh MA, Khodadadeian F, Khoubnasabjafari M, Afshar Mogaddam MR (2021) Development of a deep eutectic solvent-based ultrasound-assisted homogenous liquid-liquid microextraction method for simultaneous extraction of daclatasvir and sofosbuvir from urine samples. J Pharm Biomed Anal 204:114254. https://doi.org/10.1016/j.jpba.2021.114254
74. Hosseini MH, Rezaee M, Akbarian S, Mizani F, Pourjavid MR, Arabieh M (2013) Homogeneous liquid–liquid microextraction via flotation assistance for rapid and efficient determination of polycyclic aromatic hydrocarbons in water samples. Anal Chim Acta 762:54–60. https://doi.org/10.1016/j.aca.2012.10.030
75. Rezaee M, Saberyan K, Tajer-Mohammad-Ghazvini P (2019) Determination of malathion by homogeneous liquid-liquid micro extraction via flotation assistance combined with gas chromatography in water samples. Bull Chem Soc Ethiop 33:1–10. https://doi.org/10.4314/bcse.v33i1.1
76. Xu XY, Ye JQ, Nie J, Li ZG, Lee MR (2015) A new liquid-liquid microextraction method by ultrasound assisted salting-out for determination of triazole pesticides in water samples

coupled by gas chromatography-mass spectrometry. Anal Methods 7:1194–1199. https://doi.org/10.1039/c4ay02448f

77. Abdallah IA, Hammad SF, Bedair A, Abdelaziz MA, Danielson ND, Elshafeey AH, Mansour FR (2022) A gadolinium-based magnetic ionic liquid for supramolecular dispersive liquid–liquid microextraction followed by HPLC/UV for the determination of favipiravir in human plasma. Biomed Chromatogr 36:1–10. https://doi.org/10.1002/bmc.5365
78. Abdelaziz MA, Mansour FR, Danielson ND (2021) A gadolinium-based magnetic ionic liquid for dispersive liquid–liquid microextraction. Anal Bioanal Chem 413:205–214. https://doi.org/10.1007/s00216-020-02992-z
79. Mabrouk MM, Soliman SM, El-Agizy HM, Mansour FR (2020) Ultrasound-assisted dispersive liquid–liquid microextraction for determination of three gliflozins in human plasma by HPLC/DAD. J Chromatogr B 1136:121932. https://doi.org/10.1016/j.jchromb.2019.121932
80. Silvestre CIC, Santos JLM, Lima JLFC, Zagatto EAG (2009) Liquid-liquid extraction in flow analysis: A critical review. Anal Chim Acta 652:54–65. https://doi.org/10.1016/j.aca.2009.05.042
81. Calderilla C, Maya F, Leal LO, Cerdà V (2018) Recent advances in flow-based automated solid-phase extraction. TrAC - Trends Anal Chem 108:370–380. https://doi.org/10.1016/j.trac.2018.09.011
82. Vakh C, Falkova M, Timofeeva I, Moskvin A, Moskvin L, Bulatov A (2016) Flow analysis: a novel approach for classification. Crit Rev Anal Chem 46:374–388. https://doi.org/10.1080/10408347.2015.1087301
83. Pochivalov A, Vakh C, Garmonov S, Moskvin L, Bulatov A (2020) An automated in-syringe switchable hydrophilicity solvent-based microextraction. Talanta 209:120587. https://doi.org/10.1016/j.talanta.2019.120587
84. Vakh C, Pochivalov A, Andruch V, Moskvin L, Bulatov A (2016) A fully automated effervescence-assisted switchable solvent-based liquid phase microextraction procedure: liquid chromatographic determination of ofloxacin in human urine samples. Anal Chim Acta 907:54–59. https://doi.org/10.1016/J.ACA.2015.12.004
85. Cherkashina K, Vakh C, Lebedinets S, Pochivalov A, Moskvin L, Lezov A, Bulatov A (2018) An automated salting-out assisted liquid-liquid microextraction approach using 1-octylamine: on-line separation of tetracycline in urine samples followed by HPLC-UV determination. Talanta 184:122–127. https://doi.org/10.1016/j.talanta.2018.02.112
86. Shishov A, Volodina N, Nechaeva D, Gagarinova S, Bulatov A (2019) An automated homogeneous liquid-liquid microextraction based on deep eutectic solvent for the HPLC-UV determination of caffeine in beverages. Microchem J 144:469–473. https://doi.org/10.1016/j.microc.2018.10.014
87. Kalra A, Tugcu N, Cramer SM, Garde S (2001) Salting-in and salting-out of hydrophobic solutes in aqueous salt solutions. J Phys Chem B 105:6380–6386. https://doi.org/10.1021/jp010568+
88. Hyde AM, Zultanski SL, Waldman JH, Zhong YL, Shevlin M, Peng F (2017) General principles and strategies for salting-out informed by the Hofmeister Series. Org Process Res Dev 21:1355–1370. https://doi.org/10.1021/acs.oprd.7b00197
89. Görgényi M, Dewulf J, Van Langenhove H, Héberger K (2006) Aqueous salting-out effect of inorganic cations and anions on non-electrolytes. Chemosphere 65:802–810. https://doi.org/10.1016/j.chemosphere.2006.03.029
90. Jafari SA, Entezari MH (2020) Salting out in ACN/water systems: Hofmeister effects and partition of quercetin. J Mol Liq 312:113331. https://doi.org/10.1016/j.molliq.2020.113331
91. Wang Q, Yin C, Xu L (2013) Optimization of hydrophilic interaction LC by univariate and multivariate methods and its combination with salting-out liquid-liquid extraction for the determination of antihypertensive drugs in the environmental waters. J Sep Sci 36:1007–1014. https://doi.org/10.1002/jssc.201200985
92. Heydari R, Zarabi S (2014) Development of combined salt- and air-assisted liquid-liquid microextraction as a novel sample preparation technique. Anal Methods 6:8469–8475. https://doi.org/10.1039/c4ay01723d

93. Hammad SF, Abdallah IA, Bedair A, Mansour FR (2021) Salting-out induced liquid–liquid microextraction for alogliptin benzoate determination in human plasma by HPLC/UV. BMC Chem 15:1–10. https://doi.org/10.1186/s13065-020-00729-8
94. Farajzadeh MA, Mohebbi A, Mogaddam MRA, Davaran M, Norouzi M (2018) Development of salt-induced homogenous liquid-liquid microextraction based on iso-propanol/sodium sulfate system for extraction of some pesticides in fruit juices. Food Anal Methods 11:2497–2507. https://doi.org/10.1007/s12161-018-1238-6
95. Adlnasab L, Ebrahimzadeh H (2013) A novel salt-controlled homogenous ionic liquid phase microextraction based on the salting out effect and optimization of the procedure using the experimental design methodology. Anal Methods 5:5165–5171. https://doi.org/10.1039/c3ay40944a
96. Gupta M, Jain A, Verma KK (2010) Determination of amoxapine and nortriptyline in blood plasma and serum by salt-assisted liquid-liquid microextraction and high-performance liquid chromatography. J Sep Sci 33:3774–3780. https://doi.org/10.1002/jssc.201000434
97. Liu J, Jiang M, Li G, Xu L, Xie M (2010) Miniaturized salting-out liquid–liquid extraction of sulfonamides from different matrices. Anal Chim Acta 679:74–80. https://doi.org/10.1016/j.aca.2010.09.013
98. Chen MJ, Liu YT, Lin CW, Ponnusamy VK, Jen JF (2013) Rapid determination of triclosan in personal care products using new in-tube based ultrasound-assisted salt-induced liquid-liquid microextraction coupled with high performance liquid chromatography-ultraviolet detection. Anal Chim Acta 767:81–87. https://doi.org/10.1016/j.aca.2013.01.014
99. Hosseini M, Dalali N, Nejad SM (2012) A new mode of homogeneous liquid-liquid microextraction (HLLME) based on ionic liquids: In situ solvent formation microextraction (ISFME) for determination of lead. J Chinese Chem Soc 59:872–878. https://doi.org/10.1002/jccs.201100526
100. Mohamed AMI, Abdel-Wadood HM, Mousa HS (2014) Simultaneous determination of dorzolomide and timolol in aqueous humor: A novel salting out liquid-liquid microextraction combined with HPLC. Talanta 130:495–505. https://doi.org/10.1016/j.talanta.2014.06.074
101. Abdallah IA, Hammad SF, Bedair A, Mansour FR (2021) A green homogeneous liquid-liquid microextraction method for spectrophotometric determination of daclatasvir in human plasma. Sustain Chem Pharm 22:100498. https://doi.org/10.1016/j.scp.2021.100498
102. Sereshti H, Khosraviani M, Sadegh Amini-Fazl M (2014) Miniaturized salting-out liquid–liquid extraction in a coupled-syringe system combined with HPLC–UV for extraction and determination of sulfanilamide. Talanta 121:199–204. https://doi.org/10.1016/j.talanta.2014.01.005
103. Du D, Dong G, Wu Y, Wang J, Gao M, Wang X, Li Y (2014) Salting-out induced liquid-liquid microextraction based on the system of acetonitrile/magnesium sulfate for trace-level quantitative analysis of fluoroquinolones in water, food and biological matrices by high-performance liquid chromatography with a fluor. Anal Methods 6:6973–6980. https://doi.org/10.1039/c4ay01080a
104. Gupta M, Pillai AKKV, Singh A, Jain A, Verma KK (2011) Salt-assisted liquid-liquid microextraction for the determination of iodine in table salt by high-performance liquid chromatography-diode array detection. Food Chem 124:1741–1746. https://doi.org/10.1016/j.foodchem.2010.07.116
105. Gupta M, Dsouza A (2020) Salting-out homogeneous liquid-liquid microextraction for the spectrophotometric determination of iodate in food grade salt. J Food Compos Anal 87:103396. https://doi.org/10.1016/j.jfca.2019.103396
106. Chen W, Tu X, Wu D, Gao Z, Wu S, Huang S (2019) Comparison of the partition efficiencies of multiple phenolic compounds contained in propolis in different modes of acetonitrile–water-based homogenous liquid-liquid extraction. Molecules 24:442. https://doi.org/10.3390/molecules24030442
107. Zhang J, Myasein F, Wu H, El-Shourbagy TA (2013) Sugaring-out assisted liquid/liquid extraction with acetonitrile for bioanalysis using liquid chromatography-mass spectrometry. Microchem J 108:198–202. https://doi.org/10.1016/j.microc.2012.10.024

108. Yan L, Sun Y, Xiu Z (2016) Sugaring-out extraction coupled with fermentation of lactic acid. Sep Purif Technol 161:152–158. https://doi.org/10.1016/j.seppur.2016.01.049
109. Mahdavi P, Nojavan S, Asadi S (2019) Extraction system. J Chromatogr A 100:460411. https://doi.org/10.1016/j.chroma.2019.460411
110. Nugbienyo L, Malinina Y, Garmonov S, Kamencev M, Salahov I, Andruch V, Moskvin L, Bulatov A (2017) Automated sugaring-out liquid-liquid extraction based on flow system coupled with HPLC-UV for the determination of procainamide in urine. Talanta 167:709–713. https://doi.org/10.1016/j.talanta.2017.02.051
111. Zhang C, Huang K, Yu P, Liu H (2012) Sugaring-out three-liquid-phase extraction and one-step separation of Pt (IV), Pd (II) and Rh (III). Sep Purif Technol 87:127–134. https://doi.org/10.1016/j.seppur.2011.11.032
112. Dhamole PB, Mahajan P, Feng H (2010) Phase separation conditions for sugaring-out in acetonitrile—water systems. J Chem Eng Data 999:3803–3806
113. Timofeeva I, Shishov A, Kanashina D, Dzema D, Bulatov A (2017) On-line in-syringe sugaring-out liquid-liquid extraction coupled with HPLC-MS/MS for the determination of pesticides in fruit and berry juices. Talanta 167:761–767. https://doi.org/10.1016/j.talanta.2017.01.008
114. Shi Z, Li Z, Qiu L, Sun M, Zhang DAN, Zhang H (2015) Sugaring-out assisted liquid/liquid extraction coupled with HPLC for the analysis of Honokiol and Magnolol in traditional Chinese herbal formula Huoxiang-Zhengqi Oral Liquid. J Liq Chromatogr Relat Technol 38:722–728. https://doi.org/10.1080/10826076.2014.962147
115. Smith F, Mara C, Soares F, Nogueira M, De CR, Miranda M, Bezerra E, Lucena R (2019) Simultaneous concentration and chromatographic detection of water pesticides traces using aqueous two-phase system composed of tetrahydrofuran and fructose. Microchem J 147:303–310. https://doi.org/10.1016/j.microc.2019.03.033
116. Abdallah IA, Hammad SF, Bedair A, Elshafeey AH, Mansour FR (2022) Determination of favipiravir in human plasma using homogeneous liquid–liquid microextraction followed by HPLC/UV. Bioanalysis 14:205–216. https://doi.org/10.4155/bio-2021-0219
117. Mansour FR, Danielson ND (2012) Separation methods for captopril in pharmaceuticals and biological fluids. J Sep Sci 35:1213–1226. https://doi.org/10.1002/jssc.201200057
118. Hamad A, Elshahawy M, Negm A, Mansour FR (2020) Analytical methods for determination of glutathione and glutathione disulfide in pharmaceuticals and biological fluids. Rev Anal Chem 38:20190019. https://doi.org/10.1515/revac-2019-0019
119. Mansour FR, Wei W, Danielson ND (2013) Separation of carnitine and acylcarnitines in biological samples: a review. Biomed Chromatogr 27:1339–1353. https://doi.org/10.1002/bmc.2995
120. Alkan C, Çabuk H (2022) Matrix-induced sugaring-out liquid-liquid microextraction coupled with high-performance liquid chromatography for the determination of organophosphorus pesticides in fruit jams. Sep Sci Plus 5:416–423. https://doi.org/10.1002/sscp.202200039
121. Cai B, Ye E, Yuan B, Feng Y (2015) Sequential solvent induced phase transition extraction for profiling of endogenous phytohormones in plants by liquid chromatography-mass spectrometry. J Chromatogr B 1004:23–29. https://doi.org/10.1016/j.jchromb.2015.09.031
122. Abdallah IA, Hammad SF, Bedair A, Mansour FR (2022) Menthol-assisted homogenous liquid-liquid microextraction for HPLC/UV determination of favipiravir as an antiviral for COVID-19 in human plasma. J Chromatogr B 1189:123087. https://doi.org/10.1016/j.jchromb.2021.123087
123. Corazza G, Oenning AL, Bernardi G, Merib J, Carasek E (2021) Exploring the use of switchable hydrophilicity solvents as extraction phase for the determination of food-packaging contaminants in coconut water samples by gas chromatography-mass spectrometry. Food Anal Methods 14:319–330. https://doi.org/10.1007/s12161-020-01876-3
124. Jessop PG, Mercer SM, Heldebrant DJ (2012) CO2-triggered switchable solvents, surfactants, and other materials. Energy Environ Sci 5:7240. https://doi.org/10.1039/c2ee02912j
125. Vanderveen JR, Durelle J, Jessop PG (2014) Design and evaluation of switchable-hydrophilicity solvents. Green Chem 16:1187–1197. https://doi.org/10.1039/C3GC42164C

126. Li X, Yang Z, Sui H, Jain A, He L (2018) A hybrid process for oil-solid separation by a novel multifunctional switchable solvent. Fuel 221:303–310. https://doi.org/10.1016/j.fuel.2018.02.081
127. Lasarte-Aragonés G, Lucena R, Cárdenas S, Valcárcel M (2015) Use of switchable hydrophilicity solvents for the homogeneous liquid-liquid microextraction of triazine herbicides from environmental water samples. J Sep Sci 38:990–995. https://doi.org/10.1002/jssc.201401224
128. Behpour M, Nojavan S, Asadi S, Shokri A (2020) Combination of gel-electromembrane extraction with switchable hydrophilicity solvent-based homogeneous liquid-liquid microextraction followed by gas chromatography for the extraction and determination of antidepressants in human serum, breast milk and was. J Chromatogr A 1621:461041. https://doi.org/10.1016/j.chroma.2020.461041
129. Yıldız E, Çabuk H (2022) Determination of the synthetic antioxidants butylated hydroxyanisole (BHA) and butylated hydroxytoluene (BHT) by matrix acidity-induced switchable hydrophilicity solvent-based homogeneous liquid-liquid microextraction (MAI-SHS-HLLME) and high-performance. Anal Lett 55:480–494. https://doi.org/10.1080/00032719.2021.1941072
130. Hassan M, Erbas Z, Alshana U, Soylak M (2020) Ligandless reversed-phase switchable-hydrophilicity solvent liquid–liquid microextraction combined with flame-atomic absorption spectrometry for the determination of copper in oil samples. Microchem J 156:104868. https://doi.org/10.1016/j.microc.2020.104868
131. Lv H, Jin X, Zhang Z, Chen Y, Zhu G, Li Z, Lee M (2022) Ultrasound-assisted switchable hydrophilic solvent-based homogeneous liquid–liquid microextraction for the determination of triazole fungicides in environmental water by GC-MS. Anal Methods 14:1187–1193. https://doi.org/10.1039/D1AY02109E
132. Rahimi Kakavandi N, Ezoddin M, Abdi K, Ghazi-Khansari M, Amini M, Shahtaheri SJ (2017) Ion-pair switchable-hydrophilicity solvent-based homogeneous liquid-liquid microextraction for the determination of paraquat in environmental and biological samples before high-performance liquid chromatography. J Sep Sci 40:3703–3709. https://doi.org/10.1002/jssc.201700222
133. Shahraki S, Ahmar H, Nejati-Yazdinejad M (2018) Electrochemical determination of nitrazepam by switchable solvent based liquid-liquid microextraction combined with differential pulse voltammetry. Microchem J 142:229–235. https://doi.org/10.1016/j.microc.2018.07.003
134. Wang X, Gao M, Zhang Z, Gu H, Liu T, Yu N, Wang X, Wang H (2018) Development of CO2-mediated switchable hydrophilicity solvent-based microextraction combined with HPLC-UV for the determination of bisphenols in foods and drinks. Food Anal Methods 11:2093–2104. https://doi.org/10.1007/s12161-018-1187-0
135. Torbati M, Farajzadeh MA, Torbati M, Nabil AAA, Mohebbi A, Afshar Mogaddam MR (2018) Development of salt and pH–induced solidified floating organic droplets homogeneous liquid–liquid microextraction for extraction of ten pyrethroid insecticides in fresh fruits and fruit juices followed by gas chromatography-mass spectrometry. Talanta 176:565–572. https://doi.org/10.1016/j.talanta.2017.08.074
136. Hassan M, Alshana U (2019) Switchable-hydrophilicity solvent liquid–liquid microextraction of non-steroidal anti-inflammatory drugs from biological fluids prior to HPLC-DAD determination. J Pharm Biomed Anal 174:509–517. https://doi.org/10.1016/j.jpba.2019.06.023
137. Asadi T, Rahimi Kakavandi N, Nili Ahmadabadi A, Heshmati A, Ranjbar A, Abdi K, Ezoddin M (2022) Development of effervescence-assisted switchable polarity solvent homogeneous liquid-phase microextraction for the determination of permethrin and deltamethrin in water samples prior to gas chromatography–flame ionization detection. Biomed Chromatogr 36:1–9. https://doi.org/10.1002/bmc.5304
138. Reclo M, Yilmaz E, Soylak M, Andruch V, Bazel Y (2017) Ligandless switchable solvent based liquid phase microextraction of nickel from food and cigarette samples prior to its

micro-sampling flame atomic absorption spectrometric determination. J Mol Liq 237:236–241. https://doi.org/10.1016/j.molliq.2017.04.066

139. Hassan M, Uzcan F, Nasrullah S, Alshana U, Soylak M (2021) Switchable-hydrophilicity solvent liquid-liquid microextraction for sample cleanup prior to dispersive magnetic solid-phase microextraction for spectrophotometric determination of quercetin in food samples. Sustain Chem Pharm 22:100480. https://doi.org/10.1016/j.scp.2021.100480
140. Çabuk H, Köktürk M, Ata Ş (2014) PH-assisted homogeneous liquid–liquid microextraction using dialkylphosphoric acid as an extraction solvent for the determination of chlorophenols in water samples. J Sep Sci 37:1343–1351. https://doi.org/10.1002/JSSC.201400158
141. Kiani S, Hossein M, Ahmar H (2018) Development of a new pH assisted homogeneous liquid-liquid microextraction by a solvent with switchable hydrophilicity: application for GC-MS determination of methamphetamine. Talanta 184:103–108. https://doi.org/10.1016/j.talanta.2018.02.115

MIX
Papier aus verantwortungsvollen Quellen
Paper from responsible sources
FSC® C105338

If you have any concerns about our products,
you can contact us on
ProductSafety@springernature.com

In case Publisher is established outside the EU,
the EU authorized representative is:
Springer Nature Customer Service Center GmbH
Europaplatz 3, 69115 Heidelberg, Germany

Printed by Libri Plureos GmbH
in Hamburg, Germany